$5 = 1^2 \quad 4 - 3^2$

$5 \cdot 4^2 \, 5^2$

Probabilistic
Methods
of Signal and
System Analysis

Probabilistic
Methods
of Signal and
System Analysis

SECOND EDITION

GEORGE R. COOPER

CLARE D. McGILLEM
Purdue University

HOLT, RINEHART AND WINSTON
New York   Chicago   San Francisco   Philadelphia
Montreal   Toronto   London   Sydney
Tokyo   Mexico City   Rio de Janeiro   Madrid

DEDICATED TO

Lisa Cooper and Ann McGillem for their
encouragement and patience, without which
this revision would never have been possible

Acquisitions Editor: *Deborah L. Moore*
Production Manager: *Paul Nardi*
Project Editors: *Robert Greiner*
 Kathleen Nevils
Design Supervisor: *Robert Kopelman*
Interior Design: *Denise C. Schiff*

Library of Congress Cataloging in Publication Data

Cooper, George R.
 Probabilistic methods of signal and system analysis.

 Includes bibliographies and index.
 1. Signal processing. 2. System analysis.
3. Probabilities. I. McGillem, Clare D. II. Title.
TK5102.5.C67 1986 620'.0042 85-24768
ISBN 0-03-070614-9

CBS College Publishing
Holt, Rinehart and Winston
The Dryden Press
Saunders College Publishing

Preface

This is the second edition of a book that presents an introduction to probability theory, statistics, random processes, and the analysis of systems with random inputs. It is written at a level that is suitable for junior and senior engineering students and presumes that the student is familiar with conventional methods of system analysis such as convolution and transform techniques. However, it may also serve graduate students and engineers as a concise review of material that they previously encountered in widely scattered sources.

This edition differs from the first in several respects. In the first place, new material on statistics has been added to provide practical applications for some of the probability concepts developed in the first three chapters of the book. Explanations of the more difficult concepts have been expanded throughout the book, and many more examples to illustrate these concepts have been provided. Furthermore, there are now more exercises incorporated within the text to provide the reader with the opportunity to test his or her mastery of the concepts discussed in each section. Finally, the problems at the end of each chapter are entirely new and illustrate a wider range of applications than the previous edition.

Since this is an engineering text, the treatment is heuristic rather than rigorous, and the student will find many examples of the application of these concepts to engineering problems. However, it is not completely devoid of the mathematical subtleties, and considerable attention has been devoted to pointing out some of the difficulties that make a more advanced study of the subject essential if one is to master it. The authors believe that the educational process is best served by repeated exposure to difficult subject matter; this text is intended to be the first exposure to probability and random processes and, we hope, not the last. Thus, the book is not comprehensive, but deals selectively with those topics that the authors have found most useful in the solution of engineering problems.

A brief discussion of some of the significant features of this book will help set the stage for a discussion of the various ways it can be used. Elementary concepts of discrete probability are introduced in Chapter 1; first from the intuitive standpoint of the relative-frequency approach and then from the more rigorous standpoint of axiomatic probability. Simple examples illustrate all these concepts and are more meaningful to engineers than are the traditional examples of selecting red and white balls from urns.

The concept of a random variable is introduced in Chapter 2 along with the ideas of probability distribution and density functions, mean values, and conditional probability. A significant feature of this chapter is an extensive discussion of many different probability density functions and the physical situations in which they may occur. Chapter 3 extends the random variable concept to situations involving two or more random variables and introduces the concepts of statistical independence and correlation.

Entirely new material on statistics appears in Chapter 4. Sampling theory, as applied to statistical estimation, is considered in some detail and a thorough discussion of sample mean and sample variance is given. The distribution of the sample is described and the use of confidence intervals in making statistical decisions is both considered and illustrated by many examples of hypothesis testing. The problem of fitting smooth curves to experimental data is analyzed, and the use of linear regression is illustrated by practical examples.

A general discussion of random processes and their classification is given in Chapter 5. The emphasis here is on selecting probability models that are useful in solving engineering problems. Accordingly, a great deal of attention is devoted to the physical significance of the various process classifications, with no attempt at mathematical rigor. A unique feature of this chapter, which is continued in subsequent chapters, is an introduction to the practical problem of estimating the mean of a random process from an observed sample function.

Properties and applications of autocorrelation and crosscorrelation functions are discussed in Chapter 6. Many examples are presented in an attempt to develop some insight into the nature of correlation functions. The important problem of estimating autocorrelation functions is discussed in some detail.

Chapter 7 turns to a frequency-domain representation of random processes by introducing the concept of spectral density. Unlike most texts, which simply define spectral density as a Fourier transform of the correlation function, a more fundamental approach is adopted here in order to bring out the physical significance of the concept. This chapter is the most difficult one in the book, but the authors believe the material should be presented in this way. Instructors who wish to by-pass some of the more fundamental problems may omit Section 7–2 and bridge the gap by defining spectral density simply as the Fourier transform of the correlation function.

Chapter 8 utilizes the concepts of correlation functions and spectral density to analyze the response of linear systems to random inputs. In a sense, this chapter is a culmination of all that preceded it, and is particularly significant to engineers who must use these concepts. Hence, it contains a great many examples that are relevant to engineering problems and emphasizes the need for mathematical models that are both realistic and manageable.

Chapter 9 extends the concepts of systems analysis to consider systems that are optimum in some sense. Both the classical matched filter for known signals and the Wiener filter for random signals are considered from an elementary standpoint.

In a more general vein, each chapter contains references that the reader may use to extend his or her knowledge. There is also a wide selection of problems at the end of each chapter. A solution manual for these problems is available to the instructor. Tables of functions, integrals, and other useful information that will aid the reader in solving the problems appear in a number of appendices at the end of the book.

As an additional aid to learning and using the concepts and methods discussed in this text, there are exercises at the end of each major section. The reader should consider these exercises as part of the reading assignment and should make every effort to solve each one before going on to the next section. Answers are provided so that the reader may know when his or her efforts have been successful. It should be noted, however, that the answers to each exercise may not be listed in the same order as the questions. This is intended to provide an additional challenge. The presence of these exercises should substantially reduce the number of additional problems that need to be assigned by the instructor.

The material in this text is used at Purdue University in a one-semester, three-credit course offered in the Fall semester of the junior year. Not all sections of the text are used in this course, but at least 90% of it is covered in reasonable detail. The sections usually omitted include 3–6, 5–6, 6–4, 6–9, 7–9, and 9–6; but other choices may be made at the discretion of the instructor. There are, of course, many other ways in which the text material could be utilized. For example, a one-semester course with a more relaxed pace could be given by omitting all of Chapter 9 in addition to the sections noted above. For those schools on a quarter system, the material noted above could be covered in a four-credit course. Alternatively, if a three-credit course were desired, it is suggested that, in addition to the omissions noted above, Sections 1–5, 1–6, 1–7, 1–9, 2–6, 3–5, 7–2, 7–8, 7–10, 8–9, and all of Chapter 9 can be omitted if the instructor supplies a few explanatory words to bridge the gaps. Obviously, there are also many other possibilities that are open to the experienced instructor.

It is a pleasure for the authors to acknowledge the very substantial aid and encouragement that they have received from their colleagues and students. A complete list is too lengthy to include here, but it is appropriate to mention a few individuals who made valuable suggestions. In connection with the first edition, these individuals included Professors J. Y. S. Luh and P. A. Wintz and Dr. Lewis A. Thurman, then all at Purdue University. Furthermore, the careful and perceptive reading of the preliminary manuscript of the first edition by Professor J. E. Kemmerly of the California State University at Fullerton and Professor James L. Massey at the Swiss Federal Institute of Technology is gratefully acknowledged. Their many suggestions greatly improved that edition.

In connection with the second edition, many valuable suggestions have been received from Professor E. W. Chandler at Marquette University and from the reviewers commissioned by our editor, Deborah Moore: Richard H. Williams, University of New Mexico; Richard Christiansen, Brigham Young University; Donald Healy, Georgia Institute of Technology; Hugh Van Landingham, Virginia Polytechnic; and Soheil A. Dianat, Rochester Institute of Technology. Special thanks are due to Dr. C. P. Cheng for preparing the solutions manual and for proofreading the manuscript. Last, but not least, we acknowledge the contributions made by hundreds of students who have used and criticized the first edition of this text.

February 1986

George R. Cooper
Clare D. McGillem

Table of Contents

Chapter 7   Spectral Density   230

Chapter 8   Response of Linear Systems to Random Inputs   284

Probabilistic
Methods
of Signal and
System Analysis

CHAPTER *1*

Introduction to Probability

1–1   Engineering Applications of Probability

Before embarking on a study of elementary probability theory, it is essential to motivate such a study by considering why probability theory is useful in the solution of engineering problems. This can be done in two different ways. The first is to suggest a viewpoint, or philosophy, concerning probability that emphasizes its universal physical reality rather than treating it as another mathematical discipline which may be useful occasionally. The second is to note some of the many different types of situations that arise in normal engineering practice in which the use of probability concepts is indispensable.

A characteristic feature of probability theory is that it concerns itself with situations that involve uncertainty in some form. The popular conception of this relates probability to such activities as tossing dice, drawing cards, and spinning roulette wheels. Because the rules of probability are not widely known, and because such situations can become quite complex, the prevalent attitude is that probability theory is a mysterious and esoteric branch of mathematics that is accessible only to trained mathematicians and is of limited value in the real world. Since probability theory does deal with uncertainty, another prevalent attitude is that a probabilistic treatment of physical problems is an inferior substitute for a more desirable exact analysis and is forced on the analyst by a lack of complete information. *Both of these attitudes are false*.

Regarding the alleged difficulty of probability theory, it is doubtful there is any other branch of mathematics or analysis that is so completely based on such a small number of easily understood basic concepts. Subsequent discussion reveals that the major body of probability theory can be deduced from only three axioms that are almost self-evident. Once these axioms and their applications are understood, the remaining concepts follow in a logical manner.

The attitude that regards probability theory as a substitute for exact analysis stems from the current educational practice of presenting physical laws as deterministic, immutable, and strictly true under all circumstances. Thus, a law that describes the response of a dynamic system is supposed to predict that response precisely if the system excitation is known precisely. For example, Ohm's law

$$v(t) = Ri(t)$$

is assumed to be exactly true at every instant of time, and, on a macroscopic basis, this assumption may be well justified. On a microscopic basis, however, this assumption is patently false—a fact that is immediately obvious to anyone who has tried to connect a large resistor to the input of a high-gain amplifier and listened to the resulting noise.

In the light of modern physics and our emerging knowledge of the nature of matter, the viewpoint that natural laws are deterministic and exact is untenable. They are, at best, a representation of the average behavior of nature. In many important cases this average behavior is close enough to that actually observed so that the deviations are unimportant. In such cases, the deterministic laws are extremely valuable because they make it possible to predict system behavior with a minimum of effort. In other equally important cases, the random deviations may be significant—perhaps even more significant than the deterministic response. For these cases, analytic methods derived from the concepts of probability are essential.

From the above discussion, it should be clear that the so-called exact solution is not exact at all, but, in fact, represents an idealized special case that actually never arises in nature. The probabilistic approach, on the other hand, far from being a poor substitute for exactness, is actually the method that most nearly represents physical reality. Furthermore, it includes the deterministic result as a special case.

It is now appropriate to discuss the types of situations in which probability concepts arise in engineering. The examples presented here emphasize situations that arise in systems studies; but they do serve to illustrate the essential point that engineering applications of probability tend to be the rule rather than the exception.

Random input signals.   In order for a physical system to perform a useful task, it is usually necessary that some sort of forcing function (the input signal) be applied to it. Input signals that have simple mathematical representations are convenient for pedagogical purposes or for certain types of system analysis, but they seldom arise in actual applications. Instead, the input signal is more likely to involve a certain amount of uncertainty and unpredictability that justifies treating it as a *random* signal. There are many examples of this: speech and music signals that serve as inputs to communication systems; random digits applied to a computer; random command signals applied to an aircraft flight control system; random signals derived from measuring some characteristic of a manufactured product, and used as inputs to a process control system; steering wheel movements in an automobile power-steering system; the sequence in which the call and operating buttons of an elevator are pushed; the number of vehicles passing various checkpoints in a traffic control system; outside and inside temperature fluctuations as inputs to a building heating and airconditioning system; and many others.

Random disturbances. Many systems have unwanted disturbances applied to their input or output in addition to the desired signals. Such disturbances are almost always random in nature and call for the use of probabilistic methods even if the desired signal does not. A few specific cases serve to illustrate several different types of disturbances. If, for a first example, the output of a high-gain amplifier is connected to a loudspeaker, one frequently hears a variety of snaps, crackles, and pops. This random noise arises from thermal motion of the conduction electrons in the amplifier input circuit or from random variations in the number of electrons (or holes) passing through the transistors. It is obvious that one cannot hope to calculate the value of this noise at every instant of time since this value represents the combined effects of literally billions of individual moving charges. It is possible, however, to calculate the average power of this noise, its frequency spectrum, and even the probability of observing a noise value larger than some specified value. As a practical matter, these quantities are more important in determining the quality of the amplifier than is a knowledge of the instantaneous waveforms.

As a second example, consider a radio or television receiver. In addition to noise generated within the receiver by the mechanisms noted, there is random noise arriving at the antenna. This results from distant electrical storms, manmade disturbances, radiation from space, or thermal radiation from surrounding objects. Hence, even if perfect receivers and amplifiers were available, the received signal would be combined with random noise. Again, the calculation of such quantities as average power and frequency spectrum may be more significant than the determination of instantaneous value.

A different type of system is illustrated by a large radar antenna, which may be pointed in any direction by means of an automatic control system. The wind blowing on the antenna produces random forces that must be compensated for by the control system. Since the compensation is never perfect, there is always some random fluctuation in the antenna direction; it is important to be able to calculate the effective value and frequency content of this fluctuation.

A still different situation is illustrated by an airplane flying in turbulent air, a ship sailing in stormy seas, or an army truck traveling over rough terrain. In all these cases, random disturbing forces, acting on complex mechanical systems, interfere with the proper control or guidance of the system. It is essential to determine how the system responds to these random input signals.

Random system characteristics.   The system itself may have characteristics that are unknown and that vary in a random fashion from time to time. Some typical examples are: aircraft in which the load (that is, the number of passengers or the weight of the cargo) varies from flight to flight; troposcatter communication systems in which the path attenuation varies radically from moment to moment; an electric power system in which the load (that is, the amount of energy being used) fluctuates randomly; and a telephone system in which the number of users changes from instant to instant.

There are also many electronic systems in which the parameters may be random. For example, it is customary to specify the properties of many solid-state devices such as diodes, transistors, digital gates, shift registers, flip-flops, etc. by listing a range of values for the more important items. The actual value of the parameters are random quantities that lie somewhere in this range but are not known *a priori*.

System reliability.   All systems are composed of many individual elements, and one or more of these elements may fail, thus causing the entire system, or part of the system, to fail. The times at which such failures will occur are unknown, but it is often possible to determine the probability of failure for the individual elements and from these to determine the ''mean time to failure'' for the system. Such reliability studies are deeply involved with probability and are extremely important in engineering design. As systems become more complex, more costly, and contain larger numbers of elements, the problems of reliability become more difficult and take on added significance.

Quality control.   An important method of improving system reliability is to improve the quality of the individual elements, and this can often be done by an inspection process. As it may be too costly to inspect every element after every step during its manufacture, it is necessary to develop rules for inspecting elements selected at random. These rules are based on probabilistic concepts and serve the valuable purpose of maintaining the quality of the product with the least expense.

Information theory.   A major objective of information theory is to provide a quantitative measure for the information content of messages such as printed pages, speech, pictures, graphical data, numerical data, or physical observations of temperature, distance, velocity, radiation intensity, and rainfall. This quantitative measure is necessary in order to be able to provide communication channels that are both adequate and efficient for conveying this information from one place to another. Since such messages and observations are almost invariabley unknown in advance and random in nature, they can be described only in terms of probability. Hence, the appropriate information measure is a probabilistic one. Furthermore, the communication channels are subject to random disturbances (noise) that limit their ability to convey information, and again a probabilistic description is required.

It should be clear from the above partial listing that almost any engineering endeavor involves a degree of uncertainty or randomness that makes the use of probabilistic concepts an essential tool for the present-day engineer. In the case of system analysis, it is necessary to have some description of random signals and disturbances. There are two general methods of describing random signals mathematically. The first, and most basic, is a probabilistic description in which the random quantity is characterized by a probability model. This method is discussed later in this chapter.

The probabilistic description of random signals cannot be used directly in system analysis since it tells very little about how the random signal varies with time or what its frequency spectrum is. It does, however, lead to the statistical description of random signals, which is useful in system analysis. In this case the random signal is characterized by a statistical model, which consists of an appropriate set of average values such as the mean, variance, correlation function, spectral density, and others. These average values represent a less precise description of the random signal than that offered by the probability model, but they are more useful for system analysis because they can be computed by using straightforward and relatively simple methods. Some of the statistical averages are discussed in subsequent chapters.

There are many steps that need to be taken before it is possible to apply the probabilistic and statistical concepts to system analysis. In order that the reader may understand that even the most elementary steps are important to the final objective, it is desirable to outline these steps briefly. The first step is to introduce the concepts of probability by considering discrete random events. These concepts are then extended to continuous random variables and subsequently to random functions of time. Finally, several of the average values associated with random signals are introduced. At this point, the tools are available to consider ways of analyzing the response of linear systems to random inputs.

1–2   Random Experiments and Events

The concepts of *experiment* and *event* are fundamental to an understanding of elementary probability concepts. An experiment is some action that results in an *outcome*. A *random experiment* is one in which the outcome is uncertain before the experiment is performed. Although there is a precise mathematical definition of a random experiment, a better understanding may be gained by listing some examples of well-defined random experiments and their possible outcomes. This is done in Table 1–1. It should be noted, however, that the possible outcomes often may be defined in several different ways depending upon the wishes of the experimenter. The initial discussion is concerned with a single performance of a well-defined experiment. This single performance is referred to as a *trial*.

An important concept in connection with random events is that of *equally likely events*. For example, if we toss a coin we expect that the event of getting a *head* and the event of getting a *tail* are equally likely. Likewise, if we roll a die we expect that the events of getting any number from 1 to 6 are equally likely. Also, when a card is drawn from a deck, each of the 52 cards is equally likely. A term that is often used to be synonymous with the concept of equally likely events is that of *selected at random*. For example, when we say that a card is selected at random from a deck, we are implying that all cards in the deck are equally likely to have been chosen. In general, we assume that the outcomes of an experiment are equally likely unless there is some clear physical reason why they should not be. In the discussions that follow, there will be examples of events that are assumed to be equally likely and events that are *not* assumed to be equally likely. The reader should clearly understand the physical reasons for the assumptions in both cases.

It is also important to distinguish between *elementary* events and *composite* events. An elementary event is one for which there is only one outcome. Examples of elementary events include such things as tossing a coin or rolling a die when the events are defined in a specific way. When a coin is tossed, the event of getting a head or the event of getting a tail can be achieved in only one way. Likewise, when a die is rolled the event of getting any integer from 1 to 6 can be achieved in only one way. Hence, in both cases, the defined events are elementary events. On the other hand, it is possible to define events associated with rolling a die that are not elementary. For example, let one event be that of obtaining an even number while another event is that of obtaining an odd number. In this case, each event can be achieved in three different ways and, hence, these events are composite.

There are many different random experiments in which the events can be defined to be either elementary or composite. For example, when a card is selected at random from a deck of 52 cards, there are 52 elementary events corresponding

Table 1–1.   Possible Experiments and Their Outcomes.

Experiment	Possible Outcomes
Flipping a coin	Heads (H), tails (T)
Throwing a die	1, 2, 3, 4, 5, 6
Drawing a card	Any of the 52 possible cards
Observing a voltage	Greater than 0, less than 0
Observing a voltage	Greater than V, less than V
Observing a voltage	Between V_1 and V_2, not between V_1 and V_2

to the selection of each of the cards. On the other hand, the event of selecting a heart is a composite event containing 13 different outcomes. Likewise, the event of selecting an ace is a composite event containing 4 outcomes. Clearly, there are many other ways in which composite events could be defined.

When the number of outcomes of an experiment are countable (that is, they can be put in one-to-one correspondence with the integers), the outcomes are said to be *discrete*. All of the examples discussed above represent discrete outcomes. However, there are many experiments in which the outcomes are not countable. For example, if a random voltage is observed, and the outcome taken to be the value of the voltage, there may be an infinite and noncountable number of possible values that can be obtained. In this case, the outcomes are said to form a *continuum*. The concept of an elementary event does not apply in this case.

It is also possible to conduct more complicated experiments with more complicated sets of events. The experiment may consist of tossing ten coins, and it is apparent in this case that there are many different possible outcomes, each of which may be an event. Another situation, which has more of an engineering flavor, is that of a telephone system having 10,000 telephones connected to it. At any given time, a possible event is that 2000 of these telephones are in use. Obviously, there are a great many other possible events.

If the outcome of an experiment is uncertain before the experiment is performed, the possible outcomes are random events. To each of these events it is possible to assign a number, called the probability of that event, and this number is a measure of how likely that event is. Usually, these numbers are assumed, the assumed values being based on our intuition about the experiment. For example, if we toss a coin, we would expect that the possible outcomes of *heads* and *tails* would be equally likely. Therefore, we would assume the probabilities of these two events to be the same.

1—3 Definitions of Probability

One of the most serious stumbling blocks in the study of elementary probability is that of arriving at a satisfactory definition of the term "probability." There are, in fact, four or five different definitions for probability that have been proposed and used with varying degrees of success. They all suffer from deficiencies in concept or application. Ironically, the most successful "definition" leaves the term probability *undefined*.

Of the various approaches to probability, the two that appear to be most useful are the *relative-frequency* approach and the *axiomatic* approach. The relative-frequency approach is useful because it attempts to attach some physical significance to the concept of probability and, thereby, makes it possible to relate probabilistic concepts to the real world. Hence, the application of probability to engineering problems is almost always accomplished by invoking the concepts of relative frequency, even when the engineer may not be conscious that he is doing so.

The limitation of the relative-frequency approach is the difficulty of using it to deduce the appropriate mathematical structure for situations that are too complicated to be analyzed readily by physical reasoning, This is not to imply that this approach cannot be used in such situations, for it can, but it does suggest that there may be a much easier way to deal with these cases. The easier way turns out to be the axiomatic approach.

The axiomatic approach treats the probability of an event as a number that satisfies certain postulates but is otherwise undefined. Whether or not this number relates to anything in the real world is of no concern in developing the mathematical structure that evolves from these postulates. Engineers may object to this approach as being too artificial and too removed from reality, but they should remember that the whole body of circuit theory was developed in essentially the same way. In the case of circuit theory, the basic postulates are Kirchhoff's laws and the conservation of energy. The same mathematical structure emerges regardless of what physical quantities are identified with the abstract symbols—or even if *no* physical quantities are associated with them. It is the task of the engineer to relate this mathematical structure to the real world in a way that is admittedly not exact, but that leads to useful solutions to real problems.

From the above discussion, it appears that the most useful approach to probability for engineers is a two-pronged one, in which the relative-frequency concept is employed in order to relate simple results to physical reality, and the axiomatic approach is employed to develop the appropriate mathematics for more complicated situations. It is this philosophy that is presented here.

1—4   The Relative-Frequency Approach

As its name implies, the relative-frequency approach to probability is closely linked to the frequency of occurrence of the defined events. For any given event, the frequency of occurrence is used to define a number called the *probability* of that event and this number is a measure of how likely that event is. Usually, these numbers are assumed, the assumed values being based on our intuition about the experiment or on the assumption that the events are equally likely.

In order to make this concept more precise, consider an experiment that is performed N times and for which there are four possible outcomes that are considered to be the elementary events A, B, C, and D. Let N_A be the number of times that event A occurs, with a similar notation for the other events. It is clear that

$$N_A + N_B + N_C + N_D = N \tag{1-1}$$

We now define the relative frequency of A, $r(A)$ as

$$r(A) = \frac{N_A}{N}$$

From (1–1) it is apparent that

$$r(A) + r(B) + r(C) + r(D) = 1 \qquad (1\text{–}2)$$

Now imagine that N increases without limit. When a phenomenon known as *statistical regularity* applies, the relative frequency $r(A)$ tends to stabilize and approach a number, Pr (A), that can be taken as the probability of the elementary event A. That is

$$\text{Pr } (A) = \lim_{N \to \infty} r(A) \qquad (1\text{–}3)$$

From the relation given above, it follows that

$$\text{Pr } (A) + \text{Pr } (B) + \text{Pr } (C) + \cdots + \text{Pr } (M) = 1 \qquad (1\text{–}4)$$

and we can conclude that the sum of the probabilities of all of the mutually exclusive events associated with a given experiment must be unity.

These concepts can be summarized by the following set of statements:

1. $0 \leq \text{Pr } (A) \leq 1$.
2. $\text{Pr } (A) + \text{Pr } (B) + \text{Pr } (C) + \cdots + \text{Pr } (M) = 1$, for a complete set of mutually exclusive events.
3. An impossible event is represented by $\text{Pr } (A) = 0$.
4. A certain event is represented by $\text{Pr } (A) = 1$.

In order to make some of these ideas more specific, consider the following hypothetical example. Assume that a large bin contains an assortment of resistors of different sizes, which are thoroughly mixed. In particular, let there be 100 resistors having a marked value of 1 Ω, 500 resistors marked 10 Ω, 150 resistors marked 100 Ω, and 250 resistors marked 1000 Ω. Someone reaches into the bin and pulls out one resistor at random. There are now four possible outcomes corresponding to the value of the particular resistor selected. We desire to determine the probability of each of these events. In order to do this, we assume that the probability of each event is proportional to the number of resistors in the bin corresponding to that event. Since there are 1000 resistors in the bin all together, the resulting probabilities are

$$\text{Pr } (1 \ \Omega) = \frac{100}{1000} = 0.1 \qquad \text{Pr } (10 \ \Omega) = \frac{500}{1000} = 0.5$$

$$\text{Pr } (100 \ \Omega) = \frac{150}{1000} = 0.15 \qquad \text{Pr } (1000 \ \Omega) = \frac{250}{1000} = 0.25$$

Note that these probabilities are all positive, less than 1, and do add up to 1.

Many times one is interested in more than one event at a time. If a coin is tossed twice, one may wish to determine the probability that a head will occur on both tosses. Such a probability is referred to as a *joint probability*. In this particular case, one assumes that all four possible outcomes (*HH, HT, TH,* and *TT*) are equally likely and, hence, the probability of each is one quarter. In a more general case the situation is not this simple, so it is necessary to look at a more complicated situation in order to deduce the true nature of joint probability. The notation employed is Pr (A,B) and signifies the probability of the joint occurrence of events A and B.

Consider again the bin of resistors and specify that in addition to having different resistance values, they also have different power ratings. Let the different power ratings be 1 W, 2 W, and 5 W; the number having each rating is indicated in Table 1–2.

Before using this example to illustrate joint probabilities, consider the probability (now referred to as a *marginal probability*) of selecting a resistor having a given power rating without regard to its resistance value. From the totals given in the right-hand column, it is clear that these probabilities are

$$\text{Pr (1 W)} = \frac{440}{1000} = 0.44 \quad \text{Pr (2 W)} = \frac{200}{1000} = 0.20$$

$$\text{Pr (5 W)} = \frac{360}{1000} = 0.36$$

We now ask what the joint probability is of selecting a resistor of 10 Ω having a 5-W power rating. Since there are 150 such resistors in the bin, this joint probability is clearly

$$\text{Pr (10 }\Omega\text{, 5 W)} = \frac{150}{1000} = 0.15$$

The eleven other joint probabilities can be determined in a similar way. Note that some of the joint probabilities are zero [for example, Pr (1 Ω, 5 W) = 0] simply because a particular combination of resistance and power does not exist.

It is necessary at this point to relate the joint probabilities to the marginal probabilities. In the example of tossing a coin two times, the relationship is simply a product. That is,

$$\text{Pr }(H,H) = \text{Pr }(H)\,\text{Pr }(H) = \frac{1}{2} \times \frac{1}{2} = \frac{1}{4}$$

Table 1–2.   Resistance Values and Power Ratings.

Power Rating	Resistance Values				
	1 Ω	10 Ω	100 Ω	1000 Ω	Totals
1 W	50	300	90	0	440
2 W	50	50	0	100	200
5 W	0	150	60	150	360
Totals	100	500	150	250	1000

But this relationship is obviously not true for the resistor bin example. Note that

$$Pr (5 W) = \frac{360}{1000} = 0.36$$

and it was previously shown that

$$Pr (10 \, \Omega) = 0.5$$

Thus,

$$Pr (10 \, \Omega) \, Pr (5 W) = 0.5 \times 0.36 = 0.18 \neq Pr (10 \, \Omega, 5 W) = 0.15$$

and the joint probability is not the product of the marginal probabilities.

In order to clarify this point, it is necessary to introduce the concept of *conditional probability*. This is the probability of one event A, given that another event B has occurred; it is designated as $Pr (A \mid B)$. In terms of the resistor bin, consider the conditional probability of selecting a 10-Ω resistor when it is already known that the chosen resistor is 5 W. Since there are 360 5-W resistors, and 150 of these are 10 Ω, the required conditional probability is

$$Pr (10 \, \Omega \mid 5 W) = \frac{150}{360} = 0.417$$

Now consider the product of this conditional probability and the marginal probability of selecting a 5-W resistor.

$$Pr (10 \, \Omega \mid 5 W) \, Pr (5 W) = 0.417 \times 0.36 = 0.15 = Pr (10 \, \Omega, 5 W)$$

It is seen that this product is indeed the joint probability.

The same result can also be obtained another way. Consider the conditional probability

$$Pr (5 W \mid 10 \, \Omega) = \frac{150}{500} = 0.30$$

since there are 150 5-W resistors out of the 500 10-Ω resistors. Then form the product

$$Pr (5 W \mid 10 \, \Omega) \, Pr (10 \, \Omega) = 0.30 \times 0.5 = Pr (10 \, \Omega, 5 W) \qquad \text{(1–5)}$$

Again, the product is the joint probability.

The foregoing ideas concerning joint probability can be summarized in the general equation

$$Pr (A,B) = Pr (A \mid B) \, Pr (B) = Pr (B \mid A) \, Pr (A) \qquad \text{(1–6)}$$

which indicates that the joint probability of two events can always be expressed as the product of the marginal probability of one event and the conditional probability of the other event given the first event.

We now return to the coin-tossing problem, in which it is indicated that the joint probability can be obtained as the product of two marginal probabilities. Under what conditions will this be true? From equation (1–6) it appears that this can be true if

$$\Pr (A \mid B) = \Pr (A) \quad \text{and} \quad \Pr (B \mid A) = \Pr (B)$$

These statements imply that the probability of event A does not depend upon whether or not event B has occurred. This is certainly true in coin tossing, since the outcome of the second toss cannot be influenced in any way by the outcome of the first toss. Such events are said to be *statistically independent*. More precisely, two random events are statistically independent if and only if

$$\Pr (A,B) = \Pr (A) \Pr (B) \tag{1–7}$$

The preceding paragraphs provide a very brief discussion of many of the basic concepts of discrete probability. They have been presented in a heuristic fashion without any attempt to justify them mathematically. Instead, all of the probabilities have been formulated by invoking the concepts of relative frequency and equally likely events in terms of specific numerical examples. It is clear from these examples that it is not difficult to assign reasonable numbers to the probabilities of various events (by employing the relative-frequency approach) when the physical situation is not very involved. It should also be apparent, however, that such an approach might become unmanageable when there are many possible outcomes to any experiment and many different ways of defining events. This is particularly true when one attempts to extend the results for the discrete case to the continuous case. It becomes necessary, therefore, to reconsider all of the above ideas in a more precise manner and to introduce a measure of mathematical rigor that provides a more solid footing for subsequent extensions.

Exercise 1–4.1

a) A box contains 25 transistors, of which 8 are known to be bad. A transistor is selected at random and tested. What is the probability that it is bad?

b) If the first transistor tests bad, what is the probability that a second transistor selected at random will also be bad?

c) If the first transistor tests good, what is the probability that a second transistor selected at random will be bad?

Answers: 1/3, 8/25, 7/24

(Note: In the exercise above, and in others throughout the book, answers are not necessarily given in the same order as the questions.)

Exercise 1–4.2

A traffic survey on a busy highway reveals that one out of every three vehicles is a truck. This survey also established that one-tenth of all the automobiles are unsafe to drive and one-twentieth of all the trucks are unsafe to drive.

a) What is the probability that the next vehicle to pass a given point will be an unsafe truck?

b) What is the probability that the next vehicle will be a truck, given that it is unsafe?

c) What is the probability that the next vehicle that passes a given point will be a truck, given that the previous vehicle was an automobile?

    Answers: 1/5, 1/3, 1/60

1–5 Elementary Set Theory

The more precise formulation mentioned in Section 1–4 is accomplished by putting the ideas introduced in that section into the framework of the axiomatic approach. In order to do this, however, it is first necessary to review some of the elementary concepts of set theory.

A set is a collection of objects known as *elements*. It will be designated as

$$A = \{\alpha_1, \alpha_2, \ldots, \alpha_n\} \tag{1–8}$$

where the set is A and the elements are $\alpha_1, \ldots, \alpha_n$. The set A may consist of the integers from 1 to 6 so that $\alpha_1 = 1, \alpha_2 = 2, \ldots, \alpha_6 = 6$ are the elements. A subset of A is any set all of whose elements are also elements of A. $B = \{1,2,3\}$ is a subset of the set $A = \{1, 2, 3, 4, 5, 6\}$. The general notation for indicating that B is a subset of A is $B \subset A$. Note that every set is a subset of itself.

All sets of interest in probability theory have elements taken from the largest set called a *space* and designated as S. Hence, all sets will be subsets of the space S. The relation of S and its subsets to probability will become clear shortly, but in the meantime, an illustration may be helpful. Suppose that the elements of a space consist of the six faces of a die, and that these faces are designated as 1, 2, . . . ,6. Thus,

$$S = \{1, 2, 3, 4, 5, 6\}$$

There are many ways in which subsets might be formed, depending upon the number of elements belonging to each subset. In fact, if one includes the *null set* or *empty set,* which has no elements in it and is denoted by $\emptyset$, there are $64 = 2^6$ subsets and they may be denoted as

$$\emptyset, \{1\}, \ldots, \{6\}, \{1, 2\}, \{1, 3\}, \ldots, \{5, 6\}, \{1, 2, 3\}, \ldots, S$$

In general, if S contains n elements, then there are 2^n subsets. The proof of this is left as an exercise for the student.

One of the reasons for using set theory to develop probability concepts is that the important operations are already defined for sets and have simple geometric representations that aid in visualizing and understanding these operations. The geometric representation is the *Venn diagram* in which the space S is represented by a square and the various sets are represented by closed plane figures. For example, the Venn diagram shown in Figure 1–1 shows that B is a subset of A and that C is a subset of B (and also of A). The various operations are now defined and represented by Venn diagrams.

Equality.   Set A equals set B *iff* (if and only if) every element of A is an element of B and every element of B is an element of A. Thus

$$A = B \quad \textit{iff} \quad A \subset B \quad \textit{and} \quad B \subset A$$

The Venn diagram is obvious and will not be shown.

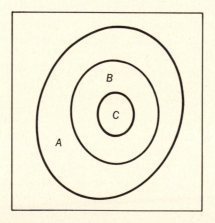

Figure 1–1.   Venn diagram for $C \subset B \subset A$.

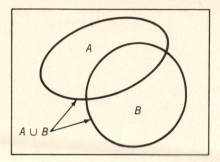

Figure 1–2. The sum of two sets, $A \cup B$.

Sums. The *sum* or *union* of two sets is a set consisting of all the elements that are elements of A or of B or of both. This is shown in Figure 1–2. Since the associative law holds, the sum of more than two sets can be written without parentheses. That is

$$(A \cup B) \cup C = A \cup (B \cup C) = A \cup B \cup C$$

The commutative law also holds, so that

$$A \cup B = B \cup A$$
$$A \cup A = A$$
$$A \cup \varnothing = A$$
$$A \cup S = S$$
$$A \cup B = A, \text{ if } B \subset A$$

Products. The *product* or *intersection* of two sets is the set consisting of all the elements that are common to both sets. It is designated as $A \cap B$ and is illustrated in Figure 1–3. A number of results apparent from the Venn diagram are

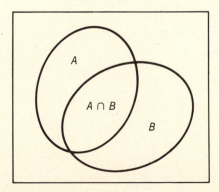

Figure 1–3. The intersection of two sets. $A \cap B$.

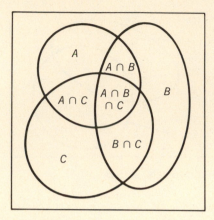

Figure 1–4. Intersections for three sets.

$$A \cap B = B \cap A \quad \text{(Commutative law)}$$
$$A \cap A = A$$
$$A \cap \varnothing = \varnothing$$
$$A \cap S = A$$
$$A \cap B = B, \text{ if } B \subset A$$

If there are more than two sets involved in the product, the Venn diagram of Figure 1–4 is appropriate. From this it is seen that

$$(A \cap B) \cap C = A \cap (B \cap C) = A \cap B \cap C$$
$$A \cap (B \cup C) = (A \cap B) \cup (A \cap C) \quad \text{(Associative law)}$$

Two sets A and B are *mutually exclusive* or *disjoint* if $A \cap B = \varnothing$. Representations of such sets in the Venn diagram do not overlap.

Complement. The complement of a set A is a set containing all the elements of S that are *not* in A. It is denoted $\bar{A}$ and is shown in Figure 1–5. It is clear that

$$\bar{\varnothing} = S$$
$$\bar{S} = \varnothing$$
$$\overline{(\bar{A})} = A$$
$$A \cup \bar{A} = S$$
$$A \cap \bar{A} = \varnothing$$
$$\bar{A} \subset \bar{B}, \qquad \text{if } B \subset A$$
$$\bar{A} = \bar{B}, \qquad \text{if } A = B$$

Two additional relations that are usually referred to as *DeMorgan's laws* are

$$\overline{(A \cup B)} = \bar{A} \cap \bar{B}$$
$$\overline{(A \cap B)} = \bar{A} \cup \bar{B}$$

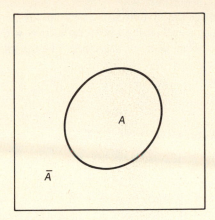

Figure 1–5. The complement of A.

Differences. The difference of two sets, $A - B$, is a set consisting of the elements of A that are not in B. This is shown in Figure 1–6. The difference may also be expressed as

$$A - B = A \cap \overline{B} = A - (A \cap B)$$

The notation $(A - B)$ is often read as "A take away B." The following results are also apparent from the Venn diagram:

$$(A - B) \cup B \neq A$$
$$(A \cup A) - A = \varnothing$$
$$A \cup (A - A) = A$$
$$A - \varnothing = A$$
$$A - S = \varnothing$$
$$S - A = \overline{A}$$

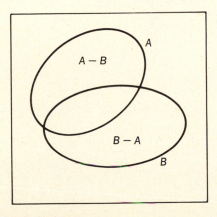

Figure 1–6. The difference of two sets.

Note that when differences are involved, the parentheses *cannot* be omitted.

It is desirable to illustrate all of the above operations with a specific example. In order to do this, let the elements of the space S be the integers from 1 to 6, as before:

$$S = \{1, 2, 3, 4, 5, 6\}$$

and define certain sets as

$$A = \{2, 4, 6\}, \quad B = \{1, 2, 3, 4\}, \quad C = \{1, 3, 5\}$$

From the definitions just presented, it is clear that

$(A \cup B) = \{1, 2, 3, 4, 6\}, \quad (B \cup C) = \{1, 2, 3, 4, 5\}$
$A \cup B \cup C = \{1, 2, 3, 4, 5, 6\} = S = A \cup C$
$A \cap B = \{2, 4\}, \quad B \cap C = \{1, 3\}, \quad A \cap C = \varnothing$
$A \cap B \cap C = \varnothing, \quad \overline{A} = \{1, 3, 5\} = C, \quad \overline{B} = \{5, 6\}$
$\overline{C} = \{2, 4, 6\} = A, \quad A - B = \{6\}, \quad B - A = \{1, 3\}$
$A - C = \{2, 4, 6\} = A, \quad C - A = \{1, 3, 5\} = C, \quad B - C = \{2, 4\}$
$C - B = \{5\}, \quad (A - B) \cup B = \{1, 2, 3, 4, 6\}$

The student should verify these results.

Exercise 1–5.1

If A and B are subsets in the same space S, find:

a) $(A - B) \cap (B - A)$

b) $(A - B) \cap \overline{B}$

c) $(A - B) \cup (A \cap B)$

Answers: A, $(A - B)$, 0

Exercise 1–5.2

A space $S = \{a, b, c, d, e, f\}$ has two subsets defined as $A = \{a, c, e\}$ and $B = \{c, d, e, f\}$. Find:

a) $A \cup B$

b) $A \cap B$

c) $(A - B)$

d) $\overline{A} \cap B$

e) $A \cap \overline{B}$

f) $(B - A) \cup A$.

Answers: $\{a, c, d, e, f\}$, $\{a\}$, $\{d, f\}$, $\{a, c, d, e, f\}$, $\{a\}$, $\{c, e\}$

1—6   The Axiomatic Approach

It is now necessary to relate probability theory to the set concepts that have just been discussed. This relationship is established by defining a *probability space* whose elements are all the outcomes (of a possible set of outcomes) from an experiment. For example, if an experimenter chooses to view the six faces of a die as the possible outcomes, then the probability space associated with throwing a die is the set

$$S = \{1, 2, 3, 4, 5, 6\}$$

The various subsets of S can be identified with the *events*. For example, in the case of throwing a die, the event $\{2\}$ corresponds to obtaining the outcome 2, while the event $\{1, 2, 3\}$ corresponds to the outcomes of either 1, or 2, or 3. Since at least one outcome must be obtained on each trial, the space S corresponds to the *certain event* and the empty set $\varnothing$ corresponds to the *impossible event*. Any event consisting of a single element is called an *elementary event*.

The next step is to assign to each event a number called, as before, the *probability of the event*. If the event is denoted as A, the probability of event A is denoted as Pr (A). This number is chosen so as to satisfy the following three conditions or *axioms*:

$$\text{Pr } (A) \geq 0 \tag{1-9}$$
$$\text{Pr } (S) = 1 \tag{1-10}$$
$$\text{If } A \cap B = \varnothing, \text{ then Pr } (A \cup B) = \text{Pr } (A) + \text{Pr } (B) \tag{1-11}$$

The whole body of probability can be deduced from these axioms. It should be emphasized, however, that axioms are postulates and, as such, it is meaningless to try to prove them. The only possible test of their validity is whether the resulting theory adequately represents the real world. The same is true of any physical theory.

A large number of corollaries can be deduced from these axioms and a few are developed here. First, since

$$S \cap \varnothing = \varnothing \quad and \quad S \cup \varnothing = S$$

it follows from (1–9) that

$$\Pr (S \cup \varnothing) = \Pr (S) = \Pr (S) + \Pr (\varnothing)$$

Hence,

$$\Pr (\varnothing) = 0 \tag{1–12}$$

Next, since

$$A \cap \overline{A} = \varnothing \quad \text{and} \quad A \cup \overline{A} = S$$

it also follows from (1–9) that

$$\Pr (A \cup \overline{A}) = \Pr (A) + \Pr (\overline{A}) = \Pr (S) = 1 \tag{1–13}$$

From this and from (1–7)

$$\Pr (A) = 1 - \Pr (\overline{A}) \leq 1 \tag{1–14}$$

Therefore, the probability of an event must be a number between 0 and 1.

If A and B are not mutually exclusive, then (1–11) usually does not hold. A more general result can be obtained, however. From the Venn diagram of Figure 1–3 it is apparent that

$$A \cup B = A \cup (\overline{A} \cap B)$$

and that A and $\overline{A} \cap B$ are mutually exclusive. Hence, from (1–11) it follows that

$$\Pr (A \cup B) = \Pr (A \cup \overline{A} \cap B) = \Pr (A) + \Pr (\overline{A} \cap B)$$

From the same figure it is also apparent that

$$B = (A \cap B) \cup (\overline{A} \cap B)$$

and that $A \cap B$ and $\overline{A} \cap B$ are mutually exclusive. From (1–9)

$$\Pr (B) = \Pr [(A \cap B) \cup (\overline{A} \cap B)] = \Pr (A \cap B) + \Pr (\overline{A} \cap B) \tag{1–15}$$

Upon eliminating $\Pr (\overline{A} \cap B)$, it follows that

$$\Pr (A \cup B) = \Pr (A) + \Pr (B) - \Pr (A \cap B) \leq \Pr (A) + \Pr (B) \tag{1–16}$$

which is the desired result.

Now that the formalism of the axiomatic approach has been established, it is desirable to look at the problem of constructing probability spaces. First consider the case of throwing a single die and the associated probability space of $S = \{1, 2, 3, 4, 5, 6\}$. The elementary events are simply the integers associated with the upper face of the die and these are clearly mutually exclusive. If the elementary events are *assumed* to be equally probable, then the probability associated with each is simply

$$\Pr \{\alpha_i\} = \frac{1}{6}, \quad \alpha_i = 1, 2, \ldots, 6$$

Note that this assumption is consistent with the relative-frequency approach, but within the framework of the axiomatic approach it is only an assumption, and any number of other assumptions could have been made.

For this same probability space, consider the event $A = \{1, 3\} = \{1\} \cup \{3\}$. From (1–11)

$$Pr\ (A) = Pr\ \{1\} + Pr\ \{3\} = \frac{1}{6} + \frac{1}{6} = \frac{1}{3}$$

and this can be interpreted as the probability of throwing *either* a 1 or a 3. A somewhat more complex situation arises when $A = \{1, 3\}$, $B = \{3, 5\}$ and it is desired to determine $Pr\ (A \cup B)$. Since A and B are not mutually exclusive, the result of (1–16) must be used. From the calculation above, it is clear that $Pr\ (A) = Pr\ (B) = \frac{1}{3}$. However, since $A \cap B = \{3\}$, an elementary event, it must be that $Pr\ (A \cap B) = \frac{1}{6}$. Hence, from (1–16)

$$Pr\ (A \cup B) = Pr\ (A) + Pr\ (B) - Pr\ (A \cap B) = \frac{1}{3} + \frac{1}{3} - \frac{1}{6} = \frac{1}{2}$$

An alternative approach is to note that $A \cup B = \{1, 3, 5\}$ which is composed of three mutually exclusive elementary events. Using (1–11) twice leads immediately to

$$Pr\ (A \cup B) = Pr\ \{1\} + Pr\ \{3\} + Pr\ \{5\} = \frac{1}{6} + \frac{1}{6} + \frac{1}{6} = \frac{1}{2}$$

Note that this can be interpreted as the probability of *either* A occurring or B occurring *or both* occurring.

Exercise 1–6.1

A dodecahedron is a solid with twelve sides and is often used to display the twelve months of the year. When this object is rolled, let the outcome be taken as the month appearing on the upper face. Also let $A = \{January\}$, $B = \{Any\ month\ with\ 30\ days\}$, and $C = \{Any\ month\ with\ 31\ days\}$. Find

   a) $Pr\ (A \cup C)$

   b) $Pr\ (A \cap C)$

c)   Pr $(B \cup C)$

d)   Pr $(A \cap B)$.

Answers: 11/12, 0, 1/12, 7/12

Exercise 1–6.2

Draw a Venn diagram showing three subsets that are *not* mutually exclusive. Using this diagram derive an expression for Pr $(A \cup B \cup C)$.

Answer: Pr (A) + Pr (B) + Pr (C) − Pr $(A \cap B)$ − Pr $(A \cap C)$ − Pr $(B \cap C)$ + Pr $(A \cap B \cap C)$

1–7   Conditional Probability

The concept of conditional probability was introduced in Section 1–3 on the basis of the relative frequency of one event when another event is specified to have occurred. In the axiomatic approach, conditional probability is a defined quantity. If an event B is assumed to have a nonzero probability, then the conditional probability of an event A, given B, is defined as

$$\text{Pr}\,(A \mid B) = \frac{\text{Pr}\,(A \cap B)}{\text{Pr}\,(B)} \qquad \text{Pr}\,(B) > 0 \tag{1–17}$$

where Pr $(A \cap B)$ is the probability of the event $A \cap B$. In the previous discussion, the numerator of (1–17) was written as Pr (A,B) and was called the joint probability of events A and B. This interpretation is still correct if A and B are elementary events, but in the more general case the proper interpretation must be based on the set theory concept of the product, $A \cap B$, of two sets. Obviously, if A and B are mutually exclusive, then $A \cap B$ is the empty set and Pr $(A \cap B)$ = 0. On the other hand, if A is contained in B (that is, $A \subset B$), then $A \cap B = A$ and

$$\text{Pr}\,(A \mid B) = \frac{\text{Pr}\,(A)}{\text{Pr}\,(B)} \geq \text{Pr}\,(A)$$

Finally, if $B \subset A$, then $A \cap B = B$ and

$$\text{Pr}\,(A \mid B) = \frac{\text{Pr}\,(B)}{\text{Pr}\,(B)} = 1$$

In general, however, when neither $A \subset B$ nor $B \subset A$, nothing can be asserted regarding the relative magnitudes of Pr (A) and Pr $(A \mid B)$.

So far it has not yet been shown that conditional probabilities are really probabilities in the sense that they satisfy the basic axioms. In the relative-frequency approach they are clearly probabilities in that they could be defined as ratios of the numbers of favorable occurrences to the total number of trials, but in the axiomatic approach conditional probabilities are defined quantities; hence, it is necessary to verify independently their validity as probabilities.

The first axiom is

$$\text{Pr } (A \mid B) \geq 0$$

and this is obviously true from the definition (1–17) since both numerator and denominator are positive numbers. The second axiom is

$$\text{Pr } (S \mid B) = 1$$

and this is also apparent since $B \subset S$ so that $S \cap B = B$ and Pr $(S \cap B) = $ Pr (B). In order to verify that the third axiom holds, consider another event, C, such that $A \cap C = \varnothing$ (that is, A and C are mutually exclusive). Then

$$\text{Pr } [(A \cup C) \cap B] = \text{Pr } [(A \cap B) \cup (C \cap B)] = \text{Pr } (A \cap B) + \text{Pr } (C \cap B)$$

since $(A \cap B)$ and $(C \cap B)$ are also mutually exclusive events and (1–11) holds for such events. So, from (1–17)

$$\text{Pr } [(A \cup C) \mid B] = \frac{\text{Pr } [(A \cup C) \cap B]}{\text{Pr } (B)} = \frac{\text{Pr } (A \cap B)}{\text{Pr } (B)} + \frac{\text{Pr } (C \cap B)}{\text{Pr } (B)}$$
$$= \text{Pr } (A \mid B) + \text{Pr } (C \mid B)$$

Thus the third axiom does hold, and it is now clear that conditional probabilities are valid probabilities in every sense.

Before extending the topic of conditional probabilities, it is desirable to consider an example in which the events are not elementary events. Let the experiment be the throwing of a single die and let the outcomes be the integers from 1 to 6. Then define event A as $A = \{1,2\}$, that is, the occurrence of a 1 or a 2. From previous considerations it is clear that Pr $(A) = \frac{1}{6} + \frac{1}{6} = \frac{1}{3}$. Define B as the event of obtaining an even number. That is, $B = \{2,4,6\}$ and Pr $(B) = \frac{1}{2}$ since it is composed of three elementary events. The event $A \cap B$ is $A \cap B = \{2\}$, from which Pr $(A \cap B) = \frac{1}{6}$. The conditional probability, Pr $(A \mid B)$, is now given by

$$\text{Pr } (A \mid B) = \frac{\text{Pr } (A \cap B)}{\text{Pr } (B)} = \frac{\dfrac{1}{6}}{\dfrac{1}{2}} = \frac{1}{3}$$

This indicates that the conditional probability of throwing a 1 or a 2, given that the outcome is even, is $\frac{1}{3}$.

On the other hand, suppose it is desired to find the conditional probability of throwing an even number given that the outcome was a 1 or a 2. This is

$$\Pr\,(B \mid A) = \frac{\Pr\,(A \cap B)}{\Pr\,(A)} = \frac{\frac{1}{6}}{\frac{1}{3}} = \frac{1}{2}$$

a result that is intuitively correct.

One of the uses of conditional probability is in the evaluation of *total probability*. Suppose there are n mutually exclusive events $A_1, A_2, \ldots, A_n$ and an arbitrary event B as shown in the Venn diagram of Figure 1–7. The events A_i occupy the entire space, S, so that

$$A_1 \cup A_2 \cup \cdots \cup A_n = S \tag{1–18}$$

Since A_i and A_j $(i \neq j)$ are mutually exclusive, it follows that $B \cap A_i$ and $B \cap A_j$ are also mutually exclusive. Further,

$$B = B \cap (A_1 \cup A_2 \cup \cdots \cup A_n) = (B \cap A_1) \cup (B \cap A_2) \cup \cdots \cup (B \cap A_n)$$

because of (1–18). Hence, from (1–11),

$$\Pr\,(B) = \Pr\,(B \cap A_1) + \Pr\,(B \cap A_2) + \cdots + \Pr\,(B \cap A_n) \tag{1–19}$$

But from (1–17)

$$\Pr\,(B \cap A_i) = \Pr\,(B \mid A_i)\,\Pr\,(A_i)$$

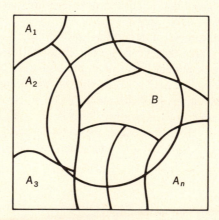

Figure 1–7.   Venn diagram for total probability.

Table 1–3. Resistance Values.

Ohms	Bin Numbers						
	1	2	3	4	5	6	Total
10 Ω	500	0	200	800	1200	1000	3700
100 Ω	300	400	600	200	800	0	2300
1000 Ω	200	600	200	600	0	1000	2600
Total	1000	1000	1000	1600	2000	2000	8600

Substituting into (1–19) yields

$$\Pr(B) = \Pr(B \mid A_1)\Pr(A_1) + \Pr(B \mid A_2)\Pr(A_2)$$
$$+ \cdots + \Pr(B \mid A_n)\Pr(A_n) \quad \text{(1–20)}$$

The quantity $\Pr(B)$ is the *total probability* and is expressed in (1–20) in terms of its various conditional probabilities.

An example serves to illustrate an application of total probability. Consider a resistor carrousel containing six bins. Each bin contains an assortment of resistors as shown in Table 1–3. If one of the bins is selected at random,[1] and a single resistor drawn from that bin at random, what is the probability that the resistor chosen will be 10 Ω? The A_i events in (1–20) can be associated with the bin chosen so that

$$\Pr(A_i) = \frac{1}{6}, \quad i = 1, 2, 3, 4, 5, 6$$

since it is *assumed* that the choices of bins are equally likely. The event B is the selection of a 10-Ω resistor and the conditional probabilities can be related to the *numbers* of such resistors in each bin. Thus

$$\Pr(B \mid A_1) = \frac{500}{1000} = \frac{1}{2} \qquad \Pr(B \mid A_2) = \frac{0}{1000} = 0$$

$$\Pr(B \mid A_3) = \frac{200}{1000} = \frac{2}{10} \qquad \Pr(B \mid A_4) = \frac{800}{1600} = \frac{1}{2}$$

$$\Pr(B \mid A_5) = \frac{1200}{2000} = \frac{6}{10} \qquad \Pr(B \mid A_6) = \frac{1000}{2000} = \frac{1}{2}$$

Hence, from (1–20) the total probability of selecting a 10-Ω resistor is

---

[1]The phrase "at random" is usually interpreted to mean "with equal probability."

$$\Pr(B) = \frac{1}{2} \times \frac{1}{6} + 0 \times \frac{1}{6} + \frac{2}{10} \times \frac{1}{6} + \frac{1}{2} \times \frac{1}{6} + \frac{6}{10} \times \frac{1}{6} + \frac{1}{2} \times \frac{1}{6}$$

$$= 0.3833$$

It is worth noting that the concepts of equally likely events and relative frequency have been used in assigning values to the conditional probabilities above, but that the basic relationships expressed by (1–20) is derived from the axiomatic approach.

The probabilities $\Pr(A_i)$ in (1–20) are often referred to as *a priori probabilities* because they are the ones that describe the probabilities of the events A_i *before* any experiment is performed. *After* an experiment is performed, and event B observed, the probabilities that describe the events A_i are the *conditional probabilities* $\Pr(A_i \mid B)$. These probabilities may be expressed in terms of those already discussed by rewriting (1–17) as

$$\Pr(A_i \cap B) = \Pr(A_i \mid B) \Pr(B) = \Pr(B \mid A_i) \Pr(A_i)$$

The last form in the above is obtained by simply interchanging the roles of B and A_i. The second equality may now be written

$$\Pr(A_i \mid B) = \frac{\Pr(B \mid A_i) \Pr(A_i)}{\Pr(B)}, \qquad \Pr(B) \neq 0 \qquad \textbf{(1–21)}$$

into which (1–20) may be substituted to yield

$$\Pr(A_i \mid B) = \frac{\Pr(B \mid A_i) \Pr(A_i)}{\Pr(B \mid A_1) \Pr(A_1) + \cdots + \Pr(B \mid A_n) \Pr(A_n)} \qquad \textbf{(1–22)}$$

The conditional probability $\Pr(A_i \mid B)$ is often called the *a posteriori probability* because it applies *after* the experiment is performed; and either (1–21) or (1–22) is referred to as *Bayes' theorem.*

The *a posteriori* probability may be illustrated by continuing the example just discussed. Suppose the resistor that is chosen from the carrousel is found to be a 10-Ω resistor. What is the probability that it came from bin three? Since B is still the event of selecting a 10-Ω resistor, the conditional probabilities $\Pr(B \mid A_i)$ are the same as tabulated before. Furthermore, the *a priori* probabilities are still $\frac{1}{6}$.

Thus, from (1–21), and the previous evaluation of $\Pr(B)$,

$$\Pr(A_3 \mid B) = \frac{\left(\frac{2}{10}\right)\left(\frac{1}{6}\right)}{0.3833} = 0.0869$$

This is the probability that the 10-Ω resistor, chosen at random, came from bin three.

Exercise 1–7.1

Using the data of Table 1–3, find the probabilities:

a)   a 1000-Ω resistor that is selected came from bin 3.

b)   a 10-Ω resistor that is selected came from bin 5.

Answers: 0.2609, 0.1067

Exercise 1–7.2

A manufacturer of electronic equipment purchases 1000 ICs from supplier A, 2000 ICs from supplier B, and 3000 ICs from supplier C. Testing reveals that the conditional probability of an IC failing during burn-in is, for devices from each of the suppliers

$$\Pr (F \mid A) = 0.1, \Pr (F \mid B) = 0.05, \Pr (F \mid C) = 0.08$$

The ICs from all suppliers are mixed together and one device is selected at random.

a)   What is the probability that it will fail during burn-in?

b)   Given that the device fails, what is the probability that the device came from supplier A?

Answers: 0.0733, 0.2273

1–8   Independence

The concept of statistical independence is a very important one in probability. It was introduced in connection with the relative-frequency approach by considering two trials of an experiment, such as tossing a coin, in which it is clear that the second trial cannot depend upon the outcome of the first trial in any way. Now that a more general formulation of events is available, this concept can be extended. The basic definition is unchanged, however. It is

Two events, A and B, are independent if and only if

$$\Pr (A \cap B) = \Pr (A) \Pr (B) \tag{1–23}$$

In many physical situations, independence of events is *assumed* because there is no apparent physical mechanism by which one event can depend upon the other. In other cases, the assumed probabilities of the elementary events lead to independence of other events defined from these. In such cases, independence may not be obvious, but can be established from (1–23).

The concept of independence can also be extended to more than two events. For example, with three events, the conditions for independence are

$$\Pr\,(A_1 \cap A_2) = \Pr\,(A_1)\,\Pr\,(A_2) \qquad \Pr\,(A_1 \cap A_3) = \Pr\,(A_1)\,\Pr\,(A_3)$$

$$\Pr\,(A_2 \cap A_3) = \Pr\,(A_2)\,\Pr\,(A_3) \quad \Pr\,(A_1 \cap A_2 \cap A_3) = \Pr\,(A_1)\,\Pr\,(A_2)\,\Pr\,(A_3)$$

Note that *four* conditions must be satisfied, and that pairwise independence is *not sufficient* for the entire set of events to be mutually independent. In general, if there are n events, it is necessary that

$$\Pr\,(A_i \cap A_j \cap \cdots \cap A_k) = \Pr\,(A_i)\,\Pr\,(A_j) \cdots \Pr\,(A_k) \qquad \textbf{(1–24)}$$

for every set of integers less than or equal to n. This implies that $2^n - (n + 1)$ equations of the form (1–24) are required to establish the independence of n events.

One important consequence of independence is a special form of (1–16), which stated

$$\Pr\,(A \cup B) = \Pr\,(A) + \Pr\,(B) - \Pr\,(A \cap B)$$

If A and B are independent events, this becomes

$$\Pr\,(A \cup B) = \Pr\,(A) + \Pr\,(B) - \Pr\,(A)\,\Pr\,(B) \qquad \textbf{(1–25)}$$

Another result of independence is

$$\Pr\,[A_1 \cap (A_2 \cup A_3)] = \Pr\,(A_1)\,\Pr\,(A_2 \cup A_3) \qquad \textbf{(1–26)}$$

if A_1, A_2, and A_3 are all independent. This is *not* true if they are only independent in pairs. In general, if A_1, A_2, . . . , A_n are independent events, then any one of them is independent of any event formed by sums, products, and complements of the others.

Examples of physical situations that illustrate independence are most often associated with two or more trials of an experiment. However, for purposes of illustration, consider two events associated with a single experiment. Let the experiment be that of rolling a pair of dice and define event A as that of obtaining a 7 and event B as that of obtaining an 11. Are these events independent? The answer is that they cannot be independent because they are mutually exclusive— if one occurs the other one cannot. Mutually exclusive events can never be statistically independent.

As a second example consider two events that are not mutually exclusive. For the pair of dice above, define event A as that of obtaining an odd number and event B as that of obtaining an 11. The event $A \cap B$ is just B since B is a subset of A. Hence, the Pr $(A \cap B)$ = Pr (B) = Pr (11) = $2/36$ = $1/18$ since there are two ways an 11 can be obtained (that is, a 5 and a 6 or a 6 and a 5). Also the Pr (A) = $1/2$ since half of all outcomes are odd. It follows then that

$$\text{Pr } (A \cap B) = 1/18 \neq \text{Pr } (A)\text{Pr } (B) = (1/2) \cdot (1/18) = 1/36$$

Thus, events A and B are *not* statistically independent. That this must be the case is obvious since if B occurs then A must also occur, although the converse is not true.

It is also possible to define events associated with a single trial that are independent, but these sets may not represent any physical situation. For example, consider throwing a single die and define two events as A = $\{1, 2, 3\}$ and B = $\{3, 4\}$. From previous results it is clear that Pr (A) = $\frac{1}{2}$ and Pr (B) = $\frac{1}{3}$. The event $(A \cap B)$ contains a single element $\{3\}$; hence, Pr $(A \cap B)$ = $\frac{1}{6}$. Thus, it follows that

$$\text{Pr } (A \cap B) = \frac{1}{6} = \text{Pr } (A) \text{ Pr } (B) = \frac{1}{2} \cdot \frac{1}{3} = \frac{1}{6}$$

and events A and B are independent, although the physical significance of this is not intuitively clear. The next section considers situations in which there is more than one experiment, or more than one trial of a given experiment, and that discussion will help clarify the matter.

Exercise 1–8.1

A card is selected at random from a standard deck of 52 cards. Let A be the event of selecting an ace, and let B be the event of selecting a spade. Are these events statistically independent? Prove your answer.

Answer: Yes

Exercise 1–8.2

In the switching circuit shown below, the switches are assumed to operate randomly and independently.

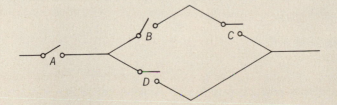

If *each* switch has a probability of 0.2 of being closed, find the probability that there is a complete path through the circuit.

Answer: 0.0464

1—9   Combined Experiments

In the discussion of probability presented thus far, the probability space, S, was associated with a single experiment. This concept is too restrictive to deal with many realistic situations, so it is necessary to generalize it somewhat. Consider a situation in which *two* experiments are performed. For example, one experiment might be throwing a die and the other one tossing a coin. It is then desired to find the probability that the outcome is, say, a "3" on the die and a "tail" on the coin. In other situations the second experiment might be simply a repeated trial of the first experiment. The two experiments, taken together, form a *combined experiment*, and it is now necessary to find the appropriate probability space for it.

Let one experiment have a space S_1 and the other experiment a space S_2. Designate the elements of S_1 as

$$S_1 = \{\alpha_1, \alpha_2, \ldots, \alpha_n\}$$

and those of S_2 as

$$S_2 = \{\beta_1, \beta_2, \ldots, \beta_m\}$$

Then form a new space, called the *cartesian product space*, whose elements are all the ordered pairs (α_1, β_1), (α_1, β_2), . . . , (α_i, β_j), . . . , (α_n, β_m). Thus, if S_1 has n elements and S_2 has m elements, the cartesian product space has mn elements. The cartesian product space may be denoted as

$$S = S_1 \times S_2$$

to distinguish it from the previous product or intersection discussed in Section 1–5.

As an illustration of the cartesian product space for combined experiments, consider the die and the coin discussed above. For the die the space is

$$S_1 = \{1, 2, 3, 4, 5, 6\}$$

while for the coin it is

$$S_2 = \{H, T\}$$

Thus, the cartesian product space has 12 elements and is

$$S = S_1 \times S_2 = \{(1, H), (1, T), (2, H), (2, T), (3, H), (3, T), (4, H),$$

$$(4, T), (5, H), (5, T), (6, H), (6, T)\}$$

It is now necessary to define the events of the new probability space. If A_1 is a subset considered to be an event in S_1, and A_2 is a subset considered to be an event in S_2, then $A = A_1 \times A_2$ is an event in S. For example, in the above illustration let $A_1 = \{1,3,5\}$ and $A_2 = \{H\}$. The event A corresponding to these is

$$A = A_1 \times A_2 = \{(1, H), (3, H), (5, H)\}$$

In order to specify the probability of event A, it is necessary to consider whether the two experiments are independent; the only cases discussed here are those in which they are independent. In such cases the probability in the product space is simply the products of the probabilities in the original spaces. Thus, if Pr (A_1) is the probability of event A_1 in space S_1, and Pr (A_2) is the probability of A_2 in space S_2, then the probability of event A in space S is

$$\text{Pr } (A) = \text{Pr } (A_1 \times A_2) = \text{Pr } (A_1) \text{ Pr } (A_2) \qquad (1\text{--}27)$$

This result may be illustrated by data from the above example. From previous results, Pr $(A_1) = \frac{1}{6} + \frac{1}{6} + \frac{1}{6} = \frac{1}{2}$ when $A_1 = \{1,3,5\}$ and Pr $(A_2) = \frac{1}{2}$ when $A_2 = \{H\}$. Thus, the probability of getting an *odd number* on the die and a *head* on the coin is

$$\text{Pr } (A) = \left(\frac{1}{2}\right)\left(\frac{1}{2}\right) = \frac{1}{4}$$

It is possible to generalize the above ideas in a straightforward manner to situations in which there are more than two experiments. However, this will be done only for the more specialized situation of repeating the same experiment an arbitrary number of times.

Exercise 1–9.1

A combined experiment is performed by rolling a die with sides numbered from 1 to 6 and a child's block with sides labeled A through F.

a) Write all of the elements in the cartesian product space.

b) Define K as the event of obtaining an even number on the die and a letter of B or C on the block and find the probability of the event K.

Answer: 1/6

Exercise 1–9.2

A combined experiment is performed by flipping a coin three times.

a) Write all of the elements in the product space by indicating them as *HHH, HTH, TTH*, etc.

b) Find the probability of obtaining exactly two heads.

c) Find the probability of obtaining more than one head.

Answers: 1/2, 3/8

1–10   Bernoulli Trials

The situation considered here is one in which the same experiment is repeated n times and it is desired to find the probability that a particular event occurs exactly k of these times. For example, what is the probability that exactly four heads will be observed when a coin is tossed ten times? Such repeated experiments are referred to as *Bernoulli trials*.

Consider some experiment for which the event A has a probability $\Pr(A) = p$. Hence, the probability that the event does not occur is $\Pr(\overline{A}) = q$, where $p + q = 1$.[2] Then repeat this experiment n times and assume that the trials are independent; that is, that the outcome of any one trial does not depend in any way upon the outcomes of any previous (or future) trials. Next determine the probability that event A occurs exactly k times in some specific order, say in the first k trials and none thereafter. Because the trials are independent, the probability of this event is

$$\underbrace{\Pr(A)\,\Pr(A)\cdots\Pr(A)}_{k \text{ of these}}\ \underbrace{\Pr(\overline{A})\,\Pr(\overline{A})\cdots\Pr(\overline{A})}_{n-k \text{ of these}} = p^k q^{n-k}$$

However, there are many other ways in which exactly k events could occur because they can arise in any order. Furthermore, because of the independence, all of these other orders have exactly the same probability as the one specified above. Hence, the event that A occurs k times in any order is the sum of the mutually exclusive events that A occurs k times in some specific order, and thus, the probability that A occurs k times is simply the above probability for a particular order multiplied by the number of different orders that can occur.

---

[2]The only justification for changing the notation from $\Pr(A)$ to p and from $\Pr(\overline{A})$ to q is that the p and q notation is traditional in discussing Bernoulli trials and most of the literature uses it.

It is necessary to digress at this point and briefly discuss the theory of combinations in order to be able to determine the number of different orders in which the event A can occur exactly k times in n trials. It is apparent that when one forms a sequence of length n, the first A can go in any one of the n places, the second A can go into any one of the remaining n-1 places, and so on, leaving $n - k + 1$ places for the kth A. Thus, the total number of different sequences of length n containing exactly k As is simply the product of these various possibilities. Thus, since the $k!$ orders of the k event places are identical

$$\frac{1}{k!}[n(n-1)(n-2)\ldots(n-k+1)] = \frac{n!}{k!(n-k)!} \qquad (1\text{--}28)$$

The quantity on the right is simply the *binomial coefficient*, which is usually denoted either as $_nC_k$ or as $\binom{n}{k}$.[3] The latter notation is employed here.

As an example of binomial coefficients, let $n = 4$ and $k = 2$. Then

$$\binom{n}{k} = \frac{4!}{2!2!} = 6$$

and there are six different sequences in which the event A occurs exactly twice. These can be enumerated easily as

$$AA\overline{AA}, \; A\overline{A}A\overline{A}, \; A\overline{AA}A, \; \overline{A}AA\overline{A}, \; \overline{A}A\overline{A}A, \; \overline{AA}AA$$

It is now possible to write the desired probability of A occurring k times as

$$p_n(k) = \text{Pr}\,\{A \text{ occurs } k \text{ times}\} = \binom{n}{k}p^k q^{n-k} \qquad (1\text{--}29)$$

As an illustration of a possible application of this result, consider a digital computer in which the binary digits (0 or 1) are organized into "words" of 32 digits each. If there is a probability of 10^{-3} that any one binary digit is incorrectly read, what is the probability that there is one error in an entire word? For this case, $n = 32$, $k = 1$, and $p = 10^{-3}$. Hence,

$$\text{Pr}\,\{\text{one error in a word}\} = p_{32}(1) = \binom{32}{1}(10^{-3})^1(0.999)^{31}$$

$$= 32(0.999)^{31}(10^{-3}) \simeq 0.031$$

It is also possible to use (1–29) to find the probability that there will be *no* error in a word. For this, $k = 0$ and $\binom{n}{0} = 1$. Thus,

---

[3] A table of binomial coefficients is given in Appendix C.

$$\text{Pr \{no error in a word\}} = p_{32}(0) = \binom{32}{0} (10^{-3})^{0}(0.999)^{32}$$

$$= (0.999)^{32} \simeq 0.9685$$

There are many other practical applications of Bernoulli trials. For example, if a system has n components and there is a probability p that any one of them will fail, the probability that one and only one component will fail is

$$\text{Pr \{one failure\}} = p_{n}(1) = \binom{n}{1} pq^{(n-1)}$$

In some cases, one may be interested in determining the probability that event A occurs at least k times, or the probability that it occurs no more than k times. These probabilities may be obtained by simply adding the probabilities of all the outcomes that are included in the desired event. For example, if a coin is tossed four times, what is the probability of obtaining at least two heads? For this case, $p = q = \frac{1}{2}$ and $n = 4$. From (1–29) the probability of getting two heads (that is, $k = 2$) is

$$p_{4}(2) = \binom{4}{2} \left(\frac{1}{2}\right)^{2}\left(\frac{1}{2}\right)^{2} = (6) \left(\frac{1}{4}\right)\left(\frac{1}{4}\right) = \frac{3}{8}$$

Similarly, the probability of three heads is

$$p_{4}(3) = \binom{4}{3} \left(\frac{1}{2}\right)^{3}\left(\frac{1}{2}\right)^{1} = (4) \left(\frac{1}{8}\right)\left(\frac{1}{2}\right) = \frac{1}{4}$$

and the probability of four heads is

$$p_{4}(4) = \binom{4}{4} \left(\frac{1}{2}\right)^{4}\left(\frac{1}{2}\right)^{0} = (1) \left(\frac{1}{16}\right) (1) = \left(\frac{1}{16}\right)$$

Hence, the probability of getting at least two heads is

$$\text{Pr \{at least two heads\}} = p_{4}(2) + p_{4}(3) + p_{4}(4) = \frac{3}{8} + \frac{1}{4} + \frac{1}{16} = \frac{11}{16}$$

The general formulation of problems of this kind can be expressed quite easily, but there are several different situations that arise. These may be tabulated as follows:

$$\text{Pr \{}A \text{ occurs } \textit{less} \text{ than } k \text{ times in } n \text{ trials\}} = \sum_{i=0}^{k-1} p_{n}(i)$$

$$\text{Pr \{}A \text{ occurs } \textit{more} \text{ than } k \text{ times in } n \text{ trials\}} = \sum_{i=k+1}^{n} p_{n}(i)$$

$$\text{Pr \{}A \text{ occurs } \textit{no more} \text{ than } k \text{ times in } n \text{ trials\}} = \sum_{i=0}^{k} p_{n}(i)$$

$$\text{Pr \{}A \text{ occurs } \textit{at least} \text{ } k \text{ times in } n \text{ trials\}} = \sum_{i=k}^{n} p_{n}(i)$$

A final comment in regard to Bernoulli trials has to do with evaluating $p_n(k)$ when n is large. Since the binomial coefficients and the large powers of p and q become difficult to evaluate in such cases, often it is necessary to seek simpler, but approximate, ways of carrying out the calculation. One such approximation, known as the *DeMoivre-Laplace theorem*, is useful if $npq \gg 1$ and if $|k - np|$ is on the order of or less than $\sqrt{npq}$. This approximation is

$$p_n(k) = \binom{n}{k} p^k q^{n-k} \simeq \frac{1}{\sqrt{2\pi npq}} e^{-(k-np)^2/2npq} \tag{1-30}$$

The DeMoivre-Laplace theorem has additional significance when continuous probability is considered in a subsequent chapter. However, a simple illustration of its utility in discrete probability is worthwhile. Suppose a coin is tossed 100 times and it is desired to find the probability of k heads, where k is in the vicinity of 50. Since $p = q = \frac{1}{2}$ and $n = 100$, (1–30) yields

$$p_n(k) \simeq \frac{1}{\sqrt{50\pi}} e^{-(k-50)^2/50}$$

for k values ranging (roughly) from 40 to 60. This is obviously much easier to evaluate than trying to find the binomial coefficient $\binom{100}{k}$ for the same range of k values.

Exercise 1–10.1

A pair of dice are tossed 8 times.

a) Find the probability that a 7 will occur exactly 4 times.

b) Find the probability that an 11 will occur 2 times.

c) Find the probability that a 12 will occur more than once.

Hint: Subtract the probability of a 12 occurring once or not at all from 1.0.

Answers: 0.0613, 0.0193, 0.02605

Exercise 1–10.2

A file containing 8000 characters is to be transferred from one computer to another. The probability of any one character being transferred in error is 0.001.

a) Find the probability that the file can be transferred without any errors.

b) Using the DeMoivre-Laplace theorem, find the probability that there will be exactly 10 errors in the transferred file.

c) What must the probability of error in transferring one character be in order to make the probability of transferring the entire file without error as large as .99?

Answers: 3.341×10^{-4}, 0.1099, 1.256×10^{-6}

PROBLEMS

Note that the first two digits of each problem number correspond to the section number in which the appropriate material is discussed.

1–1.1 A 6-cell storage battery having a nominal terminal voltage of 12 V is connected in series with an ammeter and a resistor labeled 6 Ω.

a) List as many random quantities as you can for this circuit.

b) If the battery voltage can have any value between 10.5 and 12.5, the resistor can have any value within 5% of its marked value, and the ammeter reads within 2% of the true current, find the range of possible ammeter readings. Neglect ammeter resistance.

c) List any nonrandom quantities you can for this circuit.

1–1.2 In determining the probability characteristics of printed English, it is common to consider a 27-letter alphabet in which the space between words is counted as a letter. Punctuation is usually ignored.

a) Count the number of times each of the 27 letters appears in this problem.

b) On the basis of this count, deduce the most probable letter, the next most probable letter, and the least probable letter (or letters).

1–2.1 For each of the following random experiments, list all of the possible outcomes and state whether these outcomes are equally likely.

a) Flipping two coins.

b) Observing the last digit of a telephone number selected at random from the directory.

c) Observing the *sum* of the last two digits of a telephone number selected at random from the directory.

1–2.2 State whether each of the following defined events is an elementary event or not.

a) Obtaining a seven when a pair of dice are rolled.

b) Obtaining two heads when three coins are flipped.

c) Obtaining an ace when a card is selected at random from a deck of cards.

d) Obtaining a two of spades when a card is selected at random from a deck of cards.

e) Obtaining a two when a pair of dice are rolled.

f) Obtaining three heads when three coins are flipped.

g) Observing a value less than ten when a random voltage is observed.

h) Observing the letter *e* sixteen times in a piece of text.

1–4.1 If a die is rolled, determine the probability of each of the following events:

a) Obtaining the number 5.

b) Obtaining a number greater than 3.

c) Obtaining an even number.

1–4.2 If a pair of dice are rolled, determine the probability of each of the following events:

a) Obtaining a sum of 11.

b) Obtaining a sum less than 5.

c) Obtaining a sum that is an even number.

1–4.3 A box of unmarked IC's contains 200 hex inverters, 100 dual 4-input positive-AND gates, 50 dual J-K flip flops, 25 decade counters, and 25 4-bit shift registers.

a) If an IC is selected at random, what is the probability that it is a dual J-K flip flop?

b) What is the probability that an IC selected at random is *not* a hex inverter?

c) If the first IC selected is found to be a 4-bit shift register, what is the probability that the second IC selected will also be a 4-bit shift register?

1–4.4 In the IC box of Problem 1–4.3 it is known that 10% of the hex inverters are bad, 15% of the dual 4-input positive-AND gates are bad, 18% of the dual J-K flip flops are bad, and 20% of the decade counters and 4-bit shift registers are bad.

a) If an IC is selected at random, what is the probability that it is both a decade counter and good?

b) If an IC is selected at random and found to be a J-K flip flop, what is the probability that it is good?

c) If an IC is selected at random and found to be good, what is the probability that it is a decade counter?

1–4.5 A company manufactures small electric motors having horse power ratings of 0.1, 0.5 or 1.0 horsepower and designed for operation with 120 V single-phase ac, 240 V single-phase ac or 240 V three-phase ac. The motor types can be distinguished only by their nameplates. A distributor has on hand 3000 motors in the quantities shown in the table below.

Horsepower	120 V ac	240 V ac	240 V 30
0.1	900	400	0
0.5	200	500	100
1.0	100	200	600

One motor is discovered without a nameplate. For this motor determine the probability of each of the following events.

a) The motor has a horsepower rating of 0.5 hp.

b) The motor is designed for 240 V single-phase operation.

c) The motor is 1.0 hp and is designed for 240 V three-phase operation.

d) The motor is 0.1 hp and is designed for 120 V operation.

1–4.6   In Problem 1–4.5, assume that 10% of the motors labeled 120 V single-phase are mismarked and that 5% of the motors marked 240 V single-phase are mismarked.

   a)   If a motor is selected at random, what is the probability that it is mismarked?

   b)   If a motor is picked at random from those marked 240 V single-phase, what is the probability that it is mismarked?

   c)   What is the probability that a motor selected at random is 0.5 hp and mismarked?

1–5.1   Prove that a space S containing n elements has 2^n subsets. Hint: Use the binomial expansion for $(1 + x)^n$.

1–5.2   A space S is defined as

$$S = \{1, 3, 5, 7, 9, 11\}$$

and three subsets as
$A = \{1, 3, 5\}, B = \{7, 9, 11\}, C = \{1, 3, 9, 11\}$
Find:

$A \cup B$	$A \cap B \cap C$	$\overline{(B \cap C)}$
$B \cup C$	$\overline{A}$	$A\text{-}C$
$A \cup C$	$\overline{B}$	$C\text{-}A$
$A \cap B$	$\overline{C}$	$A\text{-}B$
$A \cap C$	$\overline{A} \cap B$	$(A\text{-}B) \cup B$
$B \cap C$	$\overline{A \cap B}$	$(A\text{-}B) \cup C$

1–5.3   Draw and label the Venn diagram for Problem 1–4.4.

1–5.4   Using the algebra of sets show that the following relations are true.

   a) $A \cup (A \cap B) = A$

   b) $A \cup (B \cap C) = (A \cup B) \cap (A \cup C)$

   c) $A \cup (\overline{A} \cap B) = A \cup B$

   d) $(A \cap B) \cup (A \cap \overline{B}) \cup (\overline{A} \cap B) = A$

1–6.1   For the space and subspaces defined in Problem 1–5.2, assume that each element has a probability of 1/6. Find the following probabilities.

   a)   Pr (A) b)   Pr (B) c)   Pr (C)

   d)   Pr $(A \cup B)$ e)   Pr $(A \cup C)$ f)   Pr $[(A - C) \cup B]$

1–6.2   A card is drawn at random from a standard deck of 52 cards. Let A be the event that a king is drawn, B the event that a spade is drawn, and C the event that a ten of spades is drawn. Describe each of the events listed below and calculate its probability.

   a) $A \cup B$ b) $A \cap B$ c) $A \cup \bar{B}$

   d) $A \cup C$ e) $B \cup C$ f) $A \cap C$

   g) $B \cap C$ h) $(A \cap B) \cup \bar{C}$ i) $A \cap B \cap C$

1–6.3   An experiment consists of randomly drawing three cards in succession from a standard deck of 52 cards. Let A be the event of a king on the first draw, B the event of a king on the second draw, and C the event of a king on the third draw. Describe each of the events listed below and calculate its probability.

   a) $A \cap \bar{B}$ b) $A \cup B$ c) $\bar{A} \cup \bar{B}$

   d) $\bar{A} \cap \bar{B} \cap \bar{C}$ e) $(A \cap B) \cup (\bar{B} \cap C)$ f) $\bar{A} \cup B \cup C$

1–6.4   Prove that $\Pr(\bar{A} \cup \bar{B}) = 1 - \Pr(A \cap B)$.

1–6.5   Two solid-state diodes are connected in series. Each diode has a probability of 0.05 that it will fail as a short circuit and a probability of 0.1 that it will fail as an open circuit. If the diodes are independent, what is the probability that the series connection of diodes will function as a diode?

1–7.1   In a digital communication system, messages are encoded into the binary symbols 0 and 1. Because of noise in the system, the incorrect symbol is sometimes received. Suppose that the probability of a 0 being transmitted is 0.4 and the probability of a 1 being transmitted is 0.6. Further suppose that the probability of a transmitted 0 being received as a 1 is 0.08 and the probability of a transmitted 1 being received as a 0 is 0.05. Find:

   a)   The probability that a received 0 was transmitted as a 0.

   b)   The probability that a received 1 was transmitted as a 1.

   c)   The probability that any symbol is received in error.

1–7.2   A certain typist sometimes makes mistakes by hitting a key to the right or left of the intended key, each with a probability of 0.02. The letters

E, R, and T are adjacent to one another on the standard QWERTY keyboard, and in English they occur with probabilities of Pr (E) = 0.1031, Pr (R) = 0.0484 and Pr (T) = 0.0796.

a)   What is the probability with which the letter R appears in text typed by this typist?

b)   What is the probability that a letter R appearing in text typed by this typist will be in error?

1–7.3   A candy machine has ten buttons of which one never works, two work one-half the time, and the rest work all the time. A coin is inserted and a button is pushed at random.

a)   What is the probability that no candy is received?

b)   If no candy is received, what is the probability that the button that never works was the one pushed?

c)   If candy is received, what is the probability that one of the buttons that work one-half the time was the one pushed?

1–7.4   A fair coin is tossed. If it comes up heads, a single die is rolled. If it comes up tails, two dice are rolled. Given that the outcome of the dice is 3, but you do not know whether one or two dice were rolled, what is the probability that the coin came up heads?

1–7.5   A communication network has five links as shown below.

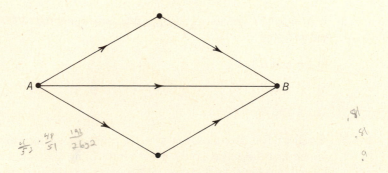

The probability that each link is working is 0.9. What is the probability of being able to transmit a message from point A to point B?

1–7.6   A manufacturer buys components in equal amounts from three different suppliers. The probability that components from supplier A are bad is 0.1, that components from supplier B are bad is 0.15, and that components from supplier C are bad is 0.05. Find:

a)   The probability that a component selected at random will be bad.

b)   If a component is found to be bad, what is the probability that it came from supplier B?

1–7.7   An electronics hobbyist has three electronic parts cabinets with two drawers each. One cabinet has NPN transistors in each drawer, while a second cabinet has PNP transistors in each drawer. The third cabinet has NPN transistors in one drawer and PNP transistors in the other drawer. The hobbyist selects one cabinet at random and withdraws a transistor from one of the drawers.

a)   What is the probability that an NPN transistor will be selected?

b)   Given that the hobbyist selects an NPN transistor, what is the probability that it came from the cabinet that contains both types?

c)   Given that an NPN transistor is selected what is the probability that it comes from the cabinet that contains only NPN transistors?

1–7.8   If the Pr (A) > Pr (B), show that Pr $(A \mid B)$ > Pr $(B \mid A)$.

1–8.1   When a pair of dice are rolled, let A be the event of obtaining a number of 6 or greater and let B be the event of obtaining a number of 6 or less. Are events A and B dependent or independent?

1–8.2   If A, B, and C are independent events, prove that the following are also independent:

a) A and $B \cup C$.

b) A and $B \cap C$.

c) A and $B - C$.

1–8.3   A pair of dice are rolled. Let A be the event of obtaining an odd number on the first die and B be the event of obtaining and odd number on the second die. Let C be the event of obtaining an odd total from both dice.

a)   Show that A and B are independent, that A and C are independent, and that B and C are independent.

b)   Show that A, B, and C are not mutually independent.

1–8.4   If A is independent of B, prove that:

    a) A is independent of $\bar{B}$.

    b) $\bar{A}$ is independent of $\bar{B}$.

1–9.1   A combined experiment is performed in which two coins are flipped and a single die is rolled. The outcomes from flipping the coins are taken to be *HH, TT,* and *HT* (which is taken to be a single outcome regardless of which coin is heads and which one is tails). The outcomes from rolling the die are the integers from one to six.

    a)   Write all of the elements in the cartesian product space.

    b)   Let A be the event of obtaining two heads *and* a number of 3 or less. Find the probability of A.

1–9.2   An electronic manufacturer uses four different types of ICs in manufacturing a particular device. The NAND gates (designated as G if good and $\bar{G}$ if bad) have a probability of 0.05 of being bad. The flip flops (F and $\bar{F}$) have a probability of 0.1 of being bad, the counters (C and $\bar{C}$) have a probability of 0.03 of being bad, and the shift registers (S and $\bar{S}$) have a probability of 0.12 of being bad.

    a)   Write all of the elements in the product space.

    b)   Determine the probability that the manufactured device will work.

    c)   If a particular device does not work, determine the probability that only the flip flops are bad.

    d)   If a particular device does not work, determine the probability that both the flip flops and the counters are bad.

1–10.1   Two men each flip a coin three times.

    a)   What is the probability that both men will get exactly two heads each?

    b)   What is the probability that one man will get no heads and the other man will get three heads?

1–10.2   In playing an opponent of equal ability, which is more probable:

    a)   To win 4 games out of 7, or to win 5 games out of 9?

    b)   To win at least 4 games out of 7, or to win at least 5 games out of 9?

1–10.3   Prove that $_nC_r$ is equal to $_{(n-1)}C_r + {}_{(n-1)}C_{(r-1)}$.

1–10.4   A football receiver, Harvey Gladiator, is able to catch two-thirds of the passes thrown to him. He must catch three passes for his team to win the game. The quarterback throws the ball to Harvey four times.

    a)   Find the probability that Harvey will drop the ball all four times.

    b)   Find the probability that Harvey will win the game.

1–10.5   Out of a group of seven EEs and five MEs, a committee consisting of three EEs and two MEs is to be formed. In how many ways can this be done if:

    a)   Any EE and any ME can be included?

    b)   One particular EE must be on the committee?

    c)   Two particular MEs cannot be on the committee?

1–10.6   In the digital communication system of Problem 1–7.1, assume that the event of an error occurring in one binary symbol is statistically independent of the event of an error occurring in any other binary symbol. Find:

    a)   The probability of receiving six successive symbols without error.

    b)   The probability of receiving six successive symbols with exactly one error.

    c)   The probability of receiving six successive symbols with more than one error.

    d)   The probability of receiving six successive symbols with one or more errors.

1–10.7   A multichannel microwave link is to provide telephone communication to a remote community having 12 subscribers, each of whom uses the link 20 percent of the time during peak hours. How many channels are needed to make the link available during peak hours to:

    a)   Eighty percent of the subscribers all of the time?

    b)   All of the subscribers 80 percent of the time?

    c)   All of the subscribers 95 percent of the time?

1–10.8   A manufacturer of electronic equipment buys 1000 ICs for which the probability of any one IC being bad is 0.01.

a) What is the probability that exactly 10 of the ICs are bad?

b) What is the probability that none of the ICs are bad?

c) What is the probability that exactly 1 of the ICs is bad?

References

All of the following texts provide coverage of the topics discussed in Chapter 1. Particularly useful and readily understandable discussions are contained in Beckmann, Drake, Gnedenko and Khinchin, Lanning and Battin, and Parzen.

Beckmann, P., *Elements of Applied Probability Theory*. New York: Harcourt, Brace and World, Inc., 1968.

> This book provides coverage of much of the material discussed in the first six chapters of the present text. The mathematical level is essentially the same as the present text but the point of view is often different, thereby providing useful amplifications or extensions of the concepts being considered. A number of interesting examples are worked out in the text.

Clarke, A. B., and R. L. Disney, *Probability and Random Processes for Engineers and Scientists*. New York: John Wiley and Sons, Inc., 1970.

> This is an undergraduate text intended for students in engineering and science. Its mathematical level is somewhat higher than that of the present text and the topical coverage is more restricted. The coverage of probability and random processes is thorough and accurate, but there is no discussion of the application of these concepts to system analysis. There is an extensive treatment of Markov processes and queueing theory, topics not usually found in an undergraduate text.

Davenport, W. B., Jr., and W. L. Root, *Introduction to Random Signals and Noise*. New York: McGraw-Hill, Inc., 1958.

> This is a graduate level text dealing with the application of probabilistic methods to the analysis of communication systems. The treatment is at an appreciably more advanced mathematical level than the present text and will require some diligent effort on the part of an undergraduate wishing to read it. However, the effort will be amply rewarded as this is the classic book in its field and is the most frequently quoted reference.

Drake, A. W., *Fundamentals of Applied Probability Theory*. New York: McGraw-Hill, Inc., 1967.

> This undergraduate text covers the elementary aspects of probability theory in a clear and readable fashion. The material relates directly to Chapters 1, 2 and 3 of the present text. Of particular interest is the use of exponential (Fourier) transforms of the probability density functions of continuous random variables and Z-transforms of the probability density functions of discrete random variables in place of the classical characteristic function procedure.

Gnedenko, B. Y. and A. Ya. Khinchin, *An Elementary Introduction to the Theory of Probability*. New York: Dover Publications, Inc., 1962.

> This small paperback book was written by two outstanding Russian mathematicians for use in high schools. It provides a very clear and easily understood introduction to many of the basic

concepts of probability that are discussed in Chapters 1 and 2 of the present text. The mathematical level is quite low and, in fact, does not go beyond simple algebra. Nevertheless, the subject matter is of fundamental importance and much useful insight into probability theory can be obtained from a study of this book.

Helstrom, C. W., *Probability and Stochastic Processes for Engineers*. New York: Macmillan, Inc., 1984.

This is an up-to-date undergraduate text written expressly for engineering students. Although somewhat more mathematical than the present text, it is straightforward and easy to read. The book emphasizes probability, random variables, and random processes, but contains very little on the application of these concepts to system analysis. A great many excellent problems are included.

Lanning, J. H., Jr. and R. H. Battin, *Random Processes in Automatic Control*. New York: McGraw-Hill, Inc., 1956.

This book is a graduate level text in the field of automatic control. However, the first half of the book provides a particularly clear and understandable treatment of probability and random processes at a level that is readily understandable by juniors or seniors in electrical engineering. A number of topics, such as random processes, are treated in greater detail than in the present text. This reference contains material relating to virtually all of the topics covered in the present text although some of the applications considered in later chapters involve more advanced mathematical concepts.

Papoulis, A., *Probability, Random Variables, and Stochastic Processes*, 2nd ed. New York: McGraw-Hill, Inc., 1984.

This is a widely used graduate level text aimed at electrical engineering applications of probability theory. Virtually all of the topics covered in the present text plus a great many more are included. The treatment is appreciably more abstract and mathematical than the present text, but a wide range of useful examples and results are given. This book provides the most readily available source for many of these results.

Parzen, E., *Modern Probability Theory and its Applications*. New York: John Wiley and Sons, Inc., 1960.

This is a standard undergraduate text on the mathematical theory of probability. The material is clearly presented and many interesting applications of probability are considered in the examples and problems.

Peebles, P. Z., *Probability, Random Variables, and Random Signal Principles*. New York: McGraw-Hill, Inc., 1980.

An undergraduate text that covers essentially the same topics as the present text, although at a slightly lower mathematical level and with fewer applications discussed. It does contain many excellent problems.

Spiegel. M. R., *Theory and Problems of Probability and Statistics*. Schaum's Outline Series in Mathematics, New York: McGraw-Hill, Inc., 1975.

This is a typical outline that might be useful for self-study when used in conjunction with the present text. Although short on discussion, it does contain all of the basic definitions and many worked-out examples. There are also many excellent problems for which answers are provided. This text is one of the few that contains material on statistics as well as probability.

Thomas, J., *An Introduction to Statistical Communication Theory*. New York: John Wiley and Sons, Inc., 1969.

> In this graduate level text, a broad coverage of probability, random processes, and their applications to the analysis of a variety of system problems is given. Most of the topics discussed in the present text are covered, although sometimes from a more advanced mathematical level. In addition, many other topics are considered. For students desiring a more detailed look at the techniques of probabilistic analysis, this text is well worth studying. The references given at the ends of the chapters are quite extensive and include journal articles as well as books.

Thomas, J. B., *An Introduction to Applied Probability and Random Processes*. New York: John Wiley and Sons, Inc., 1971.

> A graduate level text on probability and random processes. This book is more advanced than the present text but covers much of the same material. It is useful for obtaining supplementary information.

CHAPTER 2

Random Variables

2–1 Concept of a Random Variable

The previous chapter deals exclusively with situations in which the number of possible outcomes associated with any experiment is finite. Although it is never stated that the outcomes had to be finite in number (because, in fact, they do *not*), such an assumption is implied and is certainly true for such illustrative experiments as tossing coins, throwing dice, and selecting resistors from bins. There are many other experiments, however, in which the number of possible outcomes is not finite, and it is the purpose of this chapter to introduce ways of describing such experiments in accordance with the concepts of probability already established.

A good way to introduce this type of situation is to consider again the experiment of selecting a resistor from a bin. When mention is made, in the previous chapter, of selecting a 1Ω resistor, or a 10-Ω resistor, or any other value, the implied meaning is that the selected resistor is labeled "1 Ω" or "10 Ω." The actual value of resistance is expected to be close to the labeled value, but might differ from it by some unknown (but measurable) amount. The deviations from the labeled value are due to manufacturing variations and can assume any value within some specified range. Since the actual value of resistance is unknown in advance, it is a *random variable*.

To carry this illustration further, consider a bin of resistors that are all marked "100 Ω." Because of manufacturing tolerances, each of the resistors in the bin will have a slightly different resistance value. Furthermore, there are an infinite

48

number of possible resistance values, so that the experiment of selecting one resistor has an infinite number of possible outcomes. Even if it is known that all of
the resistance values lie between 9.99 Ω and 100.01 Ω, there are an infinite number of such values in this range. Thus, if one defines a particular event as the
selection of a resistor with a resistance of exactly 100.00 Ω, the probability of
this event is actually zero. On the other hand, if one were to define an event as
the selection of a resistor having a resistance between 99.9999 Ω and 100.0001
Ω, the probability of this event is nonzero. The actual value of resistance, however, is a random variable that can assume any value in a specified range of
values.

It is also possible to associate random variables with time functions, and, in
fact, most of the applications that are considered in this text are of this type.
Although Chapter 3 will deal exclusively with such random variables and random
time functions, it is worth digressing momentarily at this point, to note the relationship between the two as it provides an important physical motivation for the
present study.

A typical random time function, shown in Figure 2-1, is designated as $x(t)$. In
a given physical situation, this particular time function is only one of an infinite
number of time functions that might have occurred. The collection of all possible
time functions that might have been observed belongs to a random process, which
will be designated as $\{x(t)\}$. When the probability functions are also specified, this
collection is referred to as an *ensemble*. Any particular member of the ensemble,
say $x(t)$, is a *sample function*, and the value of the sample function at some particular time, say t_1, is a random variable which we call $X(t_1)$ or simply X_1. Thus,
$X_1 = x(t_1)$ when $x(t)$ is the particular sample function observed.

A random variable associated with a random process is a considerably more
involved concept than the random variable associated with the resistor above. In
the first place, there is a different random variable for each instant of time, although there usually is some relation between two random variables corresponding
to two different time instants. In the second place, the randomness we are concerned with is the randomness that exists from sample function to sample function
throughout the complete ensemble. There may also be randomness from time in-

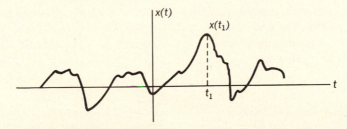

Figure 2–1 A random time function.

stant to time instant, but this is not an essential ingredient of a random process. Therefore, the probability description of the random variables being considered here is also the probability description of the random process. However, our initial discussion will concentrate on the random variables and will be extended later to the random process.

From an engineering viewpoint, a random variable is simply a numerical description of the outcome of a random experiment. Recall that the sample space $S = \{\alpha\}$ is the set of all possible outcomes of the experiment. When the outcome is α, the random variable X has a value that we might denote as $X(\alpha)$. From this viewpoint, a random variable is simply a real-valued function defined over the sample space—and in fact the fundamental definition of a random variable is simply as such a function (with a few restrictions needed for mathematical consistency). For engineering applications, however, it is usually not necessary to consider explicitly the underlying sample space. It is generally only necessary to be able to assign probabilities to various *events* associated with the random variables of interest and these probabilities can often be inferred directly from the physical situation. What events are required for a complete description of the random variable, and how the appropriate probabilities can be inferred, form the subject matter for the rest of this chapter.

If a random variable can assume any value within a specified range (possibly infinite), then it will be designated as a continuous random variable. In the following discussion all random variables will be assumed to be continuous unless stated otherwise. It will be shown, however, that discrete random variables (that is, those assuming one of a countable set of values) can also be treated by exactly the same methods.

2–2 Distribution Functions

In order to consider continuous random variables within the framework of probability concepts discussed in the last chapter, it is necessary to define the *events* to be associated with the probability space. There are many ways in which events might be defined, but the method to be described below is almost universally accepted.

Let X be a random variable as defined above and x be any allowed value of this random variable. The *probability distribution function* is defined to be the probability of the event that the observed random variable X is less than or equal to the allowed value x. That is,[1]

$$F_X(x) = \Pr(X \leq x)$$

---

[1]The subscript X denotes the random variable while the argument x could equally well be any other symbol. In much of the subsequent discussion it is convenient to suppress the subscript X when no confusion will result. Thus $F_X(x)$ will often be written $F(x)$.

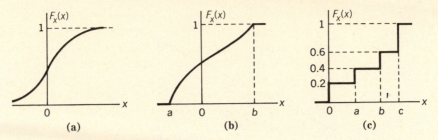

Figure 2–2 Some possible probability distribution functions.

Since the probability distribution function is a probability, it must satisfy the basic axioms and must have the same properties as the probabilities discussed in Chapter 1. However, it is also a function of x, the possible values of the random variable X, and as such must generally be defined for all values of x. Thus, the requirement that it be a probability imposes certain constraints upon the functional nature of $F_X(x)$. These may be summarized as follows:

1. $0 \leq F_X(x) \leq 1 \qquad -\infty < x < \infty$

2. $F_X(-\infty) = 0 \qquad F_X(\infty) = 1$

3. $F_X(x)$ is nondecreasing as x increases.

4. $\Pr(x_1 < X \leq x_2) = F_X(x_2) - F_X(x_1)$

Some possible distribution functions are shown in Figure 2–2. The sketch in (a) indicates a continuous random variable having possible values ranging from $-\infty$ to ∞ while (b) shows a continuous random variable for which the possible values lie between a and b. The sketch in (c) shows the probability distribution function for a discrete random variable that can assume only four possible values (that is, 0, a, b, or c). In distribution functions of this type it is important to remember that the definition for $F_X(x)$ includes the condition $X = x$ as well as $X < x$. Thus, in Figure 2–2(c), it follows (for example) that $F_X(a) = 0.4$ and not 0.2.

The probability distribution function can also be used to express the probability of the event that the observed random variable X is greater than (but not equal to) x. Since this event is simply the complement of the event having probability $F_X(x)$ it follows that

$$\Pr(X > x) = 1 - F_X(x)$$

As a specific illustration, consider the probability distribution function shown in Figure 2–3. Note that this function satisfies all of the requirements listed above. It is easy to see from the figure that the following statements (among many other possible statements) are true:

$$\Pr(X \leq -5) = 0.25$$

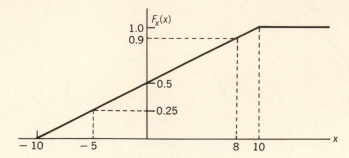

Figure 2–3 A specific probability distribution function.

$$\Pr\,(X > -5) = 1 - 0.25 = 0.75$$
$$\Pr\,(X > 8) = 1 - 0.9 = 0.1$$
$$\Pr\,(-5 < X \le 8) = 0.9 - 0.25 = 0.65$$
$$\Pr\,(X > 0) = 1 - \Pr\,(X \le 0) = 1 - 0.5 = 0.5$$

   In the example above, all of the variation of the probability distribution function takes place between finite limits. This is not always the case, however. Consider, for example, a probability distribution function defined by

$$F_X(x) = \frac{1}{2}\left(1 + \frac{2}{\pi}\tan^{-1}\frac{x}{5}\right) \qquad -\infty < x < \infty \qquad \text{(2–1)}$$

and shown in Figure 2–4. Again, there are many different statements that can be made concerning the probability that the random variable X lies in certain regions. For example, it is straightforward to verify that all of the following are true:

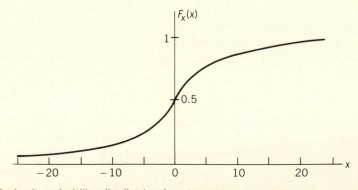

Figure 2–4 A probability distribution function with infinite range.

Pr $(X \leq -5) = 0.25$

Pr $(X > -5) = 1 - 0.25 = 0.75$

Pr $(X > 8) = 1 - 0.8222 = 0.1778$

Pr $(-5 < X \leq 8) = 0.8222 - 0.25 = 0.5722$

Pr $(X > 0) = 1 - $ Pr $(X \leq 0) = 0.5$

Exercise 2–2.1

A random experiment consists of flipping six coins and taking the random variable to be the number of heads.

a) Sketch the distribution function for this random variable.

b) What is the probability that the random variable is less than 3.5?

c) What is the probability that the random variable is greater than 2.5?

d) What is the probability that the random variable is greater than 1.5 and less than or equal to 5.0?

Answers: 0.6563, 0.875, 0.6563

Exercise 2–2.2

A particular random variable has a probability distribution function given by

$$F_X(x) = 0 \qquad -\infty < x \leq 0$$
$$= 1 - e^{-x} \qquad 0 \leq x < \infty$$

Find

a) the probability that $X > 0.5$

b) the probability that $X \leq 0.25$

c) the probability that $0.3 < X \leq 0.7$.

Answers: 0.2212, 0.6065, 0.2442

2–3   Density Functions

Although the distribution function is a complete description of the probability model for a single random variable, it is not the most convenient form for many calculations of interest. For these, it may be preferable to use the derivative of $F(x)$ rather than $F(x)$ itself. This derivative is called the *probability density function* and, when it exists, it is defined by[2]

$$f_X(x) = \lim_{e \to 0} \frac{F_X(x + e) - F_X(x)}{e} = \frac{dF_X(x)}{dx}$$

The physical significance of the probability density function is best described in terms of the probability element, $f_X(x)\, dx$. This may be interpreted as

$$f_X(x)\, dx = \Pr (x < X \leq x + dx) \tag{2-2}$$

Equation (2–2) simply states that the probability element, $f_X(x)\, dx$, is the probability of the event that the random variable X lies in the range of possible values between x and $x + dx$.

Since $f_X(x)$ is a density function and not a probability, it is not necessary that its value be less than 1; it may have any nonnegative value.[3] Its general properties may be summarized as follows:

1. $f_X(x) \geq 0 \quad -\infty < x < \infty$

2. $\displaystyle\int_{-\infty}^{\infty} f_X(x)\, dx = 1$

3. $\displaystyle F_X(x) = \int_{-\infty}^{x} f_X(u)\, du$

4. $\displaystyle\int_{x_1}^{x_2} f_X(x)\, dx = \Pr (x_1 < X \leq x_2)$

As examples of probability density functions, those corresponding to the distribution functions of Figure 2–2 are shown in Figure 2–5. Note particularly that the density function for a discrete random variable consists of a set of delta functions, each having an area equal to the magnitude of the corresponding discontinuity in the distribution function. It is also possible to have density functions that contain both a continuous part and one or more delta functions.

There are many different mathematical forms that might be probability density

---

[2]Again, the subscript denotes the random variable and when no confusion results, it may be omitted. Thus, $f_X(x)$ will often be written as $f(x)$.

[3]Because $F_X(x)$ is nondecreasing as x increases.

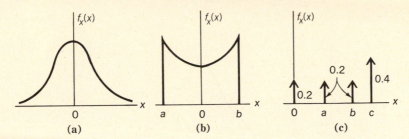

Figure 2–5 Probability density functions corresponding to the distribution functions of Figure 2–2.

functions, but only a very few of these arise to any significant extent in the analysis of engineering systems. Some of these are considered in subsequent sections and a table containing numerous density functions is given in Appendix B.

Before considering the more important probability density functions, however, let us look at the density functions that are associated with the probability distribution functions described in the previous section. It is clear from Figure 2–3 that the probability density function associated with this random variable must be zero for $x \leq -10$ and $x > 10$. Furthermore, in the interval between -10 and 10 it must have a constant value since the slope of the distribution function is constant. Thus:

$$f_X(x) = 0 \qquad x \leq -10$$
$$= 0.05 \qquad -10 < x \leq 10$$
$$= 0 \qquad x > 10$$

This is sketched in Figure 2–6.

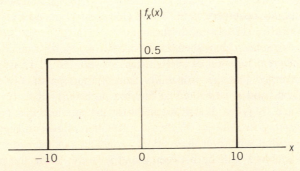

Figure 2–6 Probability density function corresponding to the distribution function of Figure 2–3.

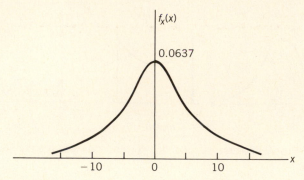

Figure 2–7 Probability density function corresponding to the distribution function of Figure 2–4.

The probability density function corresponding to the distribution function of Figure 2–4 can be obtained by differentiating the distribution function of (2–1). Thus,

$$f_X(x) = \frac{dF_X(x)}{dx} = \frac{d}{dx}\left[\frac{1}{2} + \frac{1}{\pi}\tan^{-1}\frac{x}{5}\right] = \frac{5}{\pi}\left(\frac{1}{x^2 + 25}\right) \quad -\infty < x < \infty \tag{2–3}$$

This probability density function is displayed in Figure 2–7.

A situation that frequently occurs in the analysis of engineering systems is that in which one random variable is functionally related to another random variable whose probability density function is known and it is desired to determine the probability density function of the first random variable. For example, it may be desired to find the probability density function of a power variable when the probability density function of the corresponding voltage or current variable is known. Or it may be desired to find the probability density function after some nonlinear operation is performed on a voltage or current. Although a complete discussion of this problem is not necessary here, a few elementary concepts can be presented and will be useful in subsequent discussions.

In order to formulate the mathematical framework, let the random variable Y be a single-valued, real function of another random variable X. Thus, $Y = g(X)$[4], in which it is assumed that the probability density function of X is known and is denoted by $f_X(x)$, and it is desired to find the probability density function of Y, which is denoted by $f_Y(y)$. If it is assumed for the moment that $g(X)$ is a monotonically increasing function of X, then the situation shown in Figure 2–8(a) applies. It is clear that whenever the random variable X lies between x and $x + dx$, the random variable Y will lie between y and $y + dy$. Since the probabilities of these events are $f_X(x)\,dx$ and $f_Y(y)\,dy$, one can immediately write

---

[4]This also implies that the possible values of X and Y are related by $y = g(x)$.

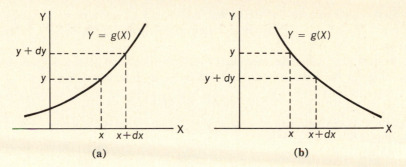

Figure 2–8 Transformation of variables.

$$f_Y(y)\ dy\ =\ f_X(x)\ dx$$

from which the desired probability density function becomes

$$f_Y(y)\ =\ f_X(x)\ \frac{dx}{dy} \qquad\qquad (2\text{–}4)$$

Of course, in the right side of (2–4), x must be replaced by its corresponding function of y.

When $g(X)$ is a monotonically decreasing function of X, as shown in Figure 2–8(b), a similar result is obtained except that the derivative is negative. Since probability density functions must be positive, and also from the geometry of the figure, it is clear that what is needed in (2–4) is simply the absolute value of the derivative. Hence, for either situation

$$f_Y(y)\ =\ f_X(x)\ \left|\frac{dx}{dy}\right| \qquad\qquad (2\text{–}5)$$

In order to illustrate the transformation of variables, consider first the problem of scaling the amplitude of a random variable. Assume that we have a random variable X whose probability density function $f_X(x)$ is known. We then consider another random variable Y that is linearly related to X by $Y = AX$. This situation arises, for example, when X is the input to an amplifier and Y is its output. Since the possible values of X and Y are related in the same way, it follows that

$$\frac{dy}{dx}\ =\ A$$

From (2–5) it is clear that the probability density function of Y is

$$f_Y(y)\ =\ \frac{1}{|A|}\ f_X\!\left(\frac{y}{A}\right)$$

Thus, it is very easy to find the probability density of any random variable that is simply a scaled, version of another random variable whose density function is known.

Consider next a specific example of the transformation of random variables by assuming that the random variable X has a density function of the form

$$f_X(x) = e^{-x}u(x)$$

where $u(x)$ is the unit step starting at $x = 0$. Now consider another random variable Y that is related to X by

$$Y = X^3$$

Since y and x are related in the same way, it follows that

$$\frac{dy}{dx} = 3x^2$$

and

$$\frac{dx}{dy} = \frac{1}{3x^2} = \frac{1}{3y^{2/3}}$$

Thus, the probability density function of Y is

$$f_Y(y) = \frac{e^{-y^{1/3}}}{3} y^{-2/3} u(y)$$

There may also be situations in which, for a given Y, $g(X)$ has regions in which the derivative is positive and other regions in which it is negative. In such cases, the regions may be considered separately and the corresponding probability densities added. An example of this sort will serve to illustrate such a transformation.

Let the functional relationship be

$$Y = X^2$$

This is shown in Figure 2–9 and represents, for example, the transformation (except for a scale factor) of a voltage random variable into a power random variable. Since the derivative, dx/dy, has an absolute value given by

$$\left|\frac{dx}{dy}\right| = \frac{1}{2\sqrt{y}}$$

and since there are two x-values for every y-value ($x = \pm\sqrt{y}$), the desired probability density function is simply

$$f_Y(y) = \frac{1}{2\sqrt{y}}[f_X(\sqrt{y}) + f_X(-\sqrt{y})] \qquad y \geq 0 \qquad \textbf{(2–6)}$$

Furthermore, since y can never be negative,

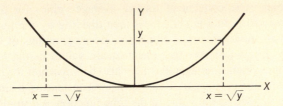

Figure 2–9 The square law transformation.

$$f_Y(y) = 0 \quad y < 0$$

Some other applications of random variable transformations are considered later.

Exercise 2–3.1

The probability density function of a random variable has the form $f_X(x) = Ke^{-2x}u(x)$, where $u(x)$ is the unit step function. Find

a) the value of K

b) the probability that $X > 1$

c) the probability that $X \le 0.5$.

   Answers: 0.1353, 0.6321, 2.0

Exercise 2–3.2

A random variable Y is related to the random variable X of Exercise 2–3.1 by

$$Y = 6X + 3$$

Find the probability density function of Y.

   Answer: $(1/3 \exp[-(y - 3)/6]$

2–4   Mean Values and Moments

One of the most important and most fundamental concepts associated with statistical methods is that of finding average values of random variables or functions of random variables. The concept of finding average values for time functions by integrating over some time interval, and then dividing by the length of the interval, is a familiar one to electrical engineers, since operations of this sort are used to find the dc component, the root-mean-square value, or the average power of the time function. Such time averages may also be important for random functions of time, but, of course, have no meaning when considering a single random variable, which is defined as the value of the time function at a single instant of time. Instead, it is necessary to find the average value by integrating over the range of possible values that the random variable may assume. Such an operation is referred to as "ensemble averaging," and the result is the mean value.

Several different notations are in standard use for the mean value but the most common ones in engineering literature are[5]

$$\overline{X} = E[X] = \int_{-\infty}^{\infty} xf(x) \, dx \tag{2-7}$$

The symbol $E[X]$ is usually read "the expected value of X" or "the mathematical expectation of X." It is shown later, that in many cases of practical interest, the mean value of a random variable is equal to the time average of any sample function from the random process to which the random variable belongs. In such cases, finding the mean value of a random voltage or current is equivalent to finding its dc component; this interpretation will be employed here for illustration.

The expected value of any function of x can also be obtained by a similar calculation. Thus,

$$E[g(X)] = \int_{-\infty}^{\infty} g(x) \, f(x) \, dx \tag{2-8}$$

A function of particular importance is $g(x) = x^n$, since this leads to the general moments of the random variable. Thus,

$$\overline{X^n} = E[X^n] = \int_{-\infty}^{\infty} x^n f(x) \, dx \tag{2-9}$$

By far the most important moments of X are those given by $n = 1$, which is the mean value discussed above, and by $n = 2$, which leads to the mean-square value.

$$\overline{X^2} = E[X^2] = \int_{-\infty}^{\infty} x^2 f(x) \, dx \tag{2-10}$$

The importance of the mean-square value lies in the fact that it may often be interpreted as being equal to the time average of the square of a random voltage

---

[5]Note that the subscript X has been omitted from $f(x)$ since there is no doubt as to what the random variable is.

or current. In such cases, the mean-square value is proportional to the average power (in a resistor) and its square root is equal to the rms or effective value of the random voltage or current.

It is also possible to define central moments, which are simply the moments of the difference between a random variable and its mean value. Thus, the nth central moment is

$$\overline{(X - \overline{X})^n} = E[(X - \overline{X})^n] = \int_{-\infty}^{\infty} (x - \overline{X})^n f(x) \, dx \qquad (2\text{-}11)$$

The central moment for $n = 1$ is, of course, zero, while the central moment for $n = 2$ is so important that it carries a special name, the *variance*, and is usually symbolized by σ^2. Thus,

$$\sigma^2 = \overline{(X - \overline{X})^2} = \int_{-\infty}^{\infty} (x - \overline{X})^2 f(x) \, dx \qquad (2\text{-}12)$$

The variance can also be expressed in an alternative form by using the rules for the expectations of sums; that is,

$$E[X_1 + X_2 + \cdots + X_m] = E[X_1] + E[X_2] + \cdots + E[X_m]$$

Thus,

$$\begin{aligned} \sigma^2 &= E[(X - \overline{X})^2] = E[X^2 - 2X\overline{X} + (\overline{X})^2] \\ &= E[X^2] - 2E[X]\overline{X} + (\overline{X})^2 \qquad (2\text{-}13) \\ &= \overline{X^2} - 2\overline{X}\,\overline{X} + (\overline{X})^2 = \overline{X^2} - (\overline{X})^2 \end{aligned}$$

and it is seen that the variance is the difference between the mean-square value and the square of the mean value. The square root of the variance, σ, is known as the *standard deviation*.

In electrical circuits, the variance can often be related to the average power (in a resistance) of the ac components of a voltage or current. The square root of the variance would be the value indicated by an ac voltmeter or ammeter of the rms type that does not respond to direct current (because of capacitive coupling, for example).

In order to illustrate some of the above ideas concerning mean values and moments, consider a random variable having a uniform probability density function as shown in Figure 2–10. A voltage waveform that would lead to such a probability density function might be a sawtooth waveform that varied linearly between 20 and 40 V. The appropriate mathematical representation for this density function is

$$\begin{aligned} f(x) &= 0 & -\infty &< x \le 20 \\ &= \frac{1}{20} & 20 &< x \le 40 \\ &= 0 & 40 &< x < \infty \end{aligned}$$

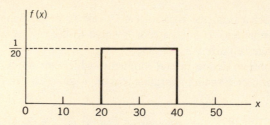

Figure 2–10 A uniform probability density function.

The mean value of this random variable is obtained by using (2–7). Thus,

$$\overline{X} = \int_{20}^{40} x\left(\frac{1}{20}\right) dx = \frac{1}{20} \cdot \frac{x^2}{2}\Big|_{20}^{40} = \frac{1}{40}(1600 - 400) = 30$$

This value is intuitively the average value of the sawtooth waveform just described. The mean-square value is obtained from (2–10) as:

$$\overline{X^2} = \int_{20}^{40} x^2\left(\frac{1}{20}\right) dx = \frac{1}{20}\frac{x^3}{3}\Big|_{20}^{40} = \frac{1}{60}(64 - 8)10^3 = 933.3$$

The variance of the random variable can be obtained from either (2–12) or (2–13). From the latter,

$$\sigma^2 = \overline{X^2} - (\overline{X})^2 = 933.3 - (30)^2 = 33.3$$

On the basis of the assumptions that will be made concerning random processes, if the sawtooth voltage were measured with a dc voltmeter, the reading would be 30 V. If it were measured with an rms-reading ac voltmeter (which did not respond to dc), the reading would be $\sqrt{33.3}$ V.

As a second illustration of the determination of the moments of a random variable, consider the probability density function

$$f(x) = kx[u(x) - u(x - 1)]$$

The value of k can be determined from the 0th moment of $f(x)$ since that is just the area of the density function and must be 1. Thus:

$$\int_0^1 kx\, dx = \frac{k}{2} = 1 \qquad \therefore k = 2$$

The mean and mean-square value of X may now be calculated readily as

$$\overline{X} = \int_0^1 x(2x)\, dx = 2/3$$

$$\overline{X^2} = \int_0^1 x^2(2x)\,dx = 1/2$$

From these two quantities the variance becomes

$$\sigma^2 = \overline{X^2} - (\overline{X})^2 = \frac{1}{2} - \left(\frac{2}{3}\right)^2 = \frac{1}{18}$$

Likewise, the 4th moment of X is

$$\overline{X^4} = \int_0^1 x^4(2x)\,dx = \frac{1}{3}$$

and the 4th central moment is given by

$$\overline{(X - \overline{X})^4} = \int_0^1 \left(x - \frac{2}{3}\right)^4 (2x)\,dx = \frac{1}{135}$$

This latter integration is facilitated by observing that

$$\left(x - \frac{2}{3}\right)^4 x = \left(x - \frac{2}{3}\right)^5 + \frac{2}{3}\left(x - \frac{2}{3}\right)^4$$

Exercise 2–4.1

For the random variable of Exercise 2–3.1, find

   a)   the mean value of X

   b)   the mean-square value of X

   c)   the variance of X.

      Answers: 1/4, 1/2, 1/2

Exercise 2–4.2

A random variable X has a probability density function of the form

$$f_X(x) = 1/4[u(x + 2) - u(x - 2)]$$

For the random variable $Y = X^2$, find

   a)   the mean value

b) the mean-square value

c) the variance.

Answers: 4/3, 64/45, 16/5

2–5   The Gaussian Random Variable

Of the various density functions that we shall study, the most important by far is the *Gaussian* or *normal* density function. There are many reasons for its importance, some of which are:

1. It provides a good mathematical model for a great many different physically observed random phenomena. Furthermore, the fact that it should be a good model can be justified theoretically in many cases.
2. It is one of the few density functions that can be extended to handle an arbitrarily large number of random variables conveniently.
3. Linear combinations of Gaussian random variables lead to new random variables that are also Gaussian. This is not true for most other density functions.
4. The random process from which Gaussian random variables are derived can be completely specified, in a statistical sense, from a knowledge of all first and second moments only. This is not true for other processes.
5. In system analysis, the Gaussian process is often the only one for which a complete statistical analysis can be carried through in either the linear or the nonlinear situation.

The mathematical representation of the Gaussian density function is

$$f(x) = \frac{1}{\sqrt{2\pi}\,\sigma} \exp\left[\frac{-(x - \overline{X})^2}{2\sigma^2}\right] \qquad -\infty < x < \infty \qquad (2\text{–}14)$$

where $\overline{X}$ and σ^2 are the mean and variance, respectively. The corresponding distribution function cannot be written in closed form. The shapes of the density function and distribution function are shown in Figure 2–11. There are a number of points in connection with these curves that are worth noting. These are:

1. There is only one maximum and it occurs at the mean value.
2. The density function is symmetrical about the mean value.
3. The width of the density function is directly proportional to the *standard deviation, σ*. The width of 2σ occurs at the points where the height is 0.607 of the maximum value. These are also the points of maximum absolute slope.

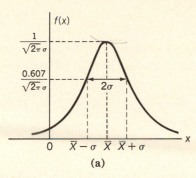
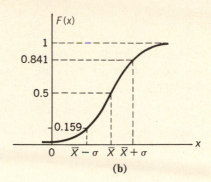

Figure 2–11 The Gaussian random variable: (a) density function and (b) distribution function.

4. The maximum value of the density function is inversely proportional to the standard deviation σ. Since the density function has an area of unity, it can be used as a representation of the impulse or delta function by letting σ approach zero. That is

$$\delta(x - \overline{X}) = \lim_{\sigma \to 0} \frac{1}{\sqrt{2\pi}\,\sigma} \exp\left[\frac{-(x - \overline{X})^2}{2\sigma^2}\right] \tag{2–15}$$

This representation of the delta function has an advantage over some others of being infinitely differentiable.

The Gaussian distribution function cannot be expressed in closed form in terms of elementary functions. It can, however, be expressed in terms of functions that are commonly tabulated. From the relation between density and distribution functions it follows that the general Gaussian distribution function is

$$F(x) = \int_{-\infty}^{x} f(u)\,du = \frac{1}{\sqrt{2\pi}\,\sigma} \int_{-\infty}^{x} \exp\left[-\frac{(u - \overline{X})^2}{2\sigma^2}\right] du \tag{2–16}$$

The function that is usually tabulated is the distribution function for a Gaussian random variable that has a mean value of zero and a variance of unity (that is, $\overline{X} = 0$, $\sigma = 1$). This distribution function is often designated by $\Phi(x)$ and is defined by

$$\Phi(x) = \frac{1}{\sqrt{2\pi}} \int_{-\infty}^{x} \exp\left(-\frac{u^2}{2}\right) du \tag{2–17}$$

By means of a simple change of variable it is easy to show that the general Gaussian distribution function of (2–14) can be expressed in terms of $\Phi(x)$ by

$$F(x) = \Phi\left(\frac{x - \overline{X}}{\sigma}\right) \tag{2-18}$$

An abbreviated table of values for $\Phi(x)$ is given in Appendix D. Since only positive values of x are tabulated, it is frequently necessary to use the additional relationship

$$\Phi(-x) = 1 - \Phi(x) \tag{2-19}$$

Another function that is closely related to $\Phi(x)$, and is often more convenient to use, is the Q-function defined by

$$Q(x) = \frac{1}{\sqrt{2\pi}} \int_x^\infty \exp\left(-\frac{u^2}{2}\right) du \tag{2-20}$$

and for which

$$Q(-x) = 1 - Q(x) \tag{2-21}$$

Upon comparing this with (2–17), it is clear that

$$Q(x) = 1 - \Phi(x)$$

Likewise, comparing with (2–18)

$$F(x) = 1 - Q\left(\frac{x - \overline{X}}{\sigma}\right)$$

A brief table of values for $Q(x)$ is given in Appendix E for small values of x.

Several alternative notations are also used in the literature for both $\Phi(x)$ and $Q(x)$. Some authors use

$$\text{erf}(x) = \Phi(x) \tag{2-22}$$

where erf(x) is called the *error function* and

$$\text{erfc}(x) = Q(x) \tag{2-23}$$

where erfc(x) is called the *complementary error function*. Still other authors define the error function by

$$\text{erf}(x) = \frac{2}{\sqrt{\pi}} \int_0^x \exp(-u^2) \, du = 2 \, \Phi\,(\sqrt{2}x) - 1 \tag{2-24}$$

This diversity of notation emphasizes the need to carefully determine the definitions being used whenever the literature is read.

Although both $\Phi(x)$ and $Q(x)$ are widely tabulated, there is an advantage in using $Q(x)$ when tables are not available or the values needed are outside the range of tables. This is because there is a relatively simple way to calculate quite accurate values of $Q(x)$ with an ordinary hand calculator. This computation procedure starts by representing $Q(x)$ as

$$Q(x) = \frac{\exp(-x^2/2)}{\sqrt{2\pi}} G(x) \qquad (2\text{--}25)$$

in which $G(x)$ is a continued fraction defined by

$$G(x) = \cfrac{1}{x + \cfrac{1}{x + \cfrac{2}{x + \cfrac{3}{x + \cfrac{4}{x + \ldots}}}}} \qquad (2\text{--}26)$$

The next step is to decide upon the number of terms to be used in evaluating $G(x)$. Increased accuracy is achieved by using more terms, but, of course, the labor involved is also increased. A rule for selecting the number of terms will be given shortly, but for the moment, let us assume that we have selected an appropriate number and have called this number n. We then calculate the following sequence of values:

$$p_n = x + \frac{n}{x}$$

$$p_{n-1} = x + \frac{n-1}{p_n}$$

$$\cdot$$
$$\cdot$$
$$\cdot$$

$$p_1 = x + \frac{1}{p_2}$$

$$G(x) = \frac{1}{p_1}$$

The last value in this sequence is the desired $G(x)$.

The number of terms required in the expansion for $G(x)$ to achieve a desired accuracy depends upon the value of x for which the computation is being made; the smaller x is, the larger n must be. As a general rule of thumb, in order to achieve six significant figures for the final value of $Q(x)$, the product of x and n should be at least 30.

The Q-function is useful in calculating the probability of events that occur very rarely. An example will serve to illustrate this application as well the technique for evaluating the Q-function. Suppose we have an IC trigger circuit that is supposed to change state whenever the input voltage exceeds 2.5 volts; that is, whenever the input goes from a "0" state to a "1" state. Assume that when the input is in the "0" state the voltage is actually 0.5 volts, but that there is Gaussian random noise superimposed on this having a variance of 0.2 volts squared. Thus, the input to the trigger circuit can be modeled as a Gaussian random variable with a mean of 0.5 and a variance of 0.2. We wish to determine the probability that the circuit will incorrectly trigger as a result of the random input exceeding 2.5. From the definition of the Q-function, it follows that the desired probability is just $Q[(2.5 - 0.5)/\sqrt{0.2}] = Q(4.472)$. Although $Q(4.472)$ is in the table in Appendix E, for purposes of illustration we will calculate it using the technique described above. From the rule of thumb, it appears that n should be 7. Thus,

$$p_7 = 4.472 + \frac{7}{4.472} = 6.037$$

$$p_6 = 4.472 + \frac{6}{6.037} = 5.466$$

$$p_5 = 4.472 + \frac{5}{5.466} = 5.387$$

.

.

.

$$p_1 = 4.472 + \frac{1}{4.868} = 4.677$$

$$G(4.472) = 0.2138$$

$$Q(4.472) = \frac{\exp(-4.472^2/2)}{\sqrt{2\pi}}(0.2138) = 3.872 \times 10^{-6}$$

Note that the probability of incorrectly triggering on any one operation is quite small. However, over a period of time in which many operations occur, the probability can become significant. The probability that false triggering does *not* occur is simply 1 minus the probability that it does occur. Thus, in n operations, the probability that false triggering occurs is

$$\text{Pr (False Triggering)} = 1 - (1 - 3.872 \times 10^{-6})^n$$

For $n = 10^5$, this probability becomes

$$\text{Pr (False Triggering)} = 0.321$$

A conclusion that can be drawn from this example is that when there is appreciable noise in a digital circuit, errors are almost certain to occur sooner or later.

Although many of the most useful properties of Gaussian random variables will become apparent only when two or more variables are considered, one that can be mentioned now is the ease with which high-order central moments can be determined. The nth central moment, which was defined in (2–11), can be expressed for a Gaussian random variable as

$$\overline{(X - \overline{X})^n} = 0 \qquad\qquad n \text{ odd} \qquad\qquad \textbf{(2–27)}$$
$$= 1 \cdot 3 \cdot 5 \cdots (n - 1)\sigma^n \quad n \text{ even}$$

As an example of the use of (2–27), if $n = 4$, the fourth central moment is $\overline{(X - \overline{X})^4} = 3\sigma^4$. A word of caution should be noted, however. The relation between the nth general moment, $\overline{X^n}$, and the nth central moment is not always as simple as it is for $n = 2$. In the $n = 4$ Gaussian case, for example,

$$\overline{X^4} = 3\sigma^4 + 6\sigma^2(\overline{X})^2 + (\overline{X})^4$$

Before leaving the subject of Gaussian density functions, it is interesting to compare the defining equation, (2–14), with the probability associated with Bernoulli trials for the case of large n as approximated in (1–30). It will be noted that, except for the fact that k and n are integers, the DeMoivre-Laplace approximation has the same form as a Gaussian density function with a mean value of np and a variance of npq. Since the Bernoulli probabilities are discrete, the exact density function for this case is a set of delta functions that increase in number as n increases, and as n becomes large the area of these delta functions follows a Gaussian law.

Another important result closely related to this is the *central limit theorem*. This famous theorem concerns the *sum* of a large number of independent random variables having the same probability density function. In particular, let the random variables be $X_1, X_2, \ldots X_n$ and assume that they all have the same mean value, m, and the same variance, σ^2. Then define a normalized sum as

$$Y = \frac{1}{\sqrt{n}} \sum_{k=1}^{n} (X_k - m) \qquad\qquad \textbf{(2–28)}$$

Under conditions that are weak enough to be realized by almost any random variable encountered in real life, the central limit theorem states that the probability density function for Y approaches a Gaussian density function as n becomes large regardless of the density function for the Xs. Furthermore, because of the normalization, the random variable Y will have zero mean and a variance of σ^2. The theorem is also true for more general conditions, but this is not the important aspect here. What is important is to recognize that a great many random phenomena that arise in physical situations result from the combined actions of many individual events. This is true for such things as: thermal agitation of electrons in a conductor, shot noise from electrons or holes in a vacuum tube or transistor, atmospheric noise, turbulence in a medium, ocean waves, and many other physical sources of random disturbances. Hence, regardless of the probability density functions of the individual components (and these density functions are usually not even known), one would expect to find that the observed disturbance has a Gaussian density function. The central limit theorem provides a theoretical justification for assuming this, and, in almost all cases experimental measurements bear out the soundness of this assumption.

Exercise 2–5.1

A Gaussian random variable has a mean value of 1 and a variance of 16. Find:

   a) the probability that the random variable has a negative value

   b) the probability that the random variable has a value between 1 and 2

   c) the probability that the random variable is greater than 4.

       Answers: 0.0987, 0.2266, 0.4013

Exercise 2–5.2

For the random variable of Exercise 2–5.1, find

   a) the fourth central moment

   b) the fourth moment

   c) the third central moment

   d) the third moment.

       Answers: 0, 49, 768, 865

2–6   Density Functions Related to Gaussian

The previous section has indicated some of the reasons for the tremendous importance of the Gaussian density function. Still another reason is that there are many other probability density functions, which arise in practical applications, that are related to the Gaussian density function and can be derived from it. The purpose of this section is to list some of these other density functions and indicate the situations under which they arise. They will not all be derived here, since in most cases insufficient background is available, but several of the more important ones will be derived as illustrations of particular techniques.

Distribution of power.   When the voltage or current in a circuit is the random variable, the power dissipated in a resistor is also a random variable that is proportional to the square of the voltage or current. The transformation that applies in this case is discussed in Section 2–3 and is used here to determine the probability density function associated with the power of a Gaussian voltage or current. In particular, let I be the random variable $I(t_1)$ and assume that $f_I(i)$ is Gaussian. The power random variable, W, is then given by

$$W = RI^2$$

and it is desired to find its probability density function $f_W(w)$. By analogy to the result in (2–6), this probability density function may be written as

$$f_W(w) = \frac{1}{2\sqrt{Rw}} \left[f_I\left(\sqrt{\frac{w}{R}} \right) + f_I\left(-\sqrt{\frac{w}{R}} \right) \right] \qquad w \geq 0$$

$$= 0 \qquad w < 0$$

$$(2\text{--}29)$$

If I is Gaussian and assumed to have zero mean, then

$$f_I(i) = \frac{1}{\sqrt{2\pi}\,\sigma_I} \exp\left(-\frac{i^2}{2\sigma_I^2} \right)$$

where σ_I^2 is the variance of I. Hence, σ_I has the physical significance of being the rms value of the current. Furthermore, since the density function is symmetrical, $f_I(i) = f_I(-i)$. Thus, the two terms of (2–29) are identical and the probability density function of the power becomes

$$f_W(w) = \frac{1}{\sigma_I \sqrt{2\pi Rw}} \exp\left(-\frac{w}{2R\sigma_I^2} \right) \qquad w \geq 0$$

$$= 0 \qquad w < 0$$

$$(2\text{--}30)$$

This density function is sketched in Figure 2–12. Straightforward calculation indicates that the mean value of the power is

$$\overline{W} = E[RI^2] = R\sigma_I^2$$

and the variance of the power is

$$\sigma_W^2 = \overline{W^2} - (\overline{W})^2 = E[R^2 I^4] - (\overline{W})^2$$

$$= 3R^2 \sigma_I^4 - (R\sigma_I^2)^2 = 2R^2 \sigma_I^4$$

It may be noted that the probability density function for the power is infinite at $w = 0$; that is, the most probable value of power is zero. This is a consequence of the fact that the most probable value of current is also zero and that the deriv-

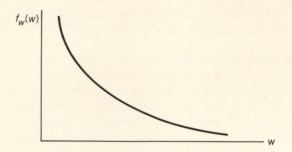

Figure 2–12 Density function for the power of a Gaussian current.

ative of the transformation (dW/dI) is zero here. It is important to note, however, that there is *not* a delta function in the probability density function.

The probability distribution function for the power can be obtained, in principle, by integrating the probability density function for the power. However, this integration does not result in a closed-form result. Nevertheless, it is possible to obtain the desired probability distribution function quite readily by employing the basic definition. Specifically, the probability that the power is less than or equal to some value w is just the same as the probability that the current is between the values of $+\sqrt{w/R}$ and $-\sqrt{w/R}$. Thus, since I is assumed to be Gaussian with zero mean and variance σ_I^2, the probability distribution function for the power becomes

$$F_W(w) = \Pr\left[i \leq \sqrt{w/R}\right] - \Pr\left[i \leq -\sqrt{w/R}\right] = \Phi\left(\frac{\sqrt{w/R}}{\sigma_I}\right) - \Phi\left(\frac{-\sqrt{w/R}}{\sigma_I}\right)$$

$$= 2\,\Phi\left(\frac{\sqrt{w/R}}{\sigma_I}\right) - 1 \quad w \geq 0$$

$$= 0 \quad w < 0$$

As an illustration of the use of the power distribution function consider the power delivered to a loudspeaker in a typical stereo system. Assume that the speaker has a resistance of 4 ohms and is rated for a maximum power of 25 watts. If the current driving the speaker is assumed to be Gaussian and at a level that provides an average power of 4 watts, what is the probability that the maximum power level of the speaker will be exceeded? Since 4 watts dissipated in 4 ohms implies a value of $\sigma_I^2 = 1$, it follows that

$$\Pr\,(W > 25) = 1 - F_W(25) = 2\left[1 - \Phi\left(\frac{\sqrt{25/4}}{1}\right)\right]$$

$$= 2(1 - 0.9798) = 0.0404$$

This probability implies that the maximum speaker power is exceeded several times per second for a Gaussian signal. The situation is probably worse than this in an actual case because the probability density function of music is not Gaussian, but tends to have peak values that are more probable than that predicted by the Gaussian assumption.

Exercise 2–6.1

A Gaussian random voltage having a mean value of zero and a standard deviation of 10 V is applied to a resistance of 4 Ω. Find:

a) the approximate probability that the power dissipated in the resistance is between 9.9 watts and 10.1 watts (use the power density function)

b) the probability that the power dissipated in the resistor is greater than 25 watts

c) the probability that the power dissipated in the resistor is less than or equal to 10 watts.

Answers: 0.00616, 0.3174, 0.472

Rayleigh distribution.   The Rayleigh probability density function arises in several different physical situations. For example, it will be shown later that the peak values (that is, the *envelope*) of a random voltage or current having a Gaussian probability density function will follow the Rayleigh density function. The original derivation of this density function (by Lord Rayleigh in 1880) was applied to the envelope of the sum of many sine waves of different frequencies. It also arises in connection with the errors associated with the aiming of firearms, missiles, and other projectiles, if the errors in each of the two rectangular coordinates have independent Gaussian probability densities. Thus, if the origin of a rectangular coordinate system is taken to be the target and the error along one axis is X and the error along the other axis is Y, the total miss distance is simply

$$R = \sqrt{X^2 + Y^2}$$

When X and Y are independent Gaussian random variables with zero mean and equal variances, σ^2, the probability density function for R is

$$f_R(r) = \frac{r}{\sigma^2} \exp\left(-\frac{r^2}{2\sigma^2}\right) \quad r \geq 0$$

$$= 0 \qquad\qquad\qquad r < 0$$

(2–31)

This is the Rayleigh probability density function and is sketched in Figure 2–13 for two different values of σ^2. Note that the maximum value of the density func-

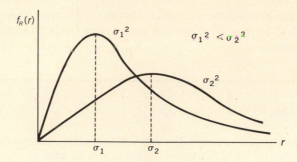

Figure 2–13   The Rayleigh probability density function.

tion is at σ, but that the density function is not symmetrical about this maximum point.

The mean value of the Rayleigh-distributed random variable is easily computed from

$$\bar{R} = \int_0^\infty r f_R(r) \, dr = \int_0^\infty \frac{r^2}{\sigma^2} \exp\left(-\frac{r^2}{2\sigma^2}\right) dr$$

$$= \sqrt{\frac{\pi}{2}}\, \sigma$$

and the mean-square value from

$$\bar{R}^2 = \int_0^\infty r^2 f_R(r) \, dr = \int_0^\infty \frac{r^3}{\sigma^2} \exp\left(-\frac{r^2}{2\sigma^2}\right) dr$$

$$= 2\sigma^2$$

The variance of R is therefore given by

$$\sigma^2_R = \bar{R}^2 - (\bar{R})^2 = \left(2 - \frac{\pi}{2}\right)\sigma^2 = 0.429\sigma^2$$

Note that this variance is *not* the same as the variance σ^2 of the Gaussian random variables that generate the Rayleigh random variable. It may also be noted that, unlike the Gaussian density function, both the mean and variance depend upon a single parameter (σ^2) and cannot be adjusted independently.

It is straightforward to find the probability distribution function for the Rayleigh random variable because the density function can be integrated readily. Thus,

$$F_R(r) = \int_0^r \frac{u}{\sigma^2} \exp\left(\frac{-u^2}{2\sigma^2}\right) du = 1 - \exp\left(\frac{-r^2}{2\sigma^2}\right) \quad r \geq 0$$

$$= 0 \qquad\qquad\qquad\qquad\qquad\qquad\qquad\qquad r < 0$$

(2–32)

As an example of the Rayleigh density function, consider an aiming problem in which an archer shoots at a target two feet in diameter and for which the bulls-eye is centered on the origin of an *XY* coordinate system. The position at which any arrow strikes the target is a random variable having an *X*-component and a *Y*-component. It is determined that the standard deviation of these components is 1/4 foot; that is, $\sigma_X = \sigma_Y = 1/4$. On the assumption that the *X* and *Y* components of the hit position are independent Gaussian random variables, the distance from the hit position to the center of the target (i.e., the *miss distance*) is a Rayleigh distributed random variable for which the probability density function is

$$f_R(r) = 16r \exp(-8r^2) \quad r \ge 0$$

Using the results obtained above, the mean value of the miss distance becomes $\overline{R} = \sqrt{\pi/2}(1/4) = 0.313$ feet and its standard deviation is $\sigma_R = \sqrt{0.429}(1/4) = 0.164$ feet. From the distribution function the probability that the target will be missed completely is

$$\Pr(\text{Miss}) = 1 - F_R(1) = 1 - \left[1 - \exp\left(-\frac{1^2}{2(0.25)^2}\right)\right]$$
$$= e^{-8} = 3.35 \times 10^{-4}$$

Similarly, if the bulls-eye is two inches in diameter, the probability of making a bulls eye is

$$\Pr(\text{Bulls-eye}) = F_R\left(\frac{1}{12}\right) = 1 - \exp\left(-\frac{8}{144}\right) = 0.0540$$

Obviously, this example describes an archer who is not very skillful, in spite of the fact that he rarely misses the entire target!

Exercise 2–6.2

An amateur marksman fires a pistol at a target 8 inches in diameter. It is determined that the probability that he will miss the target entirely is 0.01. Find the mean miss distance (from the center of the target) for all shots fired.

Answer: 1.652

Maxwell distribution.   A classical problem in thermodynamics is that of determining the probability density function of the velocity of a molecule in a perfect gas. The basic assumption is that each component of velocity is Gaussian with zero mean and a variance of $\sigma^2 = kT/m$, where k is Boltzmann's constant, T is the absolute temperature, and m is the mass of the molecule. The total velocity is, therefore,

$$V = \sqrt{V_x^2 + V_y^2 + V_z^2}$$

and is said to have a *Maxwell distribution*. The resulting probability density function can be shown to be

$$f_V(v) = \sqrt{\frac{2}{\pi}} \frac{v^2}{\sigma^3} \exp\left(-\frac{v^2}{2\sigma^2}\right) \quad v \geq 0$$

$$= 0 \qquad\qquad\qquad v < 0$$

(2–33)

The mean value of a Maxwellian-distributed random variable (the average molecule velocity) can be found in the usual way and is

$$\overline{V} = \sqrt{\frac{8}{\pi}} \sigma$$

The mean-square value and variance can be shown to be

$$\overline{V^2} = 3\sigma^2$$

$$\sigma_V^2 = \overline{V^2} - (\overline{V})^2 = \left(3 - \frac{8}{\pi}\right) \sigma^2$$

$$= 0.453\sigma^2$$

The mean kinetic energy can be obtained from $\overline{V^2}$ since

$$e = \frac{1}{2} mV^2$$

and

$$E[e] = \frac{1}{2} m\overline{V^2} = \frac{3}{2} m\sigma^2 = \frac{3}{2} m\left(\frac{kT}{m}\right) = \frac{3}{2} kT$$

which is the classical result.

The probability distribution function for the Maxwell density cannot be expressed readily in terms of elementary functions, or even in terms of tabulated functions. Thus, in most cases involving this distribution function, it is necessary to carry out the integration numerically. As an illustration of the Maxwell distribution, suppose we attempt to determine the probability that a given gas molecule will have a kinetic energy that is more than twice the mean value of kinetic energy for all the molecules. Since the kinetic energy is given by

$$e = \frac{1}{2} m V^2$$

and the mean kinetic energy is just $(3/2)m\sigma^2$, the velocity of a molecule having more than twice the mean kinetic energy is

$$V > \sqrt{6}\,\sigma$$

The probability that a molecule will have a velocity in this range is

$$\Pr\,(V > \sqrt{6}\sigma) = \int_{\sqrt{6\sigma}}^{\infty} \sqrt{\frac{2}{\pi}\frac{v^2}{\sigma^3}}\,\exp\left(-\frac{v^2}{2\sigma^2}\right)\,dv$$

This can be integrated numerically to yield

$$\Pr\,(e > 2\bar{e}) = \Pr\,(V > \sqrt{6}\sigma) = 0.1128$$

Exercise 2–6.3

In a certain gas at 300 K, it is found that the number of molecules having velocities in the vicinity of 1×10^3 meters/second is twice as great as the number of molecules having velocities in the vicinity of 5×10^3 meters/second. Find:

a) the mean velocity of the molecules

b) the mass of the molecules.

Answers: 2794.9, 1.35×10^{-27}

Chi-square distribution. A generalization of the above results arises if one defines a random variable as

$$X^2 = Y_1^2 + Y_2^2 + \cdots + Y_n^2 \tag{2-34}$$

where $Y_1, Y_2, \ldots, Y_n$ are independent Gaussian random variables with 0 mean and variance 1. The random variable X^2 is said to have a *Chi-square distribution with n degrees of freedom* and the probability density function is

$$f(x^2) = \frac{(x^2)^{n/2-1}}{2^{n/2}\Gamma(n/2)}\,\exp\left(-\frac{x^2}{2}\right) \qquad x^2 \geq 0$$
$$= 0 \qquad\qquad\qquad\qquad x^2 < 0 \tag{2-35}$$

With suitable normalization of random variables (so as to obtain unit variance), the power distribution discussed above is seen to be chi-square with $n = 1$. Likewise, in the Rayleigh distribution, the *square* of the miss-distance (R^2) is chi-square with $n = 2$; and in the Maxwell distribution, the square of the velocity

(V^2) is chi-square with $n = 3$. This latter case would lead to the probability density function of molecule *energies*.

The mean and variance of a chi-square random variable are particularly simple because of the initial assumption of unit variance for the components. Thus,

$$\overline{X^2} = n$$

$$(\sigma_{X^2})^2 = 2n$$

The chi-square distribution arises in many signal detection problems in which one is sampling an observed voltage and attempting to decide if it is just noise or if it contains a signal also. If the observed voltage is just noise, then the samples have zero mean and the chi-square distribution described above applies. If, however, there is also a signal in the observed voltage, the mean value of the samples is not zero. The random variable that results from summing the squares of the samples as in (2–34) now has a *non-central* chi-square distribution. Although detection problems of the sort described here are extremely important, further discussion of this application of the chi-square distribution is beyond the scope of this book.

Exercise 2–6.4

Ten independent samples of a Gaussian voltage are taken and each sample is found to have zero mean and a variance of 16. A new random variable is constructed by summing the squares of these samples. Find:

a)   the mean

b)   the variance of this new random variable.

Answers: 5120, 160

Log-normal distribution.   A somewhat different relationship to the Gaussian distribution arises in the case of random variables that are *defined* as the logarithms of other random variables. For example, in communication systems the attenuation of the signal power in the transmission path is frequently expressed in units of *nepers,* and is calculated from

$$A = \ln \left(\frac{W_{\text{out}}}{W_{\text{in}}} \right) \text{ nepers}$$

where W_{in} and W_{out} are the input and output signal powers respectively. An experimentally observed fact is that the attenuation A is very often quite close to being a Gaussian random variable. The question that arises, therefore, concerns the probability density function of the power ratio.

In order to generalize this result somewhat, let two random variables be related by

$$Y = \ln X$$

or, equivalently, by

$$X = e^Y$$

and assume that Y is Gaussian with a mean of $\overline{Y}$ and a variance σ_Y^2. By using (2–5) it is easy to show that the probability density function of X is

$$
\begin{aligned}
f_X(x) &= \frac{1}{\sqrt{2\pi}\,\sigma_Y x} \exp\left[-\frac{(\ln x - \overline{Y})^2}{2\sigma_Y^2} \right] \quad & x \geq 0 \\
&= 0 & x < 0
\end{aligned}
\tag{2–36}
$$

This is the *log-normal* probability density function. In engineering work base 10 is frequently used for the logarithm rather than base e, but it is simple to convert from one to the other. Some typical density functions are sketched in Figure 2–14.

The mean and variance of the log-normal random variable can be evaluated in the usual manner and become

$$\overline{X} = \exp\left(\overline{Y} + \frac{1}{2}\sigma_Y^2 \right)$$

$$\sigma_X^2 = [\exp(\sigma_Y^2) - 1]\exp 2\left(\overline{Y} + \frac{1}{2}\sigma_Y^2 \right)$$

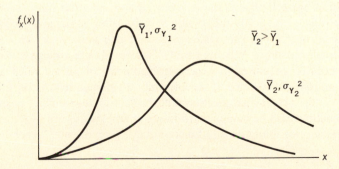

Figure 2–14 The log-normal probability density function.

The distribution function for the log-normal random variable cannot be expressed in terms of elementary functions. If calculations involving the distribution function are required, it is usually necessary to carry out the integration by numerical methods.

Exercise 2–6.5

A log-normal random variable is generated by a Gaussian random variable having a mean value of 2 and a variance of 1.

  a) Find the most probable value of the log-normal random variable.

  b) Repeat if the Gaussian random variable has a mean value of 4 and a variance of 6.

  Answers: 2.718, 0.1353

2–7 Other Probability Density Functions

In addition to the density functions that are related to the Gaussian, there are many others that frequently arise in engineering. Some of these are described here and an attempt is made to discuss briefly the situations in which they arise.

Uniform distribution. The uniform distribution was mentioned in an earlier section and used for illustrative purposes; it is generalized here. The uniform distribution usually arises in physical situations in which there is no preferred value for the random variable. For example, events that occur at random instants of time (such as the emission of radioactive particles) are often assumed to occur at times that are equally probable. The unknown phase angle associated with a sinusoidal source is usually assumed to be uniformly distributed over a range of 2π radians. The time position of pulses in a periodic sequence of pulses (such as a radar transmission) may be assumed to be uniformly distributed over an interval of one period, when the actual time position with respect to zero time is unknown. All of these situations will be employed in future examples.

The uniform probability density function may be represented generally as

$$f(x) = \frac{1}{x_2 - x_1} \quad x_1 < x \le x_2$$

$$= 0 \qquad \text{otherwise}$$

(2–37)

It is quite straightforward to show that

$$\overline{X} = \frac{1}{2}(x_1 + x_2) \tag{2-38}$$

and

$$\sigma_X{}^2 = \frac{1}{12}(x_2 - x_1)^2 \tag{2-39}$$

The probability distribution function of a uniformly distributed random variable is obtained easily from the density function by integration. The result is

$$F_X(x) = 0 \qquad x \leq x_1$$

$$= \frac{x - x_1}{x_2 - x_1} \qquad x_1 < x \leq x_2 \tag{2-40}$$

$$= 1 \qquad x > x_2$$

One of the important applications of the uniform distribution is in describing the errors associated with analog-to-digital conversion. This operation takes a continuous signal that can have any value at a given time instant and converts it into a binary number having a fixed number of binary digits. Since a fixed number of binary digits can represent only a discrete set of values, the difference between the actual value and the closest discrete value represents the error. This is illustrated in Figure 2–15. In order to determine the mean-square value of the error, it is *assumed* that the error is uniformly distributed over an interval from $-\Delta x/2$ to $\Delta x/2$ where Δx is the difference between the two closest levels. Thus, from (2–38), the mean error is zero, and from (2–39) the variance or mean-square error is $\frac{1}{12}(\Delta x)^2$.

The uniform probability density function also arises quite naturally when dealing with sinusoidal time functions in which the phase is a random variable. For example, if a sinusoidal signal is transmitted at one point and received at a distant

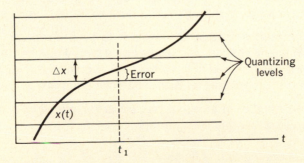

Figure 2–15 Error in analog-to-digital conversion.

point, the phase of the received signal is truly a random variable when the path over which signal travels is many wavelengths long. Since there is no physical reason for any one phase angle to be preferred over any other angle, the usual assumption is that the phase is uniformly distributed over a range of 2π. In order to illustrate this, suppose we have a time function of the form

$$x(t) = \cos(\omega t - \theta)$$

The phase angle θ is assumed to be a random variable whose probability density function is

$$f_\Theta(\theta) = \frac{1}{2\pi} \quad 0 < \theta \leq 2\pi$$

$$= 0 \quad \text{elsewhere}$$

From the previous discussion of the uniform density function, it is clear that the mean value of Θ is

$$\overline{\Theta} = \pi$$

and the variance of Θ is

$$\sigma_\theta^2 = \frac{\pi^2}{3}$$

It should also be noted that one could have just as well defined the region over which Θ exists to be $-\pi$ to $+\pi$, or any other region spanning 2π. Such a choice would not change the variance of Θ at all, but it would change the mean value.

Exercise 2–7.1

A continuous signal that can assume any value between -10 V and $+10$ V with equal probability is converted to digital form by quantizing.

a) How many discrete levels are required in order for the mean-square value of the quantizing error to be 0.01 volts squared?

b) If the number of discrete levels is to be a power of 2 in order to efficiently encode the levels into a binary number, how many levels are required to keep the mean-square value of the quantizing error not greater than 0.01 volts squared?

c) If the number of levels of part (b) are used, what is the actual mean-square quantizing error?

Answers: 0.003, 142, 256

Exponential and related distributions.   It was noted in the discussion of the uniform distribution that events occurring at random time instants are often *assumed* to occur at times that are equally probable. Thus, if the average time interval between events is denoted $\bar{\tau}$, then the probability that an event will occur in a time interval Δt that is short compared to $\bar{\tau}$ is just $\Delta t / \bar{\tau}$ regardless of where that time interval is. From this assumption it is possible to derive the probability distribution function (and, hence, the density function) for the time interval between events.

In order to carry out this derivation, consider the sketch in Figure 2–16. It is assumed that an event has occurred at time t_0, and it is desired to determine the probability that the next event will occur at a random time lying between $t_0 + \tau$ and $t_0 + \tau + \Delta t$. If the distribution function for τ is $F(\tau)$, then this probability is just $F(\tau + \Delta t) - F(\tau)$. But the probability that the event occurred in the Δt interval must also be equal to the product of the probabilities of the independent events that the event did *not* occur between t_0 and $t_0 + \tau$ and the event that it did occur between $t_0 + \tau$ and $t_0 + \tau + \Delta t$. Since

$$1 - F(\tau) = \text{probability that event did } not \text{ occur between } t_0 \text{ and } t_0 + \tau$$

$$\frac{\Delta t}{\bar{\tau}} = \text{probability that it } did \text{ occur in } \Delta t$$

it follows that

$$F(\tau + \Delta t) - F(\tau) = [1 - F(\tau)]\left(\frac{\Delta t}{\bar{\tau}}\right)$$

Upon dividing both sides by Δt and letting Δt approach zero, it is clear that

$$\lim_{\Delta t \to 0} \frac{F(\tau + \Delta t) - F(\tau)}{\Delta t} = \frac{dF(\tau)}{d\tau} = \frac{1}{\bar{\tau}}[1 - F(\tau)]$$

The latter two terms comprise a first-order differential equation that can be solved to yield

$$F(\tau) = 1 - \exp\left(\frac{-\tau}{\bar{\tau}}\right) \qquad \tau \geq 0 \qquad\qquad (2\text{–}41)$$

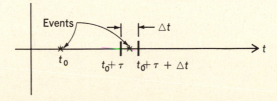

Figure 2–16   Time interval between events.

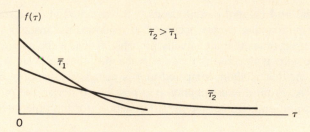

Figure 2–17   The exponential probability density function.

In evaluating the arbitrary constant, use is made of the fact that $F(0) = 0$ since τ can never be negative.

The probability density function for the time interval between events can be obtained from (2–41) by differentiation. Thus,

$$f(\tau) = \frac{1}{\bar{\tau}} \exp\left(\frac{-\tau}{\bar{\tau}}\right) \quad \tau \geq 0$$
$$= 0 \qquad\qquad \tau < 0$$

(2–42)

This is known as the exponential probability density function and is sketched in Figure 2–17 for two different values of average time interval.

As would be expected, the mean value of τ is just $\bar{\tau}$. That is,

$$E[\tau] = \int_0^\infty \frac{\tau}{\bar{\tau}} \exp\left(\frac{-\tau}{\bar{\tau}}\right) d\tau = \bar{\tau}$$

The variance turns out to be

$$\sigma_\tau^2 = (\bar{\tau})^2$$

It may be noted that this density function (like the Rayleigh) is a single-parameter one. Thus the mean and variance are uniquely related and one determines the other.

As an illustration of the application of the exponential distribution, suppose that component failures in a spacecraft occur independently and uniformly with an average time between failures of 100 days. The spacecraft starts out on a 200-day mission with all components functioning. What is the probability that it will complete the mission without a component failure? This is equivalent to asking for the probability that the time to the first failure is *greater* than 200 days; this is simply $[1 - F(200)]$ since $F(200)$ is the probability that this interval is *less* than (or equal to) 200 days. Hence, from (2–41)

$$1 - F(\tau) = 1 - \left[1 - \exp\left\{\frac{-\tau}{\bar{\tau}}\right\}\right] = \exp\left\{\frac{-\tau}{\bar{\tau}}\right\}$$

and for $\bar{\tau} = 100$, $\tau = 200$, this becomes

$$1 - F(200) = \exp\left(\frac{-200}{100}\right) = 0.1352$$

As a second example of the application of the exponential distribution consider a traveling wave tube (TWT) used as an amplifier in a satellite communication system and assume that it has a mean-time-to-failure (MTF) of 4 years. That is, the average lifetime of such a traveling wave tube is 4 years, although any particular device may fail sooner or last longer. Since the actual lifetime, T, is a random variable with an exponential distribution, we can determine the probability associated with any specified lifetime. For example, the probability that the TWT will survive for more than 4 years is

$$\Pr(T > 4) = 1 - F(4) = 1 - (1 - e^{-4/4}) = 0.368$$

Similarly, the probability that the TWT will fail within the first year is

$$\Pr(T \le 1) = F(1) = 1 - e^{-1/4} = 0.221$$

or the probability that it will fail between the 4th and the 6th years is

$$\Pr(4 < T \le 6) = F(6) - F(4) = (1 - e^{-6/4}) - (1 - e^{-4/4}) = 0.1447$$

Finally, the probability that the TWT will last as long as 10 years is

$$\Pr(T > 10) = 1 - F(10) = 1 - (1 - e^{-10/4}) = 0.0821$$

The random variable in the exponential distribution is the time interval between adjacent events. This can be generalized to make the random variable the time interval between any event and the kth following event. The probability distribution for this random variable is known as the *Erlang distribution* and the probability density function is

$$f_k(\tau) = \frac{\tau^{k-1}\exp(-\tau/\bar{\tau})}{(\bar{\tau})^k(k-1)!} \quad \tau \ge 0, k = 1, 2, 3, \ldots$$
$$= 0 \qquad\qquad\qquad \tau < 0$$

(2–43)

Such a random variable is said to be an *Erlang random variable of order k*. Note that the exponential distribution is simply the special case for $k = 1$. The mean and variance in the general case are $k\bar{\tau}$ and $k(\bar{\tau})^2$ respectively. The general Erlang distribution has a great many applications in engineering pertaining to the reliability of systems, the waiting times for users of a system (such as a telephone system or traffic system), and the number of channels required in a communication system to provide for a given number of users with random calling times and message lengths.

The Erlang distribution is also related to the *gamma distribution* by a simple change in notation. Letting $\beta = 1/\bar{\tau}$ and α be a continuous parameter that equals k for integral values, the Gamma distribution can be written as

$$f(\tau) = \frac{\beta^{\alpha}\tau^{\alpha-1}}{\Gamma(\alpha)} \exp(-\beta\tau) \qquad \tau \geq 0$$

$$\qquad\qquad = 0 \qquad\qquad\qquad\qquad \tau < 0$$

(2–44)

The mean and variance of the Gamma distribution are α/β and α/β^2 respectively.

Exercise 2–7.2

A package of 100-watt light bulbs states that the average lifetime is 750 hours. Two light bulbs from this package are installed in a lamp fixture at the same time. If it is assumed that the lifetimes of the two bulbs are independent random variables, find:

a) the probability that both bulbs will burn out before 750 hours

b) the probability that one bulb will burn out before 750 hours and the other one will burn out after 750 hours

c) the probability that both bulbs will last longer than 750 hours.

Answers: 0.1353, 0.2325, 0.3996

Delta distributions.   It was noted earlier that when the possible events could assume only a discrete set of values, the appropriate probability density function consisted of a set of delta functions. It is desirable to formalize this concept somewhat and indicate some possible applications. As an example, consider the binary waveform illustrated in Figure 2–18. Such a waveform arises in many types of communication systems or control systems since it obviously is the waveform with the greatest average power for a given peak value. It will be considered in more

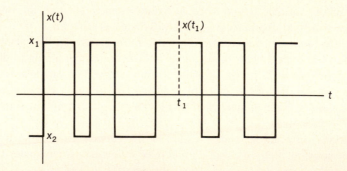

Figure 2–18   A general binary waveform.

detail throughout the study of random processes, but the present interest is in a single random variable, $X = x(t_1)$, at a specified time instant. This random variable can assume only two possible values, x_1 or x_2; it is specified that it take on value x_1 with probability p_1 and value x_2 with probability $p_2 = 1 - p_1$. Thus, the probability density function for X is

$$f(x) = p_1\, \delta(x - x_1) + p_2\, \delta(x - x_2) \qquad (2\text{–}45)$$

The mean value associated with this random variable is evaluated easily as

$$\overline{X} = \int_{-\infty}^{\infty} x[p_1\, \delta(x - x_1) + p_2\, \delta(x - x_2)]\, dx$$

$$= p_1 x_1 + p_2 x_2$$

The mean-square value is determined similarly from

$$\overline{X^2} = \int_{-\infty}^{\infty} x^2[p_1\, \delta(x - x_1) + p_2\, \delta(x - x_2)\, dx$$

$$= p_1 x_1{}^2 + p_2 x_2{}^2$$

Hence, the variance is

$$\sigma_X{}^2 = \overline{X^2} - (\overline{X})^2 = p_1 x_1{}^2 + p_2 x_2{}^2 - (p_1 x_1 + p_2 x_2)^2$$

$$= p_1 p_2 (x_1 - x_2)^2$$

in which use has been made of the fact that $p_2 = 1 - p_1$ in order to arrive at the final form.

It should be clear that similar delta distributions exist for random variables that can assume any number of discrete levels. Thus, if there are n possible levels designated as $x_1, x_2, \ldots, x_n$, and the corresponding probabilities for each level are $p_1, p_2, \ldots, p_n$, then the probability density function is

$$f(x) = \sum_{i=1}^{n} p_i\, \delta(x - x_i) \qquad (2\text{–}46)$$

in which

$$\sum_{i=1}^{n} p_i = 1$$

By using exactly the same techniques as above, the mean value of this random variable is shown to be

$$\overline{X} = \sum_{i=1}^{n} p_i x_i$$

and the mean-square value is

$$\overline{X^2} = \sum_{i=1}^{n} p_i x_i{}^2$$

From these, the variance becomes

$$\sigma_X^2 = \sum_{i=1}^{n} p_i x_i^2 - \left(\sum_{i=1}^{n} p_i x_i \right)^2$$

$$= \frac{1}{2} \sum_{i=1}^{n} \sum_{j=1}^{n} p_i p_j (x_i - x_j)^2$$

The multilevel delta distributions also arise in connection with communication and control systems, and in systems requiring analog-to-digital conversion. Typically the number of levels is an integer power of 2, so that they can be efficiently represented by a set of binary digits.

Exercise 2–7.3

When four coins are tossed, the random variable is taken to be the number of heads that result. Find:

a)   the mean value of this random variable

b)   the variance of this random variable.

Answers: 1.0, 2.0

2–8   Conditional Probability Distribution and Density Functions

The concept of conditional probability was introduced in Section 1–7 in connection with the occurrence of discrete events. In that context it was the quantity expressing the probability of one event given that another event, *in the same probability space*, had already taken place. It is desirable to extend this concept to the case of continuous random variables. The discussion in the present section will be limited to definitions and examples involving a single random variable. The case of two or more random variables is considered in Chapter 3.

The first step is to define the conditional probability distribution function for a random varaible X given that an event M has taken place. For the moment the event M is left arbitrary. The distribution function is denoted and defined by

$$F(x|M) = \Pr [X \leq x|M] \tag{2–47}$$

$$= \frac{\Pr \{X \leq x, M\}}{\Pr (M)} \qquad \Pr (M) > 0$$

where $\{X \leq x, M\}$ is the event of all outcomes ξ such that

$$X(\xi) \leq x \quad \text{and} \quad \xi \in M$$

where $X(\xi)$ is the value of the random variable X when the outcome of the experiment is ξ. Hence $\{X \leq x, M\}$ is the continuous counterpart of the set product used in the previous definition of (1–17). It can be shown that $F(x|M)$ is a valid probability distribution function and, hence, must have the same properties as any other distribution function. In particular, it has the following characteristics:

1. $0 \leq F(x|M) \leq 1 \qquad -\infty < x < \infty$
2. $F(-\infty|M) = 0 \qquad F(\infty|M) = 1$
3. $F(x|M)$ is *nondecreasing* as x increases
4. $\text{Pr } [x_1 < X \leq x_2 | M] = F(x_2|M) - F(x_1|M) \geq 0$

$$\text{for } x_1 < x_2$$

Now it is necessary to say something about the event M upon which the probability is conditioned. There are several different possibilities that arise. For example:

1. Event M may be an event that can be expressed in terms of the random variable X. Examples of this are considered in this section.
2. Event M may be an event that depends upon some other random variable, which may be either continuous or discrete. Examples of this are considered in Chapter 3.
3. Event M may be an event that depends upon both the random variable X and some other random variable. This is a more complicated situation that will not be considered at all.

As an illustration of the first possibility above, let M be the event

$$M = \{X \leq m\}$$

Then the conditional distribution function is, from (2–47),

$$F(x|M) = \text{Pr } \{X \leq x | X \leq m\} = \frac{\text{Pr } \{X \leq x, X \leq m\}}{\text{Pr } \{X \leq m\}}$$

There are now two possible situations—depending upon whether x or m is larger. If $x \geq m$, then the event that $X \leq m$ is contained in the event that $X \leq x$ and

$$\text{Pr } \{X \leq x, X \leq m\} = \text{Pr } \{X \leq m\}$$

Thus,

$$F(x|M) = \frac{\text{Pr } \{X \leq m\}}{\text{Pr } \{X \leq m\}} = 1 \qquad x \geq m$$

On the other hand, if $x \leq m$, then $\{X \leq x\}$ is contained in $\{X \leq m\}$ and

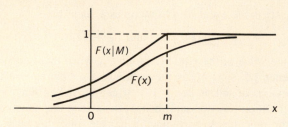

Figure 2–19   A conditional probability distribution function.

$$F(x|M) = \frac{\Pr\{X \le x\}}{\Pr\{X \le m\}} = \frac{F(x)}{F(m)}$$

The resulting conditional distribution function is shown in Figure 2–19.

The conditional probability density function is related to the distribution function in the same way as before. That is, when the derivative exists,

$$f(x|M) = \frac{dF(x|M)}{dx} \qquad (2\text{–}48)$$

This also has all the properties of a usual probability density function. That is,

1. $f(x|M) \ge 0 \quad -\infty < x < \infty$

2. $\displaystyle\int_{-\infty}^{\infty} f(x|M)\, dx = 1$

3. $F(x|M) = \displaystyle\int_{-\infty}^{x} f(u|M)\, du$

4. $\displaystyle\int_{x_1}^{x_2} f(x|M)\, dx = \Pr\,[x_1 < X \le x_2|M]$

If the example of Figure 2–19 is continued, the conditional probability density function is

$$f(x|M) = \frac{1}{F(m)}\frac{dF(x)}{dx} = \frac{f(x)}{F(m)} = \frac{f(x)}{\displaystyle\int_{-\infty}^{m} f(x)\, dx} \qquad x < m$$

$$= 0 \qquad\qquad\qquad\qquad\qquad\qquad\qquad\qquad x \ge m$$

This is sketched in Figure 2–20.

The conditional probability density function can also be used to find conditional means and conditional expectations. For example, the conditional mean is

$$E[X|M] = \int_{-\infty}^{\infty} xf(x|M)\, dx \qquad (2\text{–}49)$$

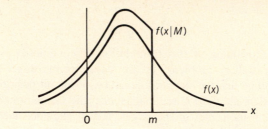

Figure 2–20 Conditional probability density function corresponding to Figure 2–19.

More generally, the conditional expectation of any $g(X)$ is

$$E[g(X)|M] = \int_{-\infty}^{\infty} g(x)f(x|M)\, dx \tag{2–50}$$

As an illustration of the conditional mean, let the $f(x)$ in the above example be Gaussian so that

$$f(x) = \frac{1}{\sqrt{2\pi}\,\sigma} \exp\left[-\frac{(x - \overline{X})^2}{2\sigma^2}\right]$$

In order to make the example simple, let $m = \overline{X}$ so that

$$F(m) = \int_{-\infty}^{m=\overline{X}} \frac{1}{\sqrt{2\pi}\,\sigma} \exp\left[-\frac{(x - \overline{X})^2}{2\sigma^2}\right] dx = \frac{1}{2}$$

Thus

$$f(x|M) = \frac{f(x)}{1/2} = \frac{2}{\sqrt{2\pi}\,\sigma} \exp\left[-\frac{(x - \overline{X})^2}{2\sigma^2}\right] \quad x < \overline{X}$$

$$= 0 \qquad\qquad\qquad\qquad\qquad x \geq \overline{X}$$

Hence, the conditional mean is

$$E[x|M] = \int_{-\infty}^{x} \frac{2x}{\sqrt{2\pi}\,\sigma} \exp\left[-\frac{(x-\overline{X})^2}{2\sigma^2}\right] dx$$

$$= \int_{-\infty}^{0} \frac{2(u + \overline{X})}{\sqrt{2\pi}\,\sigma} \exp\left(-\frac{u^2}{2\sigma^2}\right) du$$

$$= \overline{X} - \sqrt{\frac{2}{\pi}}\,\sigma$$

In words, this result says that the expected value or conditional mean of a Gaussian random variable, given that the random variable is less than its mean, is just

$$\overline{X} - \sqrt{\frac{2}{\pi}} \, \sigma$$

As a second illustration of this formulation of conditional probability, let us consider another archery problem. In this case, let the target be 12 inches in diameter and assume that the standard deviation of the hit positions is 4 inches in both the X-direction and the Y-direction. Hence, the unconditional mean value of miss distance from the center of the target, for all attempts, including those that miss the target completely, is just $\overline{R} = 4\sqrt{\pi/2} = 5.013$ inches. We now seek to find the conditional mean value of the miss distance given that the arrow strikes the target. Hence, we define the event M to be the event that the miss distance R is less than or equal to six inches. Thus, the conditional probability density function appears as

$$f(r|M] = \frac{f(r)}{F(6)}$$

Since the unconditional density function on R is

$$f(r) = \frac{r}{16} \exp\left(\frac{-r^2}{32}\right) \qquad r \geq 0$$

and the probability that R is less than or equal to 6 is

$$F(6) = 1 - e^{-6^2/32} = 0.675$$

it follows that the desired conditional density function is

$$f(r) = \frac{r}{10.806} \exp\left(-\frac{r^2}{32}\right) \qquad r \geq 0$$

Hence, the conditional mean value of the miss distance is

$$E[R|M] = \int_0^6 \frac{r^2}{10.806} \exp\left(-\frac{r^2}{32}\right) dr = 3.601 \text{ inches}$$

in which the integration has been carried out numerically. Note that this value is considerably smaller than the unconditional miss distance.

Exercise 2–8.1

A Gaussian random voltage having zero mean and a standard deviation of 10 V is connected in series with a 10–ohm resistor and an ideal diode. Find the mean value of the resulting current using the concepts of conditional probability.

Answer: 0.7979

Exercise 2–8.2

A traveling wave tube has a mean-time-to-failure of 4 years. Given that the TWT has survived for 4 years, find the conditional probability that it will fail between the 4th and 6th years.

Answer: 0.3935

2–9   Examples and Applications

The preceding sections have introduced some of the basic concepts concerning the probability distribution and density functions for a continuous random variable. Before extending these concepts to more than one variable, it is desirable to consider a few examples illustrating how they might be applied to simple engineering problems.

As a first example, consider the elementary voltage-regulating circuit shown in Figure 2–21(a). It employs a Zener diode having an idealized current-voltage characteristic as shown in Figure 2–21(b). Note that current is zero until the voltage reaches the breakdown value ($V_z = 10$) and from then on is limited by the external circuit, while the voltage across the diode remains constant. Such a circuit is often used to limit the voltage applied to solid-state devices. For example, the R_L indicated in the circuit may be a transistorized amplifier designed to work at 9 V and that is damaged if the voltage exceeds 10 V. The supply voltage, V_s, is from a power supply whose nominal voltage is 12 V, but whose actual voltage contains a sawtooth ripple and, hence, is a random variable. For purposes of this example, it will be assumed that this random variable has a uniform distribution over the interval from 9 to 15 V.

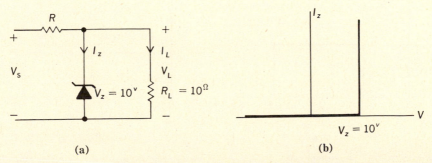

(a)   (b)

Figure 2–21   Zener diode voltage regulator: **(a)** voltage-regulating circuit and **(b)** Zener diode characteristic.

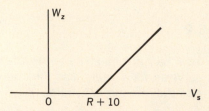

Figure 2–22   Relation between diode power dissipation and supply voltage.

Zener diodes are rated in terms of their ability to dissipate power as well as their breakdown voltage. It will be assumed that the average power rating of this diode is $W_z = 3$ W. It is then desired to find the value of series resistance, R, needed to limit the mean dissipation in the Zener diode to this rated value.

When the Zener diode is conducting, the voltage across it is $V_z = 10$, and the current through it is

$$I_z = \frac{V_s - V_z}{R} - I_L \quad \text{and} \quad V_s > \frac{V_z(R + R_L)}{R_L} = \frac{10(R + 10)}{10}$$

where the load current, I_L, is 1 A. The power dissipated in the diode is

$$W_z = V_z I_z = \frac{V_z(V_s - V_z)}{R} - I_L V_z$$

$$= \frac{10V_s - 100}{R} - 10 \qquad V_s > R + 10$$

A sketch of this power as a function of the supply voltage V_s is shown in Figure 2–22, and the probability density functions of V_s and W_z are shown in Figure 2–23. Note that the density function of W_z has a large delta function at zero, since

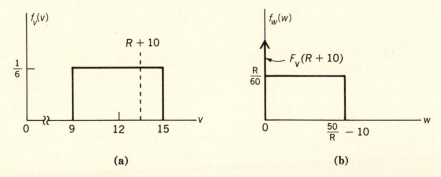

(a)   (b)

Figure 2–23   Probability density functions for supply voltage and diode power dissipation: **(a)** probability density function for V_s and **(b)** probability density function for W_z.

the diode is not conducting most of the time, but is uniform for larger values of W since W_z and V_s are linearly related in this range. From the previous discussion of transformations of density functions in Section 2–3, it is easy to show that

$$f_W(w) = F_V(R + 10)\,\delta(w) + \frac{R}{10}f_V\left(\frac{Rw}{10} + R + 10\right) \qquad 0 \le w \le \frac{50}{R} - 10$$

$$= 0 \qquad\qquad\qquad\qquad\qquad\qquad \text{elsewhere}$$

where $F_V(\cdot)$ is the distribution function of V_s. Hence, the area of the delta function is simply the probability that the supply voltage V_s is *less* than the value that causes diode conduction to start.

The mean value of diode power dissipation is now given by

$$E[W_z] = \overline{W}_z = \int_{-\infty}^{\infty} w f_W(w)\,dw$$

$$= \int_{-\infty}^{\infty} w F_V(R + 10)\,\delta(w)\,dw$$

$$+ \int_{0}^{\infty} w\left(\frac{R}{10}\right) f_V\left(\frac{Rw}{10} + R + 10\right)\,dw$$

The first integral has a value of zero (since the delta function is at $w = 0$) and the second integral can be written in terms of the uniform density function $[f_V(v) = \dfrac{1}{6}, \quad 9 < v \le 15]$, as

$$\overline{W}_z = \int_{0}^{(50/R) - 10} w\left(\frac{R}{10}\right)\left(\frac{1}{6}\right)\,dw = \frac{(5 - R)^2}{1.2R}$$

Since the mean value of diode power dissipation is to be less than or equal to 3 watts, it follows that

$$\frac{(5 - R)^2}{1.2R} \le 3 \qquad 0 < R \le 5$$

from which

$$R \ge 2.19 \ \Omega$$

It may now be concluded that any value of R greater than 2.19 Ω would be satisfactory from the standpoint of limiting the mean value of power dissipation in the Zener diode to 3 watts. The actual choice of R would be determined by the desired value of output voltage at the nominal supply voltage of 12 V. If this desired voltage is 9 V (as suggested above) then R must be

$$R = \frac{3}{\dfrac{9}{10}} = 3.33 \ \Omega$$

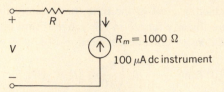

Figure 2–24 Selection of a voltmeter resistor.

which is greater than the minimum value of 2.19 Ω and, hence, would be satisfactory.

As another example, consider the problem of selecting a multiplier resistor for a dc voltmeter as shown in Figure 2–24. It will be assumed that the dc instrument produces full-scale deflection when 100 μA is passing through the coil and has a resistance of 1000 Ω. It is desired to select a multiplier resistor R such that this instrument will read full scale when 10 V is applied. Thus, the nominal value of R to accomplish this (which will be designated as R^*) is

$$R^* = \frac{10}{10^{-4}} - 1000 = 9.9 \times 10^4 \ \Omega$$

However, the actual resistor used will be selected at random from a bin of resistors marked $10^5 \ \Omega$. Because of manufacturing tolerances, the actual resistance is a random variable having a mean of 10^5 and a standard deviation of 1000 Ω. It will also be *assumed* that the actual resistance is a Gaussian random variable. (This is a customary assumption when deviations around the mean are small, even though it can never be precisely true for quantities that must be always positive, like the resistance.) On the basis of these assumptions it is desired to find the probability that the resulting voltmeter will be accurate to within 2 percent.[6]

The smallest value of resistance that would be acceptable is

$$R_{min} = \frac{10 - 0.2}{10^{-4}} - 1000 = 9.7 \times 10^4$$

while the largest value is

$$R_{max} = \frac{10 + 0.2}{10^{-4}} - 1000 = 10.1 \times 10^4$$

The probability that a resistor selected at random will fall between these two limits is

$$P_c = \text{Pr} \ [9.7 \times 10^4 < R \leq 10.1 \times 10^4] = \int_{9.7 \times 10^4}^{10.1 \times 10^4} f_R(r) \ dr \qquad \textbf{(2–51)}$$

---

[6]This is interpreted to mean that the error in voltmeter reading due to the resistor value is less than or equal to 2 percent of the full scale reading.

where $f_R(r)$ is the Gaussian probability density function for R and is given by

$$f_R(r) = \frac{1}{\sqrt{2\pi}\,(1000)} \exp\left[-\frac{(r - 10^5)^2}{2(10^6)}\right]$$

The integral in (2–51) can be expressed in terms of the standard normal distribution function, $\Phi(\cdot)$, as discussed in Section 2–5. Thus, P_c becomes

$$P_c = \Phi\left(\frac{10.1 \times 10^4 - 10^5}{10^3}\right) - \Phi\left(\frac{9.7 \times 10^4 - 10^5}{10^3}\right)$$

which can be simplified to

$$P_c = \Phi(1) - \Phi(-3)$$
$$= \Phi(1) - [1 - \Phi(3)]$$

Using the tables in Appendix D, this becomes

$$P_c = 0.8413 - [1 - 0.9987] = 0.8400$$

Thus, it appears that even though the resistors are selected from a supply that is nominally incorrect, there is still a substantial probability that the resulting instrument will be within acceptable limits of accuracy.

The third example considers an application of conditional probability. This example considers a traffic measurement system that is measuring the speed of all vehicles on an expressway and recording those speeds in excess of the speed limit of 70 miles per hour (mph). If the vehicle speed is a random variable with a Rayleigh distribution and a most probable value equal to 50 mph, it is desired to find the mean value of the excess speed. This is equivalent to finding the conditional mean of vehicle speed, given that the speed is greater than the limit, and subtracting the limit from it.

Letting the vehicle speed be S, the conditional distribution function that is sought is

$$F[s|S > 70] = \frac{\Pr\{S \le s, S > 70\}}{\Pr\{S > 70\}} \tag{2–52}$$

Since the numerator is nonzero only when $s > 70$, (2–52) can be written as

$$F[s|S > 70] = 0 \qquad\qquad s \le 70 \tag{2–53}$$
$$= \frac{F(s) - F(70)}{1 - F(70)} \qquad s > 70$$

where $F(\cdot)$ is the probability distribution function for the random variable S. The numerator of (2–53) is simply the probability that S is between 70 and s, while the denominator is the probability that S is greater than 70.

The conditional probability density function is found by differentiating (2–53) with respect to s. Thus,

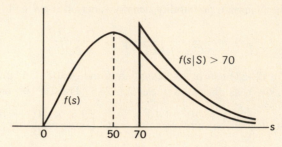

Figure 2–25   Conditional and unconditional density functions for a Rayleigh-distributed random variable.

$$f(s|S > 70) = 0 \qquad\qquad s \le 70$$

$$= \frac{f(s)}{1 - F(70)} \qquad s > 70$$

where $f(s)$ is the Rayleigh density function given by

$$f(s) = \frac{s}{(50)^2} \exp\left[-\frac{s^2}{2(50)^2} \right] \qquad s \ge 0$$

$$= 0 \qquad\qquad s < 0$$

(2–54)

These functions are sketched in Figure 2–25.

The quantity $F(70)$ is easily obtained from (2–54) as

$$F(70) = \int_0^{70} \frac{s}{(50)^2} \exp\left[-\frac{s^2}{2(50)^2} \right] ds = 1 - \exp\left[-\frac{49}{50} \right]$$

Hence,

$$1 - F(70) = \exp\left[-\frac{49}{50} \right]$$

The conditional expectation is given by

$$E[S|S > 70] = \frac{1}{\exp[-49/50]} \int_{70}^{\infty} \frac{s^2}{(50)^2} \exp - \left[\frac{s^2}{2(50)^2} \right] ds$$

$$= 70 + 50\sqrt{2\pi} \exp\left[\frac{49}{50} \right] \left\{ 1 - \Phi\left(\frac{7}{5} \right) \right\}$$

$$= 70 + 27.2$$

Thus, the mean value of the excess speed is 27.2 miles per hour. Although it is clear from this result that the Rayleigh model is not a realistic one for traffic systems (since 27.2 miles per hour excess speed is much too large for the actual situation), the above example does illustrate the general technique for finding conditional means.

The final example in this section combines the concepts of both discrete probability and continuous random variables and deals with problems that might arise in designing a satellite communication system. In such a system, the satellite normally carries a number of traveling wave tubes in order to provide more channels and to extend the useful life of the system as the tubes begin to fail. Consider a case in which the satellite is designed to carry 6 TWTs and it is desired to require that after 5 years of operation there is a probability of 0.95 that at least one of the TWTs is still good. The quantity that we need to find is the mean time to failure (MTF) for each tube in order to achieve this degree of reliability. In order to do this we need to use some of the results discussed in connection with Bernoulli trials in Sec. 1–10. In this case, let k be the number of good TWTs at any point in time and let p be the probability that any TWT is good. Since we want the probability that at least one tube is good to be 0.95, it follows that

$$\Pr(k \geq 1) = 0.95$$

or

$$\sum_{k=1}^{6} p_6(k) = 1 - p_6(0) = 1 - \binom{6}{0} p^0 (1 - p)^6 = 0.95$$

which can be solved to yield $p = 0.393$. If we assume, as usual, that the lifetime of any one TWT follows an exponential distribution, then

$$\int_5^\infty \frac{1}{T} e^{-\tau/T} \, d\tau = 0.393$$

$$T = 5.353$$

Thus, the mean time to failure for each TWT must be at least 5.353 years in order to achieve the desired reliability.

A second question that might be asked is "How many TWTs would be needed to achieve a probability of 0.99 that at least one will be still functioning after 5 years? "In this case, n is unknown but, for TWTs having the same MTF, the value of p is still 0.393. Thus,

$$1 - p_n(0) = 0.99$$

$$\binom{n}{0} p^0 (1 - p)^n = 0.01$$

This may be solved for n to yield $n = 9.22$. However, since n must be an integer, this tells us that we must use at least 10 traveling wave tubes to achieve the required reliability.

Exercise 2–9.1

The current in a semiconductor diode is often modeled by the Shockley equation

$$I = I_0[e^{\eta V} - 1]$$

in which V is the voltage across the diode, I_0 is the reverse current, η is a constant that depends upon the physical diode and the temperature, and I is the resulting diode current. For purposes of this exercise, assume that $I_0 = 10^{-9}$ and $\eta = 25$. Find the resulting mean value of current if the diode voltage is a random variable that is

a) uniformly distributed between 0 and 1

b) Gaussian with zero mean and variance of 0.07, or

c) Gaussian with zero mean and variance of 0.1. Comment on the results.

    Answers: 2.880, 3.163, 37,299

Exercise 2–9.2

A Thevenin's equivalent source has an open-circuit voltage of 18 V and a source resistance that is a random variable that is uniformly distributed between 4 ohms and 16 ohms. Find:

a) the value of load resistance that should be connected to this source in order that the mean value of the power delivered to the load is a maximum

b) the resulting mean value of power.

    Answers: 8, 9

PROBLEMS

2–1.1 For each of the following situations, list any quantities that might reasonably be considered a random variable, state whether they are continuous or discrete, and indicate a reasonable range of values for each.

a)   A weather forecast gives the prediction for July 4th as: high temper-
      ature, 84; low temperature, 67; wind, 8 mph; humidity, 75%; THI,
      72; sunrise, 5:05 am; sunset, 8:45 pm.

b)   A traffic survey on a busy street yields the following values: number
      of vehicles per minute, 26; average speed, 35 mph; ratio of cars to
      trucks, 6.81; average weight, 4000 lbs.; number of accidents per day,
      5.

c)   An electronic circuit contains 15 ICs, 12 LEDs, 43 resistors, and 12
      capacitors. The resistors are all marked 1000 ohms, the capacitors
      are all marked 0.01 microfarads, and the nominal supply voltage for
      the circuit is 5 volts.

2–1.2   State whether each of the following random variables is continuous or
         discrete and indicate a reasonable range of values for each.

a)   The outcome associated with rolling a pair of dice.

b)   The outcome resulting from measuring the voltage of a 12-volt stor-
      age battery.

c)   The outcome associated with randomly selecting a telephone number
      from the telephone directory.

d)   The outcome resulting from weighing adult males.

2–2.1   When ten coins are flipped, the event of interest is the number of heads.
         Let this number be the random variable.

a)   Plot the distribution function for this random variable.

b)   What is the probability that the random variable is between six and
      nine inclusive?

c)   What is the probability that the random variable is greater than or
      equal to eight?

2–2.2   A random variable has a probability distribution function given by

$$F_X(x) = 0 \qquad\qquad\qquad -\infty < x \le -1$$
$$= 0.5 + 0.5x \qquad -1 < x < 1$$
$$= 1 \qquad\qquad\qquad 1 \le x < \infty$$

a)   Find the probability that $x = \dfrac{1}{4}$.

b)   Find the probability that $x > \dfrac{3}{4}$.

c)   Find the probability that $-0.5 < x \le 0.5$

2–2.3   A probability distribution function for a random variable X has the form

$$F_X(x) = A\{1 - \exp[-(x-1)]\} \qquad 1 < x < \infty$$
$$= 0 \qquad\qquad\qquad\qquad -\infty < x \le 1$$

a)   For what value of A is this a valid probability distribution function?

b)   What is $F_X(2)$?

c)   What is the probability that the random variable lies in the interval $2 < X < \infty$?

d)   What is the probability that the random variable lies in the interval $1 < X \le 3$?

2–2.4   A random variable X has a probability distribution function of the form

$$F_X(x) = 0 \qquad\qquad\qquad\qquad -\infty < x \le -2$$
$$= A(1 + \cos bx) \qquad -2 < x \le 2$$
$$= 1 \qquad\qquad\qquad\qquad 2 < x < \infty$$

a)   Find the values of A and b that make this a valid probability distribution function.

b)   Find the probability that X is greater than 1.

c)   Find the probability that X is negative.

2–3.1   a)   Find the probability density function of the random variable of Problem 2–2.1 and sketch it.

b)   Using the probability density function, find the probability that the random variable is in the range between four and seven inclusive.

c)   Using the probability density function, find the probability that the random variable is less than four.

2–3.2   a)   Find the probability density function of the random variable of Problem 2–2.3 and sketch it.

b) Using the probability density function, find the probability that the random variable is in the range $2 < X \leq 3$.

c) Using the probability density function, find the probability that the random variable is less than 2.

2–3.3 a) A random variable X has a probability density function of the form

$$f_X(x) = \exp(-2|x|) \qquad -\infty < x < \infty$$

A second random variable Y is related to X by $Y = X^2$. Find the probability density function of the random variable Y.

b) Find the probability that Y is greater than 2.

2–3.4 a) A random variable Y is related to the random variable X of problem 2–3.3 by $Y = 3X - 4$. Find the probability density function of the random variable Y.

b) Find the probability that Y is negative.

c) Find the probability that Y is greater than X.

2–4.1 For the random variable of Problem 2–3.2 find:

a) The mean value of X.

b) The mean-square value of X.

c) The variance of X.

2–4.2 For the random variable X of Problem 2–2.4 find:

a) The mean value of X.

b) The mean-square value of X.

c) The third central moment of X.

d) The variance of X.

2–4.3 A random variable Y has a probability density function of the form

$$f(y) = Ky \qquad 0 < x \leq 6$$
$$= 0 \qquad \text{elsewhere}$$

a) Find the value of K for which this is a valid probability density function.

b)   Find the mean value of Y.

c)   Find the mean-square value of Y.

d)   Find the variance of Y.

e)   Find the third central moment of Y.

f)   Find the nth moment, $E[Y^n]$.

2–4.4   A power supply has five intermittent loads connected to it and each load, when in operation, draws a power of 10 W. Each load is in operation only one-quarter of the time and operates independently of all other loads.

a)   Find the mean value of the power required by the loads.

b)   Find the variance of the power required by the loads.

c)   If the power supply can provide only 40 watts, find the probability that it will be overloaded.

2–5.1   A Gaussian random voltage has a mean value of 10 and a variance of 25.

a)   What is the probability that an observed value of the voltage is greater than zero?

b)   What is the probability that an observed value of the voltage is greater than zero but less than or equal to the mean value?

c)   What is the probability that an observed value of the voltage is greater than twice the mean value?

2–5.2   For the Gaussian random variable of Problem 2–5.1 find:

a)   The fourth central moment.

b)   The fourth moment.

c)   The third central moment.

d)   The third moment.

2–5.3   A Gaussian random current has a probability of 0.5 of having value less than or equal to 1.0. It also has a probability of 0.0228 of having a value greater than 5.0.

a)   Find the mean value of this random variable.

b)   Find the variance of this random variable.

c) Find the probability that the random variable has a value less than or equal to 3.0.

2–5.4 A common method for detecting a signal in the presence of noise is to establish a threshold level and compare the value of any observation with this threshold. If the threshold is exceeded, it is decided that signal is present. Sometimes, of course, noise alone will exceed the threshold and this is known as a "false alarm." Usually, it is desired to make the probability of a false alarm very small. At the same time, we would like for any observation that does contain a signal plus the noise to exceed the threshold with a large probability. This is the probability of detection and should be as close to 1.0 as possible. Suppose we have Gaussian noise with zero mean and a variance of 1 V^2 and we set a threshold level of 5 volts.

a) Find the probability of false alarm.

b) If a signal having a value of 8 volts is observed in the presence of this noise, find the probability of detection.

2–6.1 A Gaussian random current having zero mean and a variance of 4 A^2 is passed through a resistance of 3 Ω.

a) Find the mean value of the power dissipated.

b) Find the variance of the power dissipated.

c) Find the probability that the instantaneous power will exceed 36 W.

2–6.2 A random variable X is Gaussian with zero mean and a variance of 1.0. Another random variable, Y, is defined by $Y = X^3$.

a) Write the probability density function for the random variable Y.

b) Find the mean value of Y.

c) Find the variance of Y.

2–6.3 A current having a Rayleigh probability density function is passed through a resistor having a resistance of 2π Ω. The mean value of the current is 2 A.

a) Find the mean value of the power dissipated in the resistor.

b) Find the probability that the dissipated power is less than or equal to 12 W.

c)   Find the probability that the dissipated power is greater than 72 W.

2–6.4   Marbles rolling on a flat surface have components of velocity in orthogonal directions that are independent Gaussian random variables with zero mean and a standard deviation of 3 ft/s.

a)   Find the most probable *speed* of the marbles.

b)   Find the mean value of the speed.

c)   What is the probability of finding a marble with a speed greater than 10 ft/s?

2–6.5   The average speed of a nitrogen molecule in air at 20°C is about 500 m/s. Find:

a)   The variance of molecule speed.

b)   The most probable molecule speed.

c)   The rms molecule speed.

2–6.6   Five independent observations of a Gaussian random voltage with zero mean and unit variance are made and a new random variable X^2 is formed from the *sum* of the squares of these random voltages.

a)   Find the mean value of X^2.

b)   Find the variance of X^2.

c)   What is the most probable value of X^2?

2–6.7   The log-normal density function is often expressed in terms of decibels rather than nepers. In this case, the Gaussian random variable Y is related to the log-normal random variable by $Y = 10 \log_{10} X$.

a)   Write the probabilities density function for X when this relation is used.

b)   Write an expression for the mean value of X.

c)   Write an expression for the variance of X.

2–7.1   A random variable Θ is uniformly distributed over a range of 0 to 2π. Another random variable X is related to Θ by

$$X = \cos \Theta$$

a)   Find the probability density function of X.

b)   Find the mean value of X.

c)   Find the variance of X.

d)   Find the probability that $X > 0.5$.

2–7.2   A continuous-valued random voltage ranging between -10 V and $+10$ V is to be quantized so that it can be represented by a binary sequence.

a)   If the rms quantizing error is to be less than 1% of the maximum value of the voltage, find the minimum number of quantizing levels that are required.

b)   If the number of quantizing levels is to be a power of 2, find the minimum number of quantizing levels that will still meet the requirement.

c)   How many binary digits are required to represent each quantizing level?

2–7.3   A communications satellite is designed to have a mean time to failure (MTF) of five years. If the actual time to failure is a random variable that is exponentially distributed, find:

a)   The probability that the satellite will fail sooner than five years.

b)   The probability that the satellite will survive for ten years or more.

c)   The probability that the satellite will fail during the sixth year.

2–7.4   A homeowner buys a package containing four light bulbs, each specified to have an average lifetime of 1000 hours. One bulb is placed in a single bulb table lamp and the remaining bulbs are used one after another to replace ones that burn out in this same lamp.

a)   Find the expected lifetime of the set of four light bulbs.

b)   Find the probability that the four light bulbs will last 5000 hours or more.

c)   Find the probability that the four light bulbs will all burn out in 2000 hours or less.

2–7.5   A continuous-valued signal has a probability density function that is uniform over the range from -8 V to $+8$ V. It is sampled and quantized into eight equally spaced levels ranging from -7 to $+7$.

a)   Write the probability density function for the discrete random variable representing one sample.

b) Find the mean value of this random variable.

c) Find the variance of this random variable.

2–8.1 a) For the communication satellite system of Problem 2–7.3, find the conditional probability that the satellite will survive for ten years or more given that it has survived for five years.

   b) Find the conditional mean lifetime of the system given that it has survived for three years.

2–8.2 a) For the random variable X of Problem 2–7.1, find the conditional probability density function $f(x|M)$, where M is the event $0 \le \Theta \le \frac{\pi}{2}$. Sketch this density function.

   b) Find the conditional mean $E[X|M]$, for the same event M.

2–8.3 A laser weapon is fired many times at a circular target that is 2 meters in diameter and it is found that one-tenth of the shots miss the target entirely.

   a) For those shots that hit the target, find the conditional probability that they will hit within 0.3 meter of the center.

   b) For those shots that miss the target completely, find the conditional probability that they come within 0.5 meter of the edge of the target.

2–8.4 Consider again the threshold detection system described in Problem 2–5.4.

   a) When noise only is present, find the conditional mean value of the noise that exceeds the threshold.

   b) Repeat part (a) when both the specified signal and noise are present.

2–9.1 Different types of electronic ac voltmeters produce deflections that are proportional to different characteristics of the applied waveforms. In most cases, however, the scale is calibrated so that the voltmeter correctly indicates the rms value of a *sine wave*. For other types of waveforms, the meter reading may not be equal to the rms value. Suppose the following instruments are connected to a Gaussian random voltage having zero mean and a standard deviation of 10 V. What will each read?

   a) An instrument in which the deflection is proportional to the average of the full-wave rectified waveform. That is, if $X(t)$ is applied, the deflection is proportional to $E[|X(t)|]$.

b) An instrument in which the deflection is proportional to the average of the envelope of the waveform. Remember that the envelope of a Gaussian waveform has a Rayleigh distribution.

2–9.2 In a radar system, the reflected signal pulses may have amplitudes that are Rayleigh distributed. Let the mean value of these pulses be $\sqrt{\pi/2}$. However, the only pulses that are displayed on the radar scope are those for which the pulse amplitude R is greater than some threshold r_0 in order that the effects of system noise can be supressed.

a) Determine the probability density function of the displayed pulses; that is, find $f(r|R > r_0)$. Sketch this density function.

b) Find the conditional mean of the displayed pulses if $r_0 = 0.5$.

2–9.3 A limiter has an input-output characteristic defined by

$$V_{out} = -B \qquad V_{in} < -A$$

$$= \frac{BV_{in}}{A} \qquad -A < V_{in} < A$$

$$= B \qquad V_{in} > A$$

a) If the input is a Gaussian random variable V with a mean value of $\overline{V}$ and a variance of σ_V^2, write a general expression for the probability density function of the output.

b) If $A = B = 5$ and the input is uniformly distributed from -2 to 8, find the mean value of the output.

2–9.4 Let the input to the limiter of Problem 2–9.3(b) be

$$V(t) = 10 \sin (\omega t + \Theta)$$

where Θ is a random variable that is uniformly distributed from 0 to 2π. The output of the limiter is sampled at an arbitrary time t to obtain a random varaible V_t.

a) Find the probability density function of V_t.

b) Find the mean value of V_t.

c) Find the variance of V_t.

References

See references for Chapter 1, particularly Clarke and Disney, Helstrom and Papoulis.

CHAPTER 3

Several Random Variables

3–1   Two Random Variables

All of the discussion so far has concentrated on situations involving a single random variable. This random variable may be, for example, the value of a voltage or current at a particular instant of time. It should be apparent, however, that saying something about a random voltage or current at only one instant of time is not adequate as a means of describing the nature of complete time functions. Such time functions, even if of finite duration, have an infinite number of random variables associated with them. This raises the question, therefore, of how one can extend the probabilistic description of a single random variable to include the more realistic situation of continuous time functions. The purpose of this section is to take the first step of that extension by considering *two* random variables. It might appear that this is an insignificant advance toward the goal of dealing with an infinite number of random variables, but it will become apparent later that this is really all that is needed, *provided that* the two random variables are separated in time by an arbitrary time interval. That is, if the random variables associated with *any* two instants of time can be described, then all of the information is available in order to carry out most of the usual types of systems analysis. Another situation that can arise in systems analysis is that in which it is desired to find the relation between the input and output of the system, either at the same instant of time or at two different time instants. Again, only two random variables are involved.

In order to deal with situations involving two random variables, it is necessary to extend the concepts of probability distribution and density functions that were discussed in the last chapter. Let the two random variables be designated as X and Y and define a *joint probability distribution function* as

$$F(x,y) = \Pr [X \leq x, Y \leq y]$$

Note that this is simply the probability of the event that the random variable X is less than or equal to x *and* that the random variable Y is less than or equal to y. As such, it is a straightforward extension of the probability distribution function for one random variable.

The joint probability distribution function has properties that are quite analogous to those discussed previously for a single variable. These may be summarized as follows:

1. $0 \leq F(x,y) \leq 1 \quad -\infty < x < \infty \quad -\infty < y < \infty$

2. $F(-\infty, y) = F(x, -\infty) = F(-\infty, -\infty) = 0$

3. $F(\infty, \infty) = 1$

4. $F(x,y)$ is a nondecreasing function as either x or y, or both, increase

5. $F(\infty, y) = F_Y(y) \qquad F(x, \infty) = F_X(x)$

In item 5 above, the subscripts on $F_Y(y)$ and $F_X(x)$ are introduced to indicate that these two distribution functions are not necessarily the same mathematical function of their respective arguments.

As an example of joint probability distribution functions, consider the outcomes of tossing two coins. Let X be a random variable associated with the first coin; let it have a value of 0 if a tail occurs and a value of 1 if a head occurs. Similarly let Y be associated with the second coin and also have possible values of 0 and 1. The joint distribution function, $F(x, y)$, is shown in Figure 3–1. Note that it satisfies all of the properties listed above.

It is also possible to define a *joint probability density function* by differentiating the distribution function. Since there are two independent variables, however, this differentiation must be done partially. Thus,

$$f(x, y) = \frac{\partial^2 F(x,y)}{\partial x \, \partial y} \tag{3-1}$$

and the sequence of differentiation is immaterial. The probability element is

$$f(x, y) \, dx \, dy = \Pr [x < X \leq x + dx, y < Y \leq y + dy] \tag{3-2}$$

The properties of the joint probability density function are quite analogous to those of a single random variable and may be summarized as follows:

1. $f(x, y) \geq 0 \quad -\infty < x < \infty \quad -\infty < y < \infty$

2. $\displaystyle\int_{-\infty}^{\infty} \int_{-\infty}^{\infty} f(x, y) \, dx \, dy = 1$

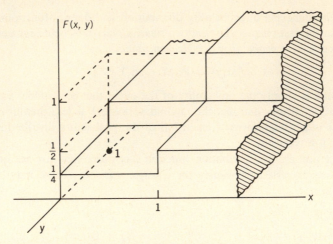

Figure 3–1   A joint probability distribution function.

3. $F(x, y) = \int_{-\infty}^{x} \int_{-\infty}^{y} f(u, v) \, dv \, du$

4. $f_X(x) = \int_{-\infty}^{\infty} f(x, y) \, dy \qquad f_Y(y) = \int_{-\infty}^{\infty} f(x, y) \, dx$

5. $\Pr[x_1 < X \le x_2, y_1 < Y \le y_2] = \int_{x_1}^{x_2} \int_{y_1}^{y_2} f(x, y) \, dy \, dx$

Note that item 2 implies that the *volume* beneath any joint probability density function must be unity.

As a simple illustration of a joint probability density function, consider a pair of random variables having a density function that is constant between x_1 and x_2 and between y_1 and y_2. Thus,

$$f(x, y) = \frac{1}{(x_2 - x_1)(y_2 - y_1)} \quad \begin{cases} x_1 < x \le x_2 \\ y_1 < y \le y_2 \end{cases}$$

$$= 0 \qquad\qquad \text{elsewhere} \qquad\qquad \textbf{(3–3)}$$

This density function and the corresponding distribution function are shown in Figure 3–2.

A physical situation in which such a probability density function could arise might be in connection with the manufacture of rectangular semiconductor substrates. Each substrate has two dimensions and the values of the two dimensions might be random variables that are uniformly distributed between certain limits.

The joint probability density function can be used to find the expected value of

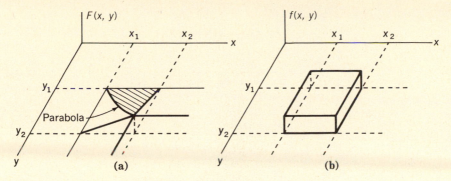

Figure 3–2 (a) Joint distribution and (b) density functions.

functions of two random variables in much the same way as with the single variable density function. In general, the expected value of any function $g(X, Y)$, can
be found from

$$E[g(X, Y)] = \int_{-\infty}^{\infty} \int_{-\infty}^{\infty} g(x, y)f(x, y) \, dx \, dy \qquad (3\text{--}4)$$

One such expected value that will be considered in great detail in a subsequent
section arises when $g(X, Y) = XY$. This expected value is known as the *correlation* and is given by

$$E[XY] = \int_{-\infty}^{\infty} \int_{-\infty}^{\infty} xyf(x, y) \, dx \, dy \qquad (3\text{--}5)$$

As a simple example of the calculation, consider the joint density function shown
in Figure 3–2(b). Since it is zero everywhere except in the specific region, (3–4)
may be written as

$$E[XY] = \int_{x_1}^{x_2} dx \int_{y_1}^{y_2} xy \left[\frac{1}{(x_2 - x_1)(y_2 - y_1)} \right] dy$$

$$= \frac{1}{(x_2 - x_1)(y_2 - y_1)} \left[\frac{x^2}{2} \Big|_{x_1}^{x_2} \right] \left[\frac{y^2}{2} \Big|_{y_1}^{y_2} \right]$$

$$= \frac{1}{4}(x_1 + x_2)(y_1 + y_2)$$

Item 4 in the above list of properties of joint probability density functions indicates that the *marginal* probability density functions can be obtained by integrating the joint density over the other variable. Thus, for the density function in
Figure 3–2(b), it follows that

$$f_X(x) = \int_{y_1}^{y_2} \frac{1}{(x_2 - x_1)(y_2 - y_1)} \, dy$$

$$= \frac{1}{(x_2 - x_1)(y_2 - y_1)} \left[y \Big|_{y_1}^{y_2} \right] \tag{3–6a}$$

$$= \frac{1}{x_2 - x_1}$$

and

$$f_Y(y) = \int_{x_1}^{x_2} \frac{1}{(x_2 - x_1)(y_2 - y_1)} \, dx$$

$$= \frac{1}{(x_2 - x_1)(y_2 - y_1)} \left[x \Big|_{x_1}^{x_2} \right] \tag{3–6b}$$

$$= \frac{1}{y_2 - y_1}$$

Exercise 3–1.1

Consider a rectangular semiconductor substrate with dimensions having mean values of 1 cm and 2 cm. Assume that the actual dimensions in both directions are uniformly distributed around the means with maximum deviations of 0.01 cm. Find:

a) the probability that both dimensions are larger than their mean values by 0.005 cm

b) the probability that the larger dimension is greater than its mean value by 0.005 cm and the smaller dimension is less than its mean value by 0.005 cm

c) the mean value of the area of the substrate.

   Answers: 1/16, 1/16, 2

Exercise 3–1.2

Two random variables X and Y have a joint probability density function given by

$$f(x,y) = Ae^{-(3x+4y)} \qquad x \geq 0, y \geq 0$$
$$= 0 \qquad x \geq 0, y \geq 0$$

Find:

a) the value of A for which this is a valid joint probability density function

b) the probability that $X > 1/2$ and $Y > 1/4$

c) the expected value of XY.

Answers: 0.0821, 12, 0.0833

3–2 Conditional Probability—Revisited

Now that the concept of joint probability for two random variables has been introduced, it is possible to extend the previous discussion of conditional probability. The previous definition of the conditional probability density function left the given event M somewhat arbitrary—although some specific examples were given. In the present discussion, the event M will be related to another random variable, Y.

There are several different ways in which the given event M can be defined in terms of Y. For example, M might be the event $Y \leq y$ and, hence, Pr (M) would be just the marginal distribution function of Y—that is, $F_Y(y)$. From the basic definition of the conditional distribution function given in (2–47) of the previous chapter, it would follow that

$$F_X(x|Y \leq y) = \frac{\text{Pr } [X \leq x, M]}{\text{Pr } (M)} = \frac{F(x,y)}{F_Y(y)} \tag{3-7}$$

Another possible definition of M is that it is the event $y_1 < Y \leq y_2$. The definition of (2–47) now leads to

$$F_X(x|y_1 < Y \leq y_2) = \frac{F(x,y_2) - F(x,y_1)}{F_Y(y_2) - F_Y(y_1)} \tag{3-8}$$

In both of the above situations, the event M has a nonzero probability—that is, Pr $(M) > 0$. However, the most common form of conditional probabiliity is one in which M is the event that $Y = y$; in almost all these cases Pr $(M) = 0$, since Y is continuously distributed. Since the conditional distribution function is defined as a ratio, it usually still exists even in these cases. It can be obtained from (3–8) by letting $y_1 = y$ and $y_2 = y + \Delta y$ and by taking a limit as Δy approaches zero. Thus,

$$F_X(x|Y = y) = \lim_{\Delta y \to 0} \frac{F(x, y + \Delta y) - F(x,y)}{F_Y(y + \Delta y) - F_Y(y)} = \frac{\partial F(x,y)/\partial y}{\partial F_Y(y)/\partial y} \tag{3-9}$$

$$= \frac{\int_{-\infty}^{x} f(u,y)\, du}{f_Y(y)}$$

The corresponding conditional density function is

$$f_X(x|Y = y) = \frac{\partial F_X(x|Y = y)}{\partial x} = \frac{f(x,y)}{f_Y(y)} \tag{3-10}$$

and this is the form that is most commonly used. By interchanging X and Y it follows that

$$f_Y(y|X = x) = \frac{f(x,y)}{f_X(x)} \tag{3-11}$$

Because this form of conditional density function is so frequently used, it is convenient to adopt a shorter notation. Thus, when there is no danger of ambiguity, the conditional density functions will be written as

$$f(x|y) = \frac{f(x,y)}{f_Y(y)} \tag{3-12}$$

$$f(y|x) = \frac{f(x,y)}{f_X(x)} \tag{3-13}$$

From these two equations one can obtain the continuous version of *Bayes' theorem,* which was given by (1–21) for the discrete case. Thus, eliminating $f(x,y)$ leads directly to

$$f(y|x) = \frac{f(x|y)f_Y(y)}{f_X(x)} \tag{3-14}$$

It is also possible to obtain the total probability from (3–12) or (3–13) by noting that

$$f_X(x) = \int_{-\infty}^{\infty} f(x,y)\, dy = \int_{-\infty}^{\infty} f(x|y)f_Y(y)\, dy \tag{3-15}$$

and

$$f_Y(y) = \int_{-\infty}^{\infty} f(x, y)\, dx = \int_{-\infty}^{\infty} f(y|x)f_X(x)\, dx \tag{3-16}$$

These equations are the continuous counterpart of (1–20), which applied to the discrete case.

A point that might be noted in connection with the above results is that the joint probability density function completely specifies both marginal density functions and both conditional density functions. As an illustration of this, consider a joint probability density function of the form

$$f(x,y) = \frac{6}{5}(1 - x^2 y) \qquad 0 \le x \le 1, 0 \le y \le 1$$

$$= 0 \qquad\qquad \text{elsewhere}$$

Integrating this function with respect to x alone and with respect to y alone yields the two marginal density functions as

$$f_X(x) = \frac{6}{5}\left(1 - \frac{x^2}{2}\right) \qquad 0 \le x \le 1$$

and

$$f_Y(y) = \frac{6}{5}\left(1 - \frac{y}{3}\right) \qquad 0 \le y \le 1$$

From (3–12) and (3–13) the two conditional density functions may now be written as

$$f(x|y) = \frac{1 - x^2 y}{1 - \dfrac{y}{3}} \qquad 0 \le x \le 1, 0 \le y \le 1$$

and

$$f(y|x) = \frac{1 - x^2 y}{1 - \dfrac{x^2}{2}} \qquad 0 \le x \le 1, 0 \le y \le 1$$

The use of conditional density functions arises in many different situations, but one of the most common (and probably the simplest) is that in which some observed quantity is the sum of two quantities—one of which is usually considered to be a signal while the other is considered to be a noise. Suppose, for example, that a signal $X(t)$ is perturbed by additive noise $N(t)$ and that the sum of these two, $Y(t)$, is the only quantity that can be observed. Hence, at some time instant, there are three random variables related by

$$Y = X + N$$

and it is desired to find the conditional probability density function of X given the observed value of Y—that is, $f(x|y)$. The reason for being interested in this is that the most probable values of X, given the observed value Y, may be a reasonable guess, or estimate, of the true value of X when X can only be observed in the presence of noise. From Bayes' theorem this conditional probability is

$$f(x|y) = \frac{f(y|x)f_X(x)}{f_Y(y)}$$

But if X is given, as implied by $f(y|x)$, then the only randomness about Y is the noise N, and it is assumed that its density function, $f_N(n)$, is known. Thus, since $N = Y - X$, and X is given,

$$f(y|x) = f_N(n = y - x) = f_N(y - x)$$

The desired conditional probability density, $f(x|y)$, can now be written as

$$f(x|y) = \frac{f_N(y - x)f_X(x)}{f_Y(y)} = \frac{f_N(y - x)f_X(x)}{\int_{-\infty}^{\infty} f_N(y - x) f_X(x) \, dx} \tag{3-17}$$

in which the integral in the denominator is obtained from (3–16). Thus, if the *a priori* density function of the signal, $f_X(x)$, and the noise density function, $f_N(n)$, are known, it becomes possible to determine the conditional density function, $f(x|y)$. When some particular value of Y is observed, say y_1, then the value of x for which $f(x|y_1)$ is a maximum is a good estimate for the true value of X.

As a specific example of the above application of conditional probability, suppose that the signal random variable, X, has an exponential density function so that

$$f_X(x) = b \exp(-bx) \qquad x \geq 0$$

$$= 0 \qquad x < 0$$

Such a density function might arise, for example, as a signal from a space probe in which the *time intervals* between counts of high-energy particles are converted to *voltage amplitudes* for purposes of transmission back to earth. The noise that is added to this signal is assumed to be Gaussian, with zero mean, so that its density function is

$$f_N(n) = \frac{1}{\sqrt{2\pi} \, \sigma_N} \exp\left(-\frac{n^2}{2\sigma_N^2}\right)$$

The marginal density function of Y, which appears in the denominator of (3–17), now becomes

$$f_Y(y) = \int_0^{\infty} \frac{b}{\sqrt{2\pi} \, \sigma_N} \exp\left[-\frac{(y - x)^2}{2\sigma_N^2}\right] \exp(-bx) \, dx$$

$$= \frac{b}{2} \exp\left(-by + \frac{b^2\sigma_N^2}{2}\right)\left[1 + \operatorname{erf}\left(\frac{y - b\sigma_N^2}{\sqrt{2} \, \sigma_N}\right)\right]^1 \tag{3-18}$$

---

[1]The *error function* is related to the normal probability distribution function and is defined as

$$\operatorname{erf}(z) = \frac{2}{\sqrt{\pi}} \int_0^z e^{-u^2} \, du = 2\Phi(\sqrt{2} \, z) - 1$$

It should be noted, however, that if one is interested only in locating the maximum of $f(x|y)$, it is not necessary to evaluate $f_Y(y)$ since it is *not* a function of x. Hence, for a given Y, $f_Y(y)$ is simply a constant.

The desired conditional density function can now be written, from (3–17), as

$$f(x|y) = \frac{b}{\sqrt{2\pi}\,\sigma_{Nf_Y}(y)}\exp\left[-\frac{(y-x)^2}{2\sigma_N^2}\right]\exp(-bx) \qquad x \geq 0$$

$$= 0 \qquad\qquad\qquad\qquad\qquad\qquad\qquad\qquad x < 0$$

This may also be written as

$$f(x|y) = \frac{b}{\sqrt{2\pi}\,\sigma_{Nf_Y}(y)}\exp\left\{-\frac{1}{2\pi_N^2}[x^2 - 2(y-b\sigma_N^2)x+y^2]\right\}x \geq 0$$

$$= 0 \qquad\qquad\qquad\qquad\qquad\qquad\qquad\qquad x < 0 \qquad\text{(3–19)}$$

and this is sketched in Figure 3–3 for two different values of y.

It was noted earlier that when a particular value of Y is observed, a reasonable estimate for the true value of X is that value of x which maximizes $f(x|y)$. Since the conditional density function is a maximum (with respect to x) when the exponent is a *minimum*, it follows that this value of x can be determined by equating the derivative of the exponent to zero. Thus

$$2x - 2(y - b\sigma_N^2) = 0$$

or

$$x = y - b\sigma_N^2 \qquad\text{(3–20)}$$

is the location of the maximum, *provided that* $y - b\sigma_N^2 > 0$. Otherwise, there is no point of zero slope on $f(x|y)$ and the largest value occurs at $x = 0$. Suppose, therefore, that the value $Y = y_1$ is observed. Then, if $y_1 > b\sigma_N^2$, the appropriate estimate for X is $\hat{X} = y_1 - b\sigma_N^2$. On the other hand, if $y_1 < b\sigma_N^2$, the appropri-

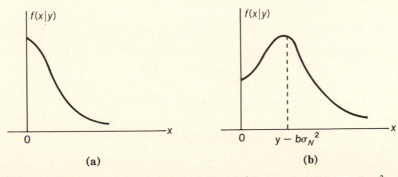

(a)   (b)

Figure 3–3 The conditional density function, $f(x|y)$: (a) case for $y < b\sigma_N^2$ and (b) case for $y > b\sigma_N^2$.

ate estimate for X is $\hat{X} = 0$. Note that as the noise gets smaller ($\sigma_N^2 \to 0$), the estimate of X approaches the observed value y_1.

Exercise 3–2.1

Two random variables, X and Y, have a joint probability density function of the form

$$f(x, y) = k(x + y) \qquad 0 \le x \le 1, 0 \le y \le 1$$
$$= 0 \qquad \text{elsewhere}$$

Find:

a) the value of k for which this is a valid joint probability density function

b) the conditional probability that X is greater than 1/2 given that $Y = 1/2$

c) the conditional probability that Y is less than, or equal to, 1/2 given that X is 1/2.

Answers: 3/8, 5/8, 1

Exercise 3–2.2

A random signal X is uniformly distributed between 6 and 10 V. It is observed in the presence of Gaussian noise N having zero mean and a standard deviation of 2 V.

a) If the observed value of signal plus noise, $(X + N)$, is 4, find the best estimate of the signal amplitude.

b) Repeat (a) if the observed value of signal plus noise is 8.

c) Repeat (a) if the observed value of signal plus noise is 12.

Answers: 6, 8, 10

3–3   Statistical Independence

The concept of statistical independence was introduced earlier in connection with discrete events, but is equally important in the continuous case. Random variables that arise from different physical sources are almost always statistically independent. For example, the random thermal voltage generated by one resistor in a circuit is in no way related to the thermal voltage generated by another resistor. Statistical independence may also exist when the random variables come from the same source but are defined at greatly different times. For example, the thermal voltage generated in a resistor tomorrow almost certainly does not depend upon the voltage today. When two random variables are statistically independent, a knowledge of one random variable gives no information about the value of the other.

The joint probability density function for statistically independent random variables can always be factored into the two marginal density functions. Thus, the relationship

$$f(x,y) = f_X(x)f_Y(y) \tag{3-21}$$

can be used as a definition for statistical independence, since it can be shown that this factorization is both a necessary and sufficient condition. As an example, this condition is satisfied by the joint density function given in (3–3). Hence, these two random variables are statistically independent.

One of the consequences of statistical independence concerns the correlation defined by (3–5). Because the joint density function is factorable, (3–5) can be written as

$$
\begin{aligned}
E[XY] &= \int_{-\infty}^{\infty} x f_X(x)\, dx \int_{-\infty}^{\infty} y f_Y(y)\, dy \\
&= E[X]E[Y] = \overline{X}\,\overline{Y}
\end{aligned}
\tag{3-22}
$$

Hence, the expected value of the product of two statistically independent random variables is simply the product of their mean values. The result will be zero, of course, if *either* random variable has zero mean.

Another consequence of statistical independence is that conditional probability density functions become marginal density functions. For example, from (3–12)

$$f(x|y) = \frac{f(x,y)}{f_Y(y)}$$

but if X and Y are statistically independent the joint density function is factorable and this becomes

$$f(x|y) = \frac{f_X(x)f_Y(y)}{p_Y(y)} = f_X(x)$$

Similarly,

$$f(y|x) = \frac{f(x,y)}{f_X(x)} = \frac{f_X(x)f_Y(y)}{f_X(x)} = f_Y(y)$$

It may be noted that the random variables described by the joint probability density function of Exercise 3–1.2 are statistically independent since the joint density function can be factored into the product of a function of x only and a function of y only. However, the random variables defined by the joint probability density function of Exercise 3–2.1 are not statistically independent since this density function cannot be factored in this manner.

Exercise 3–3.1

Two random variables, X and Y, have a joint probability density function of the form

$$f(x,y) = k(xy + 2x + y + a) \qquad 0 \le x \le 1, 0 \le y \le 1$$
$$= 0 \qquad \text{elsewhere}$$

Find:

a) the values of k and a for which the random variables X and Y are statistically independent

b) the expected value of XY.

      Answers: 4/15, 10/9, 2

Exercise 3–3.2

Two random variables, X and Y, have Gaussian probability density functions with means of 1 and 2 respectively and variances of 1 and 4 respectively. Find the probability that $XY > 0$.

      Answer: 0.7078

3–4   Correlation Between Random Variables

As noted above, one of the important applications of joint probability density functions is that of specifying the *correlation* of two random variables; that is, whether one random variable depends in any way upon another random variable.

If two random variables X and Y have possible values x and y, then the expected value of their product is known as the correlation, defined in (3–5) as

$$E[XY] = \int_{-\infty}^{\infty} \int_{-\infty}^{\infty} xyf(x,y)\ dx\ dy = \overline{XY} \tag{3–5}$$

If both of these random variables have nonzero means, then it is frequently more convenient to find the correlation with the mean values subtracted out. Thus,

$$E[(X - \overline{X})(Y - \overline{Y})] = \overline{(X - \overline{X})(Y - \overline{Y})} \tag{3–23}$$

$$= \int_{-\infty}^{\infty} \int_{-\infty}^{\infty} (x - \overline{X})(y - \overline{Y})f(x,y)\ dx\ dy$$

This is known as the *covariance*, by analogy to the variance of a single random variable.

If it is desired to express the degree to which two random variables are correlated without regard to the magnitude of either one, then the *correlation coefficient* or *normalized covariance* is the appropriate quantity. The correlation coefficient, which is denoted by ρ, is defined as

$$\rho = E\left\{ \left[\frac{X - \overline{X}}{\sigma_X} \right] \left[\frac{Y - \overline{Y}}{\sigma_Y} \right] \right\} = \int_{-\infty}^{\infty} \int_{-\infty}^{\infty} \frac{x - \overline{X}}{\sigma_X} \cdot \frac{y - \overline{Y}}{\sigma_Y} f(x,y)\ dx\ dy \tag{3–24}$$

Note that each random variable has its mean subtracted out and is divided by its standard deviation. The resulting random variable is often called the *standardized variable* and is one with zero mean and unit variance.

An alternative, and sometimes simpler, expression for the correlation coefficient can be obtained by multiplying out the terms in equation (3–24). This yields

$$\rho = \int_{-\infty}^{\infty} \int_{-\infty}^{\infty} \frac{xy - \overline{X}y - \overline{Y}x + \overline{X}\,\overline{Y}}{\sigma_X \sigma_Y} f(x,y)dx\ dy$$

Carrying out the integration leads to

$$\rho = \frac{E(XY) - \overline{X}\,\overline{Y}}{\sigma_X \sigma_Y} \tag{3–25}$$

In order to investigate some of the properties of ρ, define the standardized variables ξ and η as

$$\xi = \frac{X - \overline{X}}{\sigma_X} \qquad \eta = \frac{Y - \overline{Y}}{\sigma_Y}$$

$$\overline{\xi} = 0 \qquad \overline{\eta} = 0$$

$$\sigma_\xi^2 = 1 \qquad \sigma_\eta^2 = 1$$

Then,

$$\rho = E[\xi\eta]$$

Now look at

$$E[(\xi \pm \eta)^2] = E[\xi^2 \pm 2\xi\eta + \eta^2] = 1 \pm 2\rho + 1$$
$$= 2(1 \pm \rho)$$

Since $(\xi \pm \eta)^2$ is always positive, its expected value must also be positive, so that

$$2(1 \pm \rho) \geq 0$$

Hence, ρ can never have a magnitude greater than one and thus

$$-1 \leq \rho \leq 1$$

If X and Y are statistically independent, then

$$\rho = E[\xi\eta] = \overline{\xi}\,\overline{\eta} = 0$$

since both ξ and η are zero mean. Thus, the correlation coefficient for statistically independent random variables is always zero. The converse is not necessarily true, however. A correlation coefficient of zero does not automatically mean that X and Y are statistically independent unless they are Gaussian, as will be seen.

In order to illustrate the above properties, consider two random variables for which the joint probability density function is

$$f(x,y) = x + y \qquad 0 \leq x \leq 1, \qquad 0 \leq y \leq 1$$
$$= 0 \qquad\qquad \text{elsewhere}$$

From Property 4 pertaining to joint probability density functions, it is straightforward to obtain the marginal density functions as

$$f_X(x) = \int_0^1 (x + y)\, dy = x + \frac{1}{2} \qquad 0 \leq x \leq 1$$

and

$$f_Y(y) = \int_0^1 (x + y)\, dx = y + \frac{1}{2} \qquad 0 \leq y \leq 1$$

from which the mean values of X and Y can be obtained immediately as

$$\overline{X} = \int_0^1 x\left(x + \frac{1}{2}\right) dx = \frac{7}{12}$$

with an identical value for $E[Y]$. The variance of X is readily obtained from

$$\sigma_X{}^2 = \int_0^1 \left(x - \frac{7}{12}\right)^2 \left(x + \frac{1}{2}\right) dx = \frac{11}{144}$$

Again there is an identical value for $\sigma_Y{}^2$. Also the expected value of XY is given by

$$E[XY] = \int_0^1 \int_0^1 xy(x + y)dx\,dy = \frac{1}{3}$$

Hence, from (3–25) the correlation coefficient becomes

$$\rho = \frac{E[XY] - \bar{X}\bar{Y}}{\sigma_X \sigma_Y} = \frac{1/3 - (7/12)^2}{11/144} = -\frac{1}{11}$$

Although the correlation coefficient can be defined for any pair of random variables, it is particularly useful for random variables that are individually and jointly Gaussian. In these cases, the joint probability density function can be written as

$$f(x,y) = \frac{1}{2\pi\sigma_X\sigma_Y\sqrt{1 - \rho^2}}$$

$$\exp\left\{\frac{-1}{2(1 - \rho^2)}\left[\frac{(x - \bar{X})^2}{\sigma_X^2} + \frac{(y - \bar{Y})^2}{\sigma_Y^2} - \frac{2(x - \bar{X})(y - \bar{Y})\rho}{\sigma_X\sigma_Y}\right]\right\} \qquad (3\text{–}26)$$

Note that when $\rho = 0$, this reduces to

$$f(x,y) = \frac{1}{2\pi\sigma_X\sigma_Y}\exp\left\{-\frac{1}{2}\left[\frac{(x - \bar{X})^2}{\sigma_X^2} + \frac{(y - \bar{Y})^2}{\sigma_Y^2}\right]\right\}$$

$$= f_X(x)f_Y(y)$$

which is the form for statistically independent Gaussian random variables. Hence, $\rho = 0$ does imply statistical independence in the Gaussian case.

It is also of interest to use the correlation coefficient to express some results for general random variables. For example, from the definitions of the standardized variables it follows that

$$X = \sigma_X\xi + \bar{X} \qquad \text{and} \qquad Y = \sigma_Y\eta + \bar{Y}$$

and, hence

$$\overline{XY} = E[(\sigma_X\xi + \bar{X})(\sigma_Y\eta + \bar{Y})] = E(\sigma_X\sigma_Y\xi\eta + \bar{X}\sigma_Y\eta + \bar{Y}\sigma_X\xi + \bar{X}\bar{Y}] \qquad (3\text{–}27)$$

$$= \rho\sigma_X\sigma_Y + \bar{X}\bar{Y}$$

As a further example, consider

$$E[(X \pm Y)^2] = E[X^2 \pm 2XY + Y^2] = \overline{X^2} \pm 2\overline{XY} + \overline{Y^2}$$

$$= \sigma_X^2 + (\bar{X})^2 \pm 2\rho\sigma_X\sigma_Y \pm 2\bar{X}\bar{Y} + \sigma_Y^2 + (\bar{Y})^2$$

$$= \sigma_X^2 + \sigma_Y^2 \pm 2\rho\sigma_X\sigma_Y + (\bar{X} \pm \bar{Y})^2$$

Since the last term is just the square of the mean of $(X \pm Y)$, it follows that the variance of $(X \pm Y)$ is

$$[\sigma_{(X \pm Y)}]^2 = \sigma_X^2 + \sigma_Y^2 \pm 2\rho\sigma_X\sigma_Y \qquad (3\text{–}28)$$

Note that when random variables are uncorrelated ($\rho = 0$), the variance of sum or difference is the sum of the variances.

Exercise 3–4.1

The result expressed in equation (3–28) may be used to measure the correlation coefficient of two random signals. To illustrate this, assume that we measure the mean and variance of a random signal to be 5 and 8 respectively, and the mean and variance of another signal to be 3 and 10 respectively. We then combine the two signals and measure the mean-square value of the sum as 75. Find:

a) the variance of the sum

b) the correlation of the signals

c) the correlation coefficient.

        Answers: -0.391, 11, 11.5

Exercise 3–4.2

A random variable X has a variance of 9 and a statistically independent random variable Y has a variance of 16. Their sum is another random variable $Z = X + Y$. Without assuming that either random variable has zero mean, find:

a) the correlation coefficient for X and Z

b) the correlation coefficient for Y and Z

c) the variance of Z.

        Answers: 25, 0.6, 0.8

3–5   Density Function of the Sum of Two Random Variables

The above example illustrates that the mean and variance associated with the sum (or difference) of two random variables can be determined from a knowledge of the individual means and variances and the correlation coefficient without any regard to the probability density functions of the random variables. A more difficult question, however, pertains to the probability density function of the sum of two random variables. The only situation of this sort that is considered here is the one in which the two random variables are statistically independent. The more general case is beyond the scope of the present discussion.

Let X and Y be statistically independent random variables with density functions of $f_X(x)$ and $f_Y(y)$, and let the sum be

$$Z = X + Y$$

It is desired to obtain the probability density function of Z, $f_Z(z)$. The situation is best illustrated graphically as shown in Figure 3–4. The probability distribution function for Z is just

$$F_Z(z) = \text{Pr } (Z \leq z) = \text{Pr } (X + Y \leq z)$$

and can be obtained by integrating the joint density function, $f(x,y)$, over the region *below* the line, $x + y = z$. For every fixed y, x must be such that $-\infty < x < z - y$. Thus,

$$F_Z(z) = \int_{-\infty}^{\infty} \int_{-\infty}^{z-y} f(x,y) \, dx \, dy \tag{3–29}$$

For the special case in which X and Y are statistically independent, the joint density function is factorable and (3–29) can be written as

$$F_Z(z) = \int_{-\infty}^{\infty} \int_{-\infty}^{z-y} f_X(x)f_Y(y) \, dx \, dy$$

$$= \int_{-\infty}^{\infty} f_Y(y) \int_{-\infty}^{z-y} f_X(x) \, dx \, dy$$

The probability density function of Z is obtained by differentiating $F_Z(z)$ with respect to z. Hence

$$f_Z(z) = \frac{dF_Z(z)}{dz} = \int_{-\infty}^{\infty} f_Y(y)f_X(z - y) \, dy \tag{3–30}$$

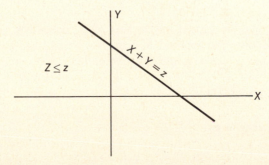

Figure 3–4 Showing the region for $X + Y = Z \leq z$.

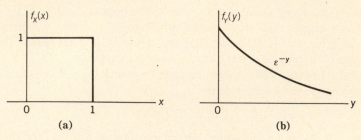

Figure 3–5 Density functions for two random variables.

since z appears only in the upper limit of the second integral. Thus, the probability density function of Z is simply the *convolution* of the density functions of X and Y.

It should also be clear that (3–29) could have been written equally well as

$$F_Z(z) = \int_{-\infty}^{\infty} \int_{-\infty}^{z-x} f(x,y) \, dy \, dx$$

and the same procedure would lead to

$$f_Z(z) = \int_{-\infty}^{\infty} f_X(x) f_Y(z - x) \, dx \qquad \textbf{(3–31)}$$

Hence, just as in the case of system analysis, there are two equivalent forms for the convolution integral.

As a simple example of this procedure, consider the two density functions shown in Figure 3–5. These may be expressed analytically as

$$f_X(x) = 1 \qquad 0 \le x \le 1$$
$$= 0 \qquad \text{elsewhere}$$

and

$$f_Y(y) = e^{-y} \quad y \ge 0$$
$$= 0 \qquad y < 0$$

The convolution must be carried out in two parts, depending on whether z is greater or less than one. The appropriate diagrams, based on (3–30), are sketched in Figure 3–6. When $0 < z \le 1$, the convolution integral becomes

$$f_Z(z) = \int_0^z (1)e^{-(z-x)} \, dx = 1 - e^{-z} \qquad 0 < z \le 1$$

When $z > 1$, the integral is

$$f_Z(z) = \int_0^1 (1)e^{-(z-x)} \, dx = (e - 1)e^{-z} \qquad 1 < z < \infty$$

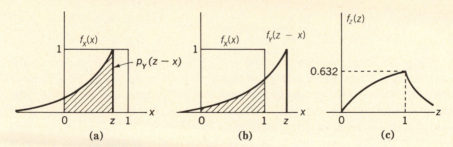

Figure 3–6 Convolution of density functions: (a) $0 < z \leq 1$, (b) $1 < z < \infty$, and (c) $f_Z(z)$.

When $z < 0$, $f_Z(z) = 0$ since both $f_X(x) = 0$, $x < 0$ and $f_Y(y) = 0$, $y < 0$. The resulting density function is sketched in Figure 3–6(c).

It is straightforward to extend the above result to the difference of two random variables. In this case let

$$Z = X - Y$$

All that is necessary in this case is to replace y by $-y$ in equation (3–30). Thus,

$$f_Z(z) = \int_{-\infty}^{\infty} f_Y(y) f_X(z + y) dy \tag{3–32}$$

There is also an alternative expression analogous to equation (3–31). This is

$$f_Z(z) = \int_{-\infty}^{\infty} f_X(x) f_Y(x - z)\, dx \tag{3–33}$$

It is also of interest to consider the case of the sum of two independent Gaussian random variables. Thus let

$$f_X(x) = \frac{1}{\sqrt{2\pi}\,\sigma_X} \exp\left[\frac{-(x - \bar{X})^2}{2\sigma_X^2}\right]$$

and

$$f_Y(y) = \frac{1}{\sqrt{2\pi}\,\sigma_Y} \exp\left[\frac{-(y - \bar{Y})^2}{2\sigma_Y^2}\right]$$

Then if $Z = X + Y$, the density function for z is [based on (3–31)]

$$f_Z(z) = \frac{1}{2\pi\sigma_X\sigma_Y} \int_{-\infty}^{\infty} \exp\left[\frac{-(x - \bar{X})^2}{2\sigma_X^2}\right] \exp\left[\frac{-(z - x - \bar{Y})^2}{2\sigma_Y^2}\right] dx$$

It is left as an exercise for the student to verify that the result of this integration is

$$f_Z(z) = \frac{1}{\sqrt{2\pi(\sigma_X^2 + \sigma_Y^2)}} \exp\left\{\frac{-[z - (\overline{X} + \overline{Y})]^2}{2(\sigma_X^2 + \sigma_Y^2)}\right\} \tag{3–34}$$

This result clearly indicates that the sum of two *independent* Gaussian random variables is still Gaussian with a mean that is the sum of the means and a variance that is the sum of the variances. It should also be apparent that by adding more random variables, the sum is still Gaussian. Thus, the sum of any number of independent Gaussian random variables is still Gaussian. Density functions that exhibit this property are said to be *reproducible;* the Gaussian case is one of a very limited class of density functions that are reproducible. Although it will not be proven here, it can likewise be shown that the sum of *correlated* Gaussian random variables is also Gaussian with a mean that is the sum of the means and a variance that can be obtained from (3–28).

The fact that sums (and differences) of Gaussian random variables are still Gaussian is a very important one in the analysis of linear systems. It can also be shown that derivatives and integrals of time functions that have a Gaussian distribution are still Gaussian. Thus, one can carry out the analysis of linear systems for Gaussian inputs with the assurance that signals everywhere in the system are Gaussian. This is analogous to the use of sinuosidal functions for carrying out steady state system analysis in which signals everywhere in the system are still sinusoids at the same frequency.

Exercise 3–5.1

Let a random variable X have a probability density function of

$$f_X(x) = 5e^{-5x} \qquad x \geq 0$$

and a statistically independent random variable Y have a probability density function of

$$f_Y(y) = 2e^{-2y} \qquad y \geq 0$$

For the random variable $Z = X + Y$ find:

a) $f_Z(0)$

b) the value of z for which $f_Z(z)$ is a maximum

c) the probability that Z is greater than 1.0.

Answers: 0, 0.221, 0.305

Exercise 3–5.2

The resistance values of a supply of resistors are independent random variables and are uniformly distributed between 100 and 120 ohms. If two resistors are selected at random and connected in series, find:

a) the most probable value of resistance for the series combination

b) the largest value of resistance for the series combination

c) the probability that the series combination will have a resistance value greater than 220 ohms.

Answers: 220, 0.5, 240

3–6 The Characteristic Function

It is shown in the previous section that the probability density function of the sum of two independent random variables can be obtained by convolving the individual density functions. When more than two random variables are summed, the resulting density function can obviously be obtained by repeating the convolution until every random variable has been taken into account. Since this is a lengthy and tedious procedure, it is natural to inquire if there is some easier way.

When convolution arises in system and circuit analysis, it is well known that transform methods can be used to simplify the computation since convolution then becomes a simple multiplication of the transforms. Repeated convolution is accomplished by multiplying more transforms together. Thus, it seems reasonable to try to use transform methods when dealing with density functions. This section discusses how to do it.

The *characteristic function* of a random variable X is defined to be

$$\phi(u) = E[e^{juX}] \tag{3-35}$$

and this expected value can be obtained from

$$\phi(u) = \int_{-\infty}^{\infty} f(x)e^{jux} \, dx \tag{3-36}$$

The right side of (3–36) is (except for a minus sign in the exponent) the Fourier transform of the density function $f(x)$. The difference in sign for the characteristic function is traditional rather than fundamental, and makes no essential difference in the application or properties of the transform. By analogy to the inverse Fourier transform, the density function can be obtained from

$$f(x) = \frac{1}{2\pi} \int_{-\infty}^{\infty} \phi(u)e^{-jux} \, du \qquad (3\text{--}37)$$

In order to illustrate one application of characteristic functions, consider once again the problem of finding the probability density function of the sum of two independent random variables X and Y, where $Z = X + Y$. The characteristic functions for these random variables are

$$\phi_X(u) = \int_{-\infty}^{\infty} f_X(x)e^{jux} \, dx$$

and

$$\phi_Y(u) = \int_{-\infty}^{\infty} f_Y(y)e^{jux} \, dy$$

Since convolution corresponds to multiplication of transforms (characteristic functions) it follows that the characteristic function of Z is

$$\phi_Z(u) = \phi_X(u)\phi_Y(u)$$

The resulting density function for Z becomes

$$f_Z(z) = \frac{1}{2\pi} \int_{-\infty}^{\infty} \phi_X(u)\phi_Y(u)e^{-juz} \, du \qquad (3\text{--}38)$$

This technique can be illustrated by reworking the example of the previous section, in which X was uniformly distributed and Y exponentially distributed. Since

$$f_X(x) = 1 \qquad 0 \le x \le 1$$
$$= 0 \qquad \text{elsewhere}$$

the characteristic function is

$$\phi_X(u) = \int_0^1 (1)e^{jux} \, dx = \left. \frac{e^{jux}}{ju} \right|_0^1$$

$$= \frac{e^{ju} - 1}{ju}$$

Likewise,

$$f_Y(y) = e^{-y} \qquad y \ge 0$$
$$= 0 \qquad y < 0$$

so that

$$\phi_Y(u) = \int_0^\infty e^{-y}e^{juy}\,dy = \frac{e^{(-1+ju)y}}{(-1+ju)}\bigg|_0^\infty = \frac{1}{1-ju}$$

Hence, the characteristic function of Z is

$$\phi_Z(u) = \phi_X(u)\phi_Y(u) = \frac{e^{ju}-1}{ju(1-ju)}$$

and the corresponding density function is

$$f_Z(z) = \frac{1}{2\pi}\int_{-\infty}^\infty \frac{e^{ju}-1}{ju(1-ju)}e^{-juz}\,du$$

$$= \frac{1}{2\pi}\int_{-\infty}^\infty \frac{e^{ju(1-z)}}{ju(1-ju)}\,du - \frac{1}{2\pi}\int_{-\infty}^\infty \frac{e^{-juz}}{ju(1-ju)}\,du$$

$$= 1 - e^{-z} \qquad \text{when} \qquad 0 < z < 1$$

$$= (e-1)e^{-z} \qquad \text{when} \qquad 1 < z < \infty$$

The integration can be carried out by standard inverse Fourier transform methods or by the use of tables.

Another application of the characteristic function is to find the moments of a random variable. Note that if $\phi(u)$ is differentiated, the result is

$$\frac{d\phi(u)}{du} = \int_{-\infty}^\infty f(x)(jx)e^{jux}\,dx$$

For $u = 0$, the derivative becomes

$$\frac{d\phi(u)}{du}\bigg|_{u=0} \equiv j\int_{-\infty}^\infty xf(x)\,dx = j\overline{X} \qquad\qquad \text{(3–39)}$$

Higher order derivatives introduce higher powers of x into the integrand so that the general nth moment can be expressed as

$$\overline{X^n} = E[X^n] = \frac{1}{j^n}\left[\frac{d^n\phi(u)}{du^n}\right]_{u=0} \qquad\qquad \text{(3–40)}$$

If the characteristic function is available, this may be much easier than carrying out the required integrations of the direct approach.

There are some fairly obvious extensions of the above results. For example, (3–38) can be extended to an arbitrary number of independent random variables. If $X_1, X_2, \ldots, X_n$ are independent and have characteristic functions of $\phi_1(u)$, $\phi_2(u), \ldots, \phi_n(u)$, and if

$$Y = X_1 + X_2 + \ldots + X_n$$

then Y has a characteristic function of

$$\phi_Y(u) = \phi_1(u)\phi_2(u) \cdots \phi_n(u)$$

and a density function of

$$f_Y(y) = \frac{1}{2\pi} \int_{-\infty}^{\infty} \phi_1(u)\phi_2(u) \cdots \phi_n(u)e^{-juy}\, du \tag{3-41}$$

The characteristic function can also be extended to cases in which random variables are not independent. For example, if X and Y have a joint density function of $f(x, y)$, then they have a joint characteristic function of

$$\phi_{X,Y}(u, v) = E[e^{j(uX + vY)}] = \int_{-\infty}^{\infty} \int_{-\infty}^{\infty} f(x, y)e^{j(ux + vy)}\, dx\, dy \tag{3-42}$$

The corresponding inversion relation is

$$f(x, y) = \frac{1}{(2\pi)^2} \int_{-\infty}^{\infty} \int_{-\infty}^{\infty} \phi_{XY}(u, v)e^{-j(ux + vy)}\, du\, dv \tag{3-43}$$

The joint characteristic function can be used to find the correlation between the random variables. Thus, for example,

$$E[XY] = \overline{XY} = -\left[\frac{\partial^2 \phi_{XY}(u, v)}{\partial u\, \partial v}\right]_{u=v=0} \tag{3-44}$$

More generally,

$$E[X^i Y^k] = \overline{X^i Y^k} = \frac{1}{j^{i+k}} \left[\frac{\partial^{i+k} \phi_{XY}(u, v)}{\partial u^i\, \partial v^k}\right]_{u=v=0} \tag{3-45}$$

The results given in (3–40), (3–43), and (3–45) are particularly useful in the case of Gaussian random variables since the necessary integrations and differentiations can always be carried out. One of the valuable properties of Gaussian random variables is that moments and correlations of all orders can be obtained from only a knowledge of the first two moments and the correlation coefficient.

Exercise 3–6.1

For the two random variables in Exercise 3–5.1, find the probability density function of $Z = X + Y$ by using the characteristic function.

Answer: Same as found in Exercise 3–5.1.

Exercise 3–6.1

A random variable X has a probability density function of the form

$$f(x) = 2e^{-4|x|} \qquad -\infty < x < \infty$$

Using the characteristic function, find the 1st and 2nd moments of this random variable.

Answers: 0, 1/8

PROBLEMS

3–1.1 Two random variables have a joint probability distribution function defined by

$$F(x, y) = 0 \qquad x < 0, y < 0$$
$$= xy \qquad 0 \le x \le 1, 0 \le y \le 1$$
$$= 1 \qquad x > 1 \, y > 1$$

a) Sketch this distribution function.

b) Find the joint probability density function and sketch it.

c) Find the joint probability of the event $X \le \dfrac{3}{4}$ and $Y > \dfrac{1}{4}$.

3–1.2 Two random variables, X and Y, have a joint probability density function given by

$$f(x,y) = kxy \qquad 0 \le x \le 1, 0 \le y \le 1$$
$$= 0 \qquad \text{elsewhere}$$

a) Determine the value of k that makes this a valid probability density function.

b) Determine the joint probability distribution function $F(x,y)$.

c) Find the joint probability of the event $X \le \dfrac{1}{2}$ and $Y > \dfrac{1}{2}$.

d) Find the marginal density function, $f_X(x)$.

3–1.3 a) For the random variables of Problem 3–1.1 find $E[XY]$.

   b) For the random variables of Problem 3–1.2 find $E[XY]$.

3–1.4 Let X be the outcome from rolling one die and Y the outcome from rolling a second die.

   a) Find the joint probability of the event $X \le 3$ and $Y > 3$.

   b) Find $E[XY]$.

   c) Find $E\left[\dfrac{X}{Y}\right]$.

3–2.1 A signal X has a Rayleigh density function and a mean value of 10 and is added to noise, N, that is uniformly distributed with a mean value of zero and a variance of 12. X and N are statistically independent and can be observed only as $Y = X + N$.

   a) Find, sketch, and label the conditional probability density function, $f(x|y)$, as a function of x for $y = 0, 6,$ and 12.

   b) If an observation yields a value of $y = 12$, what is the best estimate of the true value of X?

3–2.2 For the joint probability density function of Problem 3–1.2, find:

   a) The conditional probability density function $f(x|y)$.

   b) The conditional probability density function $f(y|x)$.

3–2.3 A dc signal having a uniform distribution over the range from -5 V to $+5$ V is measured in the presence of an independent noise voltage having a Gaussian distribution with zero mean and a variance of 2 V^2.

   a) Find, sketch, and label the conditional probability density function of the signal given the value of the measurement.

   b) Find the best estimate of the signal voltage if the measurement is 6 V.

   c) Find the best estimate of the noise voltage if the measurement is 7 V.

3–2.4   A random signal X can be observed only in the presence of independent additive noise N. The observed quantity is $Y = X + N$. The joint probability density function of X and Y is

$$f(x,y) = K \exp\left[-(x^2 + y^2 + 4xy)\right] \qquad \textit{all } x \textit{ and } y$$

a)   Find a general expression for the best estimate of X as function of the observation $Y = y$.

b)   If the observed value of Y is $y = 3$, find the best estimate of X.

3–3.1   For each of the following joint probability density functions state whether the random variables are statistically independent and find $E[XY]$.

a)
$$f(x,y) = \frac{kx}{y} \qquad 0 \le x \le 1,\, 1 \le y \le 2$$
$$= 0 \qquad \text{elsewhere}$$

b)
$$f(x,y) = k(x^2 + y^2) \qquad 0 \le x \le 1,\, 0 \le y \le 1$$
$$= 0 \qquad \text{elsewhere}$$

c)
$$f(x,y) = k(xy + 2x + 3y + 6) \qquad 0 \le x \le 1,\, 0 \le y \le 1$$
$$= 0 \qquad \text{elsewhere}$$

3–3.2   Let X and Y be statistically independent random variables. Let $W = g(X)$ and $V = h(Y)$ be any transformations with continuous derivatives on X and Y. Show that W and V are also statistically independent random variables.

3–4.1   Two random variables have zero mean and variances of 16 and 36. Their correlation coefficient is 0.5.

a)   Find the variance of their sum.

b)   Find the variance of their difference.

c)   Repeat (a) and (b) if the correlation coefficient is -0.5.

3–4.2   Two statistically independent random variables, X and Y, have variances of $\sigma_X^2 = 9$ and $\sigma_Y^2 = 25$. Two new random variables are defined by

$$U = 3X + 4Y$$
$$V = 5X - 2Y$$

   a) Find the variances of U and V.

   b) Find the correlation coefficient of U and V.

3–4.3 Let X be a zero-mean random variable having a variance of 9 and let Y be another zero-mean random variable. The sum of X and Y has a variance of 29 and the difference of X and Y has a variance of 21.

   a) Find the variance of Y.

   b) Find the correlation coefficient of X and Y.

   c) Find the variance of $U = 3X - 5Y$.

3–4.4 Three zero mean, unit variance random variables X, Y, and Z are added to form a new random variable, $W = X + Y + Z$. Random variables X and Y are uncorrelated, X and Z have a correlation coefficient of 1/2, and Y and Z have a correlation coefficient of $-1/2$.

   a) Find the variance of W.

   b) Find the correlation coefficient between W and X.

   c) Find the correlation coefficient between W and the sum of Y and Z.

3–5.1 A random variable X has a probability density function of

$$f_X(x) = 2x \qquad 0 \le x \le 1$$
$$= 0 \qquad \text{elsewhere}$$

and an independent random variable Y is uniformly distributed between -1.0 and 1.0.

   a) Find the probability density function of the random variable $Z = X + 2Y$.

   b) Find the probability that $0 < Z \le 1$.

3–5.2 A commuter attempts to catch the 8:00 am train every morning although his arrival time at the station is a random variable that is uniformly distributed between 7:55 am and 8:05 am. The train's departure time from the station is also a random variable that is uniformly distributed between 8:00 am and 8:10 am.

a) Find the probability density function of the time interval between the commuter's arrival at station and the train's departure time.

b) Find the probability that the commuter will catch the train.

c) If the commuter gets delayed 3 minutes by a traffic jam, find the probability that the train will still be at the station.

3–5.3 A sinusoidal signal has the form

$$X(t) = \cos(100t + \Theta)$$

where Θ is a random variable that is uniformly distributed between 0 and 2π. Another sinusoidal signal has the form

$$Y(t) = \cos(100t + \psi)$$

where ψ is independent of Θ and is also uniformly distributed between 0 and 2π. The sum of these two sinusoids, $Z(t) = X(t) + Y(t)$ can be expressed in terms of its magnitude and phase as

$$Z(t) = A \cos(100t + \phi)$$

a) Find the probability that $A > 1$.

b) Find the probability that $A \leq \dfrac{1}{2}$.

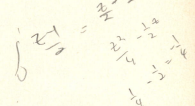

3–5.4 Many communication systems connecting computers employ a technique known as "packet transmission." In this type of system, a collection of binary digits (perhaps 1000 of them) are grouped together and transmitted as a "packet." The time interval between packets is a random variable that is usually assumed to be exponentially distributed with a mean value that is the reciprocal of the average number of packets per second that are transmitted. Under some conditions it is necessary for a user to delay transmission of a packet by a random amount that is uniformly distributed between 0 and T. If a user is generating 100 packets per second, and his maximum delay time, T, is 1 ms, find:

a) The probability density function of the time interval between packets.

b) The mean value of the time interval between packets.

3–6.1 A random variable X has a probability density function of the form

$$f_X(x) = e^{-x}u(x)$$

and an independent random variable Y has a probability density function of

$$f_Y(y) = 3e^{-3y}u(y)$$

Using characteristic functions, find the probability density function of $Z = X + Y$.

3–6.2 a) Find the characteristic function of a Gaussian random variable with zero mean and variance σ^2.

b) Using the characteristic function, verify the result in Section 2–5 for the nth central moment of a Gaussian random variable.

3–6.3 The characteristic function of the Bernoulli distribution is

$$\phi(u) = 1 - p + pe^{ju}$$

where p is the probability that the event of interest will occur at any one trial. Find:

a) The mean value of the Bernoulli random variable.

b) The mean-square value of the random variable.

c) The 3rd central moment of the random variable.

References

See references for Chapter 1, particularly Clarke and Disney, Helstrom and Papoulis.

CHAPTER 4

Elements of Statistics

4—1   Introduction

Now that we have completed an introductory study of probability and random variables, it is desirable to turn our attention to some of the important engineering applications of these concepts. One such application is in the field of statistics. Although our major objective in this text is to apply probabilistic concepts to the study of signals and systems, the field of statistics is of such importance to the engineer that it would not be appropriate to proceed without a brief discussion of the subject. Therefore, the objective of this chapter is to present a very brief introduction to some of the elementary concepts of statistics before turning all of our attention to signals and systems. It may be noted, however, that this material may be omitted without jeopardizing the understanding of subsequent chapters if time does not permit its inclusion.

Probability and statistics are often considered to be one and the same subject and they are often linked together in courses and textbooks. However, they are really two different areas of study even though statistics relies heavily upon probabilistic concepts. In fact, the usual definition of statistics makes no reference at all to probability. Instead, it defines statistics as the science of assembling, classifying, tabulating, and analyzing data or facts. In apparent agreement with this definition, a popular undergraduate textbook on statistics does not even discuss probability until the eighth chapter!

There are two general branches of statistics that are frequently designated as *descriptive statistics* and *inductive statistics* or *statistical inference*. Descriptive statistics involves collecting, grouping, and presenting data in a way that can be easily understood or assimilated. Statistical inference, on the other hand, uses the data to draw conclusions about, or estimate parameters of, the environment from which the data came.

The field of statistics is very large and includes a great many areas of specialty. For our purposes, however, it is convenient to classify them into five theoretical areas. They are:

a) *Sampling theory,* which deals with problems associated with selecting samples from some collection of data that is too large to be examined completely.

b) *Estimation theory,* which is concerned with making some estimate or prediction based on the data that is available.

c) *Hypothesis testing,* which attempts to decide which of two or more hypotheses about the data are true.

d) *Curve fitting and regression,* which attempts to find mathematical expressions that best represent the data.

e) *Analysis of variance,* which attempts to assess the significance of variations in the data and the relation of these variations to the physical situations from which the data arose.

One cannot hope to cover all of these topics in one brief chapter, so we will limit our attention to some simple concepts associated with sampling theory, a brief exposure to hypothesis testing, and a short discussion and example of linear regression.

4–2   Sampling Theory—The Sample Mean

A problem that often arises in connection with quality control of manufactured items is that of determining whether the items are meeting the desired quality standards without actually testing all of them. Usually, the number of items being manufactured is so large that it would be impractical to test every one. The alternative is to test only a few items and hope that these few are representative of all the items. Similar problems arise in connection with taking polls of public opinion, in determining the popularity of certain television programs, or in determining any sort of average about the general population.

Problems of the type listed above are solved by *sampling* the collection of items or facts that are being considered. A sufficient number of samples must be taken in order to obtain an answer in which one has reasonable confidence. Clearly, one would not predict the outcome of a presidential election by taking the result of asking the first person met on the street. Nor would one claim that one million transistors are all good or all bad on the basis of testing only one of them. On the other hand, it may be very expensive and time consuming to take samples; thus, it is important not to take more samples than are actually required. One of the purposes of this section is to determine how many samples are required for a given degree of confidence in the result.

It is necessary to introduce some terminology in connection with sampling. The collection of data that is being studied is known as the *population*. For example, if a production line is set up to make a particular device, then all of these devices that are produced in a given run become the population. If one is concerned with predicting the outcome of an election, then the population is all persons voting in that election. The number of items or pieces of data that make up the population is designated as N. This is said to be the *size* of the population. If N is not a very large number, then its value may be significant. On the other hand, if N is very large it is often convenient to assume that it is infinity. The calculations for infinite populations are somewhat easier to carry out than for finite values of N, and, as will be seen, for very large N it makes very little difference whether the actual value of N is used or if one assumes N is infinite.

A *sample*, or more precisely a *random sample*, is simply part of the population that has been selected at random. As mentioned in Chapter 1, the term "selected at random" implies that all members of the population are equally likely to be selected. This is a very important consideration and one must often go to considerable difficulty to ensure that all members of the population do have an equal probability of being selected. The number of items or pieces of data in the sample is denoted as n and is called the size of the sample.

There are a number of calculations that can be made with the members of the sample and one of the most important of these is the *sample mean*. For most engineering purposes, every item in the sample can be assigned a numerical value. Obviously, there are other types of samples, such as might arise in public opinion sampling, where numerical values cannot be assigned; we are not going to be concerned with such situations. For our purposes, let us assume that we have a sample of size n drawn from a population of size N, and that each element of the sample has a numerical value that is designated by $x_1, x_2, \ldots, x_n$. For example, if we are testing bipolar transistors these x-values might be the dc current gain, β. We also assume that we have a truly random sample so that the elements we have are truly representative of the entire population. The sample mean is simply the average of the numerical values that make up the sample. Hopefully, this average value will be close to the average value of the population from which the sample is drawn. How close it might be is one of the problems addressed here.

When one has a particular sample, the sample mean is denoted by

$$\bar{x} = \frac{1}{n} \sum_{i=1}^{n} x_i \tag{4–1}$$

where the x_i are the particular values in the sample. More generally, however, we are interested in describing the statistical properties of arbitrary random samples rather than those of any particular sample. In this case, the sample mean becomes a random variable, as do the members of the sample. Thus, it is appropriate to denote the sample mean as

$$\hat{\bar{X}} = \frac{1}{n} \sum_{i=1}^{n} X_i \qquad (4\text{--}2)$$

where the X_i are random variables from the population and each is assumed to have the population probability density function $f(x)$. Note that the notation here is consistent with that used previously in connection with random variables; capital letters being used for random variables and small letters for possible values of the random variable. This notation is used throughout this chapter and it is important to distinguish general results, which deal with random variables, from specific cases in which particular values are used.

The true mean value of the population from which the sample came is denoted by $\bar{X}$. Hopefully, the sample mean will be close to this value. Since the sample mean, in the general case, is a random variable, it also has a mean value. Thus,

$$E[\hat{\bar{X}}] = E\left[\frac{1}{n} \sum_{i=1}^{n} X_i\right]$$

$$= \frac{1}{n} \sum_{i=1}^{n} E[X_i]$$

$$= \frac{1}{n} \sum_{i=1}^{n} \bar{X} = \bar{X}$$

It is clear from this result that the mean value of the sample mean is equal to the true mean value of the population. It is said, therefore, that the sample mean is an *unbiased estimate* of the population mean. The term "unbiased estimate" is one that arises often in the study of statistics and it simply implies that the mean value of the estimate of any parameter is the same as the true mean value of the parameter.

Although it is certainly desirable for the sample mean to be an unbiased estimate of the true mean, this is not sufficient to indicate whether the sample mean is a *good* estimator of the true population mean. Since the sample mean is itself a random variable, it will have a value that fluctuates around the true population mean as different samples are drawn. Therefore, it is desirable to know something about the magnitude of this fluctuation; that is, to determine the variance of the sample mean. This is done first for the case in which the population size is very much greater than the sample size; that is, $N \gg n$. In such cases, it is reasonable to assume that the characteristics of the population do not change as the sample is drawn. It is also equivalent to assuming that $N = \infty$.

In order to calculate the variance, we look at the difference between the mean-square value of $\hat{\bar{X}}$ and the square of the mean value of $\hat{\bar{X}}$, which, as we have just seen, is the true mean of the population, $\bar{X}$. Thus,

$$\text{Var } (\hat{\bar{X}}) = E\left[\frac{1}{n^2} \sum_{i=1}^{n} \sum_{j=1}^{n} X_i X_j\right] - (\bar{X})^2$$

(4–3)

$$= \frac{1}{n^2} \sum_{i=1}^{n} \sum_{j=1}^{n} E[X_i X_j] - (\bar{X})^2$$

Since X_i and X_j are parameters of different items in the population, it is reasonable to assume that they are statistically independent random variables when $i \neq j$. Hence, it follows that

$$E[X_i X_j] = \overline{X^2} \qquad\qquad i = j$$

$$= (\bar{X})^2 \text{ or } (X)^2 \quad i \neq j$$

Using this result in (4–3) leads to

$$\text{Var } (\hat{\bar{X}}) = \frac{1}{n^2} [n\overline{X^2} + (n^2 - n)(\bar{X})^2] - (\bar{X})^2$$

(4–4)

$$= \frac{\overline{X^2} - (\bar{X})^2}{n} = \frac{\sigma^2}{n}$$

where σ^2 is the true variance of the population. Note that the variance of the sample mean can be made small by making n large. This suggests that large sample sizes lead to a better estimate of the population mean, since the expected value of the sample mean is always equal to the true population mean, regardless of sample size, but the variance of the sample mean decreases as n gets large.

As noted previously, the result given in (4–4) assumed that N was very large. There is an alternative approach to sampling that leads to the same result as assuming a large population. Recall that the basic reason for assuming that the population size is very large is to assure that the statistical characteristics of the population do not change as we withdraw the members of the sample. For example, suppose we have a population consisting of five 10-Ω resistors and five 100-Ω resistors. Withdrawing even one resistor will leave the remaining population with a significantly different proportion of the two resistor types. However, if the population consisted of one million 10-Ω resistors and one million 100-Ω resistors then withdrawing one resistor, or even a thousand resistors, is not going to alter the composition of the remaining population significantly. The same sort of freedom from changing population characteristics can be achieved by replacing an item that is withdrawn after it has been examined, tested, and recorded. Since every item is drawn from exactly the same population, the effect of having an infinite population is achieved. Of course, one may select an item that has already been examined, but if the selection is done in a truly random fashion this will make no difference to the validity of the conclusions that might be drawn. Sampling done in this manner is said to be *sampling with replacement*.

There may be situations, of course, in which one may not wish to replace a sample or may be unable to replace it. For example, if the testing to be done is a life test, or a test that involves destroying the item, replacement is not possible. Similarly, in a public opinion poll or TV program survey, one simply does not wish to question the same person twice. In such situations, it is still possible to calculate the variance of the sample mean even when the population size is quite small. The mathematical expression for this, which is simply quoted here without proof, is

$$\text{Var}\,(\hat{\bar{X}}) = \frac{\sigma^2}{n}\left(\frac{N-n}{N-1}\right) \tag{4-5}$$

Note that as N becomes very large, this expression approaches the previous one. Note also, that if $N = n$, the sample variance becomes zero. This must be the case because this condition corresponds to every item in the population being sampled and, hence, the sample mean must be *exactly* the same as the population mean. It is clear, however, that one would not do this if destructive testing were involved! Two examples serve to illustrate the above ideas. The first example considers a case in which the population size is infinite or very large. Suppose we have a random waveform such as illustrated in Figure 4–1 and we wish to estimate the mean value of this waveform, which, we shall assume, has a true mean value of 10 and a true variance of 9.

As indicated in Figure 4–1, the value of this waveform is being sampled at equally spaced time instants $t_1, t_2, \ldots, t_n$. In the general situation, these sample values are random variables and are denoted by $X_i = X(t_i)$ for $i = 1, 2, \ldots, n$. We would like to find how many samples should be taken to estimate the mean value of this waveform with a standard deviation that is only one percent of the true mean value. If we assume that the waveform lasts forever, so that the population of time samples is infinite, then from (4–4)

$$\text{Var}\,(\hat{\bar{X}}) = \frac{\sigma^2}{n} = \frac{9}{n} = (0.01 \times 10)^2 = 0.01$$

in which the two right-hand terms are the desired variance of the estimate and correspond to a standard deviation of one percent of the true mean. Thus,

$$n = \frac{9}{0.01} = 900$$

This result indicates that the sample size must be quite large in most cases of sampling an infinite population, or in sampling with replacement, if it is desired to obtain a sample mean with a small variance.

Of course, estimating the mean value of the random time function with the specified variance does not necessarily imply that the estimate is really within one percent of true mean. It is possible, however, to determine the probability that the estimate of the mean is within one percent (or any amount) of the true mean. In order to do this, the probability density function of the estimate must be known.

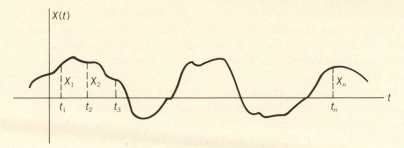

Figure 4–1 A random waveform that is being sampled.

In the case of a large sample size, the central limit theorem comes to the rescue and assures us that since the estimated mean is related to the sum of a large number of independent random variables, the sum is very nearly Gaussian regardless of the density function of the individual sample values. Thus, we can say that the probability that $\hat{\overline{X}}$ is within one percent of $\overline{X}$ is

$$\Pr\,(9.9 < \hat{\overline{X}} \le 10.1) = F(10.1) - F(9.9)$$

$$= \phi\left(\frac{10.1 - 10}{0.1}\right) - \phi\left(\frac{9.9 - 10}{0.1}\right) = \phi(1) - \phi(-1) = 2\phi(1) - 1$$

$$= 2 \times 0.8413 - 1 = 0.6826$$

Hence, there is a significant probability (0.3174) that the estimate of the population mean is actually more than one percent away from the true population mean.

The assumption of a Gaussian probability density function for sample means is quite realistic when the sample size is large, but may not be very good for small sample sizes. A method of dealing with small sample sizes is discussed in a subsequent section.

The second example considers a situation in which the population size is not large and sampling is done without replacement. In this example, there is a population of 100 bipolar transistors for which one wishes to estimate the mean value of the current gain, β. If the true population mean is $\overline{\beta} = 120$ and the true population variance is $\sigma_\beta^2 = 25$, how large a sample size is required to obtain a sample mean that has a standard deviation that is one percent of the true mean? Since the desired variance of the sample mean is

$$\text{Var}\,(\hat{\overline{\beta}}) = (0.01 \times 120)^2 = 1.44$$

it follows from (4–5) that

$$\frac{25}{n}\left(\frac{100 - n}{100 - 1}\right) = 1.44$$

This may be solved for n to yield $n = 14.92$, which implies a sample size of 15 since n must be an integer. This relatively small sample size is a consequence of having a small population size. In this case, for example, a sample size of 100 (that is, sampling every item) would result in a variance of the sample mean of exactly zero.

It is also possible to calculate the probability that the sample mean is within one percent of the true population mean, but it is not reasonable in this case to assume that the sample mean has a Gaussian density function unless, of course, the original β random variables are Gaussian. This is because the sample size of 15 is too small for the central limit theorem to be effective. As a rule of thumb, it is often assumed that a sample size of at least 30 is required to make the Gaussian assumption. A technique for dealing with smaller sample sizes is considered when sampling distributions are discussed.

Exercise 4–2.1

An endless production line is turning out solid-state diodes and every 100th diode is tested for reverse current I_{-1} and forward current I_1 at diode voltages of -1 and $+1$ respectively.

a) If the random variable I_{-1} has a true mean value of 10^{-6} and a variance of 10^{-11}, how many diodes must be tested to obtain a sample mean whose standard deviation is five percent of the true mean?

b) If the random variable I_1 has a true mean value of 0.1 and a variance of 0.0025, how many diodes must be tested to obtain a sample mean whose standard deviation is one percent of the true mean?

c) If the larger of the two numbers found in (a) and (b) is used for both tests, what will the standard deviations of the sample mean be for each test?

Answers: 5×10^{-8}, 2500, 0.00079, 4000

Exercise 4–2.2

A population of 80 resistors is to be tested without replacement to obtain a sample mean whose standard deviation is two percent of the true population mean.

a) How large must the sample size be if the true population mean is 100 Ω and the true standard deviation is 5 Ω?

b) How large must the sample size be if the true population mean is 100 Ω and the true standard deviation is 1 Ω?

c) If the sample size is 10, what is the standard deviation of the sample mean for the population of part (b)?

Answers: 1, 6, 0.1

4–3   Sampling Theory—The Sample Variance

In the previous section, we discussed estimating the mean value of a population of random variables by averaging the values in the sample taken from that population. We also determined the variance of that estimate and indicated how it influenced the sample size. However, in addition to the mean value, we may also be interested in estimating the variance of the random variables in the population. A knowledge of the variance is important because it indicates something about the spread of values around the mean. For example, it is not sufficient to test resistors and find that the sample mean is very close to the desired resistance value. If the standard deviation of the resistance values is very large, then regardless of how close the sample mean is, many of the resistors can be quite far from the desired value. Hence, it is necessary to control the variance of the population as well as its mean.

There is also another reason for wanting to estimate the variance of the population. You may recall that the population variance is needed in order to determine the sample size required to achieve a desired variance of the sample mean. Initially, one may not know the population variance and, thus, not have any idea as to how large the sample size should be. Estimating the population variance will at least provide some information as to how the sample size should be changed to achieve the desired results.

The sample variance is denoted initially by S^2, the change in notation being adopted in order to avoid undue notational complexity in distinguishing among the several variances. In terms of the random variables in the sample, $X_1, \ldots, X_n$, the sample variance may be defined as

$$S^2 = \frac{1}{n} \sum_{i=1}^{n} (X_i - \hat{\bar{X}})^2$$

$$= \frac{1}{n} \sum_{i=1}^{n} [X_i - \frac{1}{n} \sum_{j=1}^{n} X_j]^2 \qquad (4\text{–}6)$$

Note that the second summation in this expression is just the sample mean, so the entire expression represents the sample mean of the *square* of the difference between the random variables and the sample mean.

The expected value of S^2 can be obtained by expanding the squared term in (4–6) and taking the expected value of each term in the expansion. The details are tedious, but the method is straightforward and the result is

$$E[S^2] = \frac{n - 1}{n} \sigma^2 \tag{4–7}$$

where σ^2 is the true variance of the population. Note that the expected value of the sample variance is not the true variance. Thus, this is a *biased* estimate of the variance rather than an unbiased one. For most applications, one would like to have an unbiased estimate of any parameter. Hence, it is desirable to see if an unbiased estimate can be achieved readily. From (4–7), it is clear that one need only modify the original estimate by the factor $n/(n - 1)$. Therefore, an unbiased estimate of the population variance can be achieved by defining the sample variance as

$$\bar{S}^2 = \frac{n}{n - 1} S^2 \tag{4–8}$$

$$= \frac{1}{n - 1} \sum_{i=1}^{n} (X_i - \hat{\bar{X}})^2$$

Both of the above results have assumed that the population size is very large; i.e., $N = \infty$. When the population is not large, the expected value of S^2 is given by

$$E[S^2] = \frac{N}{N - 1} \cdot \frac{n - 1}{n} \sigma^2 \tag{4–9}$$

Note that this is also a biased estimate, but that the bias can be removed by defining $\bar{S}^2$ as

$$\bar{S}^2 = \frac{N - 1}{N} \cdot \frac{n}{n - 1} S^2 \tag{4–10}$$

Note that both of these results reduce to the previous ones as $N \rightarrow \infty$.

The variance of the estimates of variance can also be obtained by straightforward, but tedious, methods. For example, it can be shown that the variance of S^2 is given by

$$\text{Var}(S^2) = \frac{\mu_4 - \sigma^4}{n} \tag{4–11}$$

where μ_4 is the fourth central moment of the population and is defined by

$$\mu_4 = E[(X - \bar{X})^4] \tag{4-12}$$

The variance of $\bar{S}^2$ follows immediately from (4–7) and (4–8) as

$$\text{Var } \bar{S}^2 = \frac{n(\mu_4 - \sigma^4)}{(n - 1)^2} \tag{4-13}$$

Only the large sample size case will be considered to illustrate an application of the above results. For this purpose, consider again the random time function displayed in Figure 4–1 and for which the sample mean has been discussed. It is found in that discussion that a sample size of 900 is required to reduce the standard deviation of the sample mean to a value that is one percent of the true mean. Now suppose this same sample of size 900 is used to determine the sample variance; specifically, we will use it to calculate $\bar{S}^2$ as defined in (4–8). Recall that $\bar{S}^2$ is an unbiased estimate of the population variance. The variance of this estimate can now be evaluated from (4–13) if we know the fourth central moment. Unfortunately, the fourth central moment is not easily obtained unless we know the probability density function of the random variables. For the purpose of this discussion, let us assume that the random waveform under consideration is Gaussian and that the random variables that make up the sample are mutually statistically independent. From equation (2–27) in Section 2–5, we know that the fourth central moment of a Gaussian random variable is just $3\sigma^4$. Using this value in (4–13), and remembering that for this waveform σ^2 is 9, leads to

$$\text{Var } (\bar{S}^2) = \frac{900(3 \times 9^2 - 9^2)}{(900 - 1)^2} = 0.1804$$

This value of variance corresponds to a standard deviation of 0.4247, which is 4.72 percent of the true population variance. One conclusion that can be drawn from this example, and which turns out to be fairly true in general, is that it takes a larger sample size to achieve a given accuracy in estimating the population variance than it does to estimate the population mean.

It is also possible to determine the probability that the sample variance is within any specified region if the probability density function of $\bar{S}^2$ is known. In the large sample size case, this probability density function may be assumed Gaussian as is done in the case of the sample mean. In the small sample size case, this is not reasonable. In fact, if the original random variables are Gaussian the probability density function of $\bar{S}^2$ is chi-squared for any sample size. Another situation is discussed in a subsequent section.

Exercise 4–3.1

For the random waveform of Fig. 4–1, find the sample size that would be required to estimate the true variance of the waveform with:

a) a standard deviation of one percent of the true variance if an unbiased estimator is used

b) a standard deviation of one percent of the true variance if a biased estimator is used.

Answers: 20,000, 20,002

Exercise 4–3.2

Independent samples are taken from a random time function having a probability density function of

$$f_i(x) = e^{-x} \qquad x \geq 0$$
$$ = 0 \qquad x < 0$$

How many samples are required to estimate the variance of this time function with a standard deviation that is five percent of the true variance if an unbiased estimator is used.

Answer: 3133

4–4 Sampling Distributions and Confidence Intervals

Although the mean and variance of any estimate of a population parameter do give useful information about the population, it is not sufficient to answer questions about the probability that these estimates are within specified bounds. In order to answer these questions, it is necessary to know the probability density functions associated with parameter estimates such as the sample mean or the sample variance. A great deal of effort has been expended in the study of statistics to determine these probability density functions and many such functions are described in the literature. Only two probability density functions are discussed here and these are discussed only for sample means.

The sample mean is defined in (4–2) as

$$\hat{\bar{X}} = \frac{1}{n} \sum_{i=1}^{n} X_i$$

where n is the sample size and X_i are random variables from the population. If the X_i are Gaussian and independent, with a mean of $\bar{X}$ and a variance of σ^2, then the normalized random variable Z, defined by

$$Z = \frac{\hat{\bar{X}} - \bar{X}}{\sigma/\sqrt{n}} \tag{4–14}$$

is Gaussian with zero mean and unit variance. Thus, when the population is Gaussian, the sample mean is also Gaussian regardless of the size of the population or the size of the sample provided that the true population standard deviation is known so that it can be used in (4–14) to normalize the random variable. If the population is not Gaussian, the central limit theorem assures us that Z is asymptotically Gaussian as $n \rightarrow \infty$. Hence, for large n, the sample mean may still be assumed to be Gaussian. Also, if the true population variance is not known, the σ in (4–14) may be replaced by its estimate, $\tilde{S}$ since this estimate should be close to the true value for large n. The questions that arise in this case, however, are how large does n have to be and what does one do if n is not this large?

A rule of thumb that is often used is that the Gaussian assumption is reasonable if $n \geq 30$. If the sample size is less than 30, and if the population random variables are not Gaussian, very little can be said in general and each situation must be examined in the light of its own particular characteristics. However, if the population random variables are Gaussian and the true population variance is not known, the normalized sample mean is no longer Gaussian because the $\tilde{S}$ that is used to replace σ in (4–14) is also a random variable. It is possible to specify the probability density function of the normalized sample mean, however, and this topic is considered next.

When $n < 30$, define the normalized sample mean as

$$T = \frac{\hat{\bar{X}} - \bar{X}}{\tilde{S}/\sqrt{n}} = \frac{\hat{\bar{X}} - \bar{X}}{S/\sqrt{n-1}} \tag{4–15}$$

The random variable T is said to have a *Student's t distribution*[1] with $n - 1$ degrees of freedom.

---

[1] The Student's t distribution was discovered by William Gosset who published it using the pen name 'Student' because his employer, the Guinness Brewery, had a strict rule against their employees publishing their discoveries under their own names.

In order to define the Student's t probability density function, let $\nu = n - 1$ be denoted as the degrees of freedom. The density function then is defined by

$$f_T(t) = \frac{\Gamma\left(\frac{\nu + 1}{2}\right)}{\sqrt{\nu\pi}\,\Gamma\left(\frac{\nu}{2}\right)}\left(1 + \frac{t^2}{\nu}\right)^{-\frac{\nu+1}{2}} \tag{4-16}$$

where $\Gamma(\cdot)$ is the gamma function, some of whose essential properties are discussed below. This density function, for $\nu = 1$, is displayed in Figure 4–2, along with the normalized Gaussian density function for purposes of comparison. It may be noted that the Student's t density function has heavier tails than does the Gaussian density function. However, when $n \geq 30$ the two density functions are almost indistinguishable.

In order to evaluate the Student's t density function it is necessary to evaluate the gamma function. Fortunately, this can be done readily in this case by noting a few special relations. First, there is a recursion relation of the form

$$\Gamma(k + 1) = k\Gamma(k) \qquad \text{any } k \tag{4-17}$$
$$= k! \qquad \text{integer } k$$

Next, some special values of the gamma function are

$$\Gamma(1) = \Gamma(2) = 1, \quad \Gamma(1/2) = \sqrt{\pi}$$

Note that in evaluating the Student's t density function all arguments of the gamma function are either integers or one-half plus an integer. As an illustration of the application of (4–17), let $k = 3.5$.

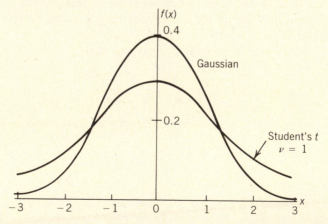

Figure 4–2   Comparison of Student's t and Gaussian probability density functions.

Table 4–1. Confidence Interval Width for a Gaussian Density Function.

q%	k
90	1.64
95	1.96
99	2.58
99.9	3.29
99.99	3.89

Thus

$$\Gamma(3.5) = 2.5 \cdot \Gamma(2.5) = 2.5 \cdot 1.5 \cdot \Gamma(1.5) = 2.5 \cdot 1.5 \cdot .5 \cdot \Gamma(.5)$$
$$= 2.5 \cdot 1.5 \cdot 5 \cdot \sqrt{\pi} = 3.323$$

The concept of a *confidence interval* is one that arises very often in the study of statistics. Although the confidence interval is most appropriately considered in connection with estimation theory, it is convenient to discuss it here as an application of the probability density function of the sample mean. The sample mean, as we defined it, is really a *point estimate* in the sense that it assigns a single value to the estimate. The alternative to a point estimate is an *interval estimate* in which the parameter being estimated is declared to lie within a certain interval with a certain probability. This interval is the confidence interval.

More specifically, a *q-percent confidence interval* is the interval within which the estimate will lie with a probability of $q/100$. The limits of this interval are the *confidence limits* and the value of q is said to be the *confidence level*.

When considering the sample mean, the q-percent confidence interval is defined as

$$\overline{X} - \frac{k\sigma}{\sqrt{n}} \leq \hat{\overline{X}} \leq \overline{X} + \frac{k\sigma}{\sqrt{n}} \qquad (4\text{–}18)$$

where k is a constant that depends upon q and the probability density function of $\hat{\overline{X}}$. Specifically,

$$q = 100 \int_{\overline{X}-k\sigma}^{\overline{X}+k\sigma} f_{\hat{\overline{X}}}(x)dx \qquad (4\text{–}19)$$

For the Gaussian density function, the values of k can be tabulated readily as a function of the confidence level. A very limited table of this sort is given in Table 4–1.

As an illustration of the use of this table, consider once again the random waveform of Figure 4–1 for which the true population mean is 10, the true population

Table 4–2.   Probability Distribution for Student's t Function ($\nu = 8$).

t	$F_T(t)$
0.262	0.60
0.706	0.75
1.397	0.90
1.860	0.95
2.306	0.975
2.896	0.99
3.355	0.995

variance is 9, and 900 samples are taken. The width of a 95 percent confidence interval is just

$$10 - \frac{1.96\sqrt{9}}{\sqrt{900}} \leq \hat{\overline{X}} \leq 10 + \frac{1.96\sqrt{9}}{\sqrt{900}}$$

$$9.804 \leq \hat{\overline{X}} \leq 10.196$$

Thus, there is a probability of 0.95 that the sample mean will lie in the interval between 9.804 and 10.196.

It is worth noting that large confidence levels correspond to wide confidence intervals. Hence, there is a small probability that an estimate will lie within a very narrow confidence interval, but a large probability that it will lie within a broad confidence interval. It follows, therefore, that a 99 percent confidence level represents a *poorer* estimate than does, say, a 90 percent confidence level when the same sample sizes are being compared.

The same information regarding confidence intervals can be obtained from the probability distribution function. Note that the integral in (4–19) can be replaced by the difference of two distribution functions. Hence, this relation could have been written as

$$q = 100[F_{\hat{\overline{X}}}(\overline{X} + k\sigma) - F_{\hat{\overline{X}}}(\overline{X} - k\sigma)] \tag{4–20}$$

It is also possible to tabulate k-values for the Student's t distribution, but a different set of values is required for each value of ν, the degrees of freedom. However, it is customary to present this information in terms of the probability distribution function. A modest table of these values is given in Appendix F, while a much smaller table for the particular case of eight degrees of freedom is given in Table 4–2 to assist in the discussion that follows.

The application of this table to several aspects of hypothesis testing is discussed in the next section.

Exercise 4–4.1

Calculate the probability density function for the Student's t density for t = 1 and for

  a) 5 degrees of freedom,

  b) 10 degrees of freedom.

     Answers: 0.2197, 0.2304

Exercise 4–4.2

A very large population of resistor values has a true mean of 100 Ω and a standard deviation of 5 Ω. The resistance values may be assumed to be Gaussian random variables. Find the confidence limits on the sample mean for a confidence level of 99 percent if it is computed from

  a) a sample size of 100,

  b) a sample size of 9.

     Answers: 94.41 to 105.59, 98.71 to 101.29

4–5 Hypothesis Testing

One of the important applications of statistics is that of making decisions about the parameters of a population. In the preceding sections we have seen how to estimate the mean value or the variance of a population and how to assign confidence intervals to these estimates for any specified level of confidence. The next step is to make some hypothesis about the population and then determine if the observed sample confirms or rejects this hypothesis. For example, a manufacturer may claim that the light bulbs he produces have an average lifetime of 1000 hours. The hypothesis is then made that the mean value of this population (i.e., the lifetimes of all light bulbs produced) is 1000 hours. Since it is not possible to run life tests on all the light bulbs produced, a small fraction is tested and the sample mean determined. The question then is: does the result of this test verify the hypothesis? To take an extreme example, suppose only two light bulbs are tested and the sample mean is found to be 900 hours. Does this prove that the hypothesis about the average lifetime of the population of all light bulbs is false? Probably not, because the sample size is too small to be able to make a reasonable decision. On the other hand, suppose the sample mean of these two light bulbs is 1000 hours. Does this prove that the hypothesis is correct? Again, the answer is probably not. The question then becomes: how does one decide to accept or reject a given hypothesis when the sample size and the confidence level are specified? We now have the background necessary to answer that question and will do so in several specific cases by means of examples.

One way of classifying hypothesis tests is based on whether they are one-sided or two-sided. In a one-sided test, one is concerned with what happens on one side of the desired value of the parameter. For example, in the light bulb situation above, we are concerned only if the average lifetime is less than 1000 hours and would be happy to have the average lifetime greater than 1000 hours by any amount. There are many other situations of a comparable nature. On the other hand, in a two-sided test we are concerned about deviations in either direction from the hypothesized value. For example, if we have a supply of 100-Ω resistors that we are testing, it is equally serious if the resistance is either too high or too low.

To consider the one-sided test first, imagine that a capacitor manufacturer claims that his capacitors have a mean value of breakdown voltage of 300 volts or greater. We test the breakdown voltage of a sample of 100 capacitors and find that the sample mean is 290 volts and the unbiased sample standard deviation, $\tilde{S}$, is 40 volts. Is the manufacturer's claim valid if a 99 percent confidence level is used? Note that this is a one-sided test since we don't care how much greater than 300 volts the mean value of breakdown voltage might be.

We start by making the hypothesis that the true mean value of the population is 300 volts and then check to see if this hypothesis is consistent with the observed data. Since the sample size is greater than 30, the Gaussian assumption may be employed here, with σ set equal to $\tilde{S}$. Thus, the value of the normalized random variable, $Z = z$, is

$$z = \frac{\bar{x} - \bar{X}}{\sigma/\sqrt{n}} = \frac{290 - 300}{40/\sqrt{100}} = -2.5$$

For a one-sided confidence level of 99 percent the critical value of z is found from that value above which the area of $F_Z(z)$ is 0.99. That is,

$$\int_{z_c}^{\infty} f_Z(z)\,dz = 1 - \Phi(z_c) = 0.99$$

from which $z_c = -2.33$. Since the observed value of z is less than z_c, we would reject the hypothesis; that is, we would say that the claim that the mean breakdown voltage is 300 volts or greater is not valid.

An often confusing point in connection with hypothesis testing is the real meaning of the decision made. In the example above, the decision means that there is a probability of 0.99 that the observed sample did *not* come from a population having a true mean of 300 volts. This seems clear enough; the confusing point, however, is that had we chosen a confidence level of 99.5 percent we would have *accepted* the hypothesis because the critical value of z for this level of confidence is -2.575 and the observed z-value is now greater than z_c. Thus, choosing a high confidence level makes it more likely that any given sample will result in accepting the hypothesis. This seems contrary to logic, but the reason is clear; a high confidence results in a wider confidence interval because a greater fraction of the probability density function must be contained in it. Conversely, selecting a small confidence level makes it less likely that any given sample will result in accepting the hypothesis and, thus, is a more severe requirement. Because the use of the term confidence level does seem to be contradictory, some statisticians prefer to use the *level of significance*, which is just the confidence level subtracted from 100 percent. Thus, a confidence level of 99 percent corresponds to a 1 percent level of significance while a confidence level of 99.5 percent is only a 0.5 percent level of significance. A larger level of significance corresponds to a more severe test of the hypothesis.

The example concerning the capacitor breakdown voltage is now reconsidered when the sample size is small. Suppose we test only 9 capacitors and find that the mean value of breakdown voltage is 290 V and the unbiased sample standard deviation is 40 V. Note that these are the same values that were obtained with a large sample size. However, since the sample size is less than 30 we will use the T random variable, which for this case is

$$ t = \frac{\bar{x} - \overline{X}}{\bar{s}/\sqrt{n}} = \frac{290 - 300}{\dfrac{40}{\sqrt{9}}} = -0.75 $$

For the Student's t density function with $\nu = n - 1 = 8$ degrees of freedom, the critical value of t for a confidence level of 99 percent is, from Table 4–2, $t_c = -2.896$. Since the observed value of t is now greater than t_c we would *accept* the hypothesis that the true mean breakdown voltage is 300 V or greater.

Note that the use of a small sample size tends to increase the value of t and, hence, makes it more likely to exceed the critical value. Furthermore, the small sample size leads to the use of the Student's t distribution, which has heavier tails than the Gaussian distribution and, thus, leads to a smaller value of t_c. Both of these factors together make small sample size tests less reliable than large sample size tests.

The next example considers a two-sided hypothesis test. Suppose that a manufacturer of Zener diodes claims that a certain type has a mean breakdown voltage of 10 V. Since a Zener diode is used as a voltage regulator, deviations from the desired value of breakdown voltage in either direction are equally undesirable.

Hence, we hypothesize that the true mean value of the population is 10 V and then seek a test that either accepts or rejects this hypothesis and utilizes the fact that deviations on either side of 10 are of concern.

Considering a large sample size test first, suppose we test 100 Zener diodes and find that the sample mean is 10.3 V and the unbiased sample standard deviation is 1.2 V. Is the claim valid if a 95 percent confidence level is used? Since the sample size is greater than 30, we can use the Gaussian random variable, Z, which for this sample is

$$z = \frac{10.3 - 10}{1.2/\sqrt{100}} = 2.5$$

For a 95 percent confidence level, the critical values of the Gaussian random variable are, from Table 4–1, ± 1.96. Thus, in order to accept the hypothesis it is necessary for z to lie in the region $-1.96 \leq z \leq 1.96$. Since $z = 2.5$ does not lie in this interval, the hypothesis is *rejected;* that is, the manufacturer's claim is not valid since the observed sample could not have come from a population having a mean value of 10 with a probability of 0.95.

This same test is now repeated with a small sample size. Suppose that 9 Zener diodes are tested and it is found that the mean value of their breakdown voltages is again 10.3 V and the unbiased sample standard deviation is 1.2 V. The Student's t random variable now has a value of

$$t = \frac{\bar{x} - \overline{X}}{\bar{s}/\sqrt{n}} = \frac{10.3 - 10}{1.2/\sqrt{9}} = 0.75$$

The critical values of t can be obtained from Table 4–2 since there are once again 8 degrees of freedom. Since Table 4–2 lists the distribution function for the Student's t random variable and we are interested in finding the interval around zero that contains 95 percent of the area, there will be 2.5 percent of the area above t_c and 2.5 percent below t_c. Thus, the value that we need from the table is that corresponding to 0.975. This is seen easily by noting that

$$\Pr\left[-t_c < T \leq t_c\right] = F_T(t_c) - F_T(-t_c) = 2F_T(t_c) - 1 = 0.95$$

Therefore

$$F_T(t_c) = \frac{1.95}{2} = 0.975$$

From Table 4–2 the required value is $t_c = 2.306$. In order to accept the hypothesis, it is necessary that the observed value of t lie in the range $-2.306 < t \leq 2.306$. Since $t = 0.75$ does lie in this range, the hypothesis is accepted and the manufacturer's claim is considered to be valid. Again we see that a small sample test is not as severe as a large sample test.

Exercise 4–5.1

A certain type of bipolar transistor is claimed to have a mean value of current gain of $\beta \geq 200$. A sample of these transistors is tested and the sample mean value of current gain is found to be 190 and the unbiased sample standard deviation is 40. If a 95 percent confidence level is employed, is this claim valid if

a) the sample size is 100?

b) the sample size is 20?

Answers: $z = -2.5$, $z_c = -1.645$, no;
$t = -1.118$, $t_c = -1.729$, yes

Exercise 4–5.2

A certain type of bipolar transistor is claimed to have mean collector current of 4 mA. A sample of these transistors is tested and the sample mean value of collector current is found to be 4.2 mA and the unbiased sample standard deviation is 0.8 mA. If a 95 percent confidence level is employed, is this claim valid if

a) the sample size is 100?

b) the sample size is 20?

Answers: $z = 2.5$, $z_c = \pm 1.96$, no
$t = 1.118$, $t_c = \pm 2.09$, yes

4–6   Curve Fitting and Linear Regression

The topic considered in this section is considerably different from those in previous sections, but it does represent an important application of statistics in engineering problems. Frequently, statistical data reveals a relationship between two or more variables and it is desired to express this relationship in mathematical form by determining an equation that connects the variables. For example, one might collect data on the lifetime of light bulbs as a function of the applied voltage. Such data might be presented in the form of a *scatter diagram*, such as shown in Figure 4–3, in which each observed lifetime and the corresponding operating voltage are plotted as a point on a two-dimensional plane.

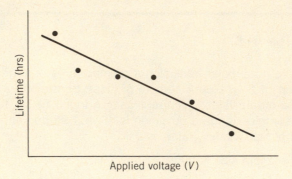

Figure 4–3   Scatter diagram of light bulb lifetimes and applied voltage.

Also shown in Figure 4–3 is a solid curve that represents, in some sense, the best fit between the data points and a mathematical expression that relates the two variables. The objective of this section is to show one way of obtaining such a mathematical relationship.

For purposes of discussion, it is convenient to consider the two variables as x and y. Since the data consists of specific numerical values, in keeping with our previously adopted notation, this data is represented by lower case letters. Thus, for a sample size of n we would have values of one variable denoted as x_1, x_2, . . . , x_n and corresponding values of the other variable as y_1, y_2, . . . , y_n. For example, for the data displayed in Figure 4–3 each x-value might be an applied voltage and each y-value the corresponding lifetime.

The general problem of finding a mathematical relationship to represent the data is called *curve fitting*. The resulting curve is called a *regression curve* and the mathematical equation is the *regression equation*. In order to find a "best" regression equation it is first necessary to establish a criterion that will be used to define what is meant by "best." Consider the scatter diagram and regression curve shown in Figure 4–4.

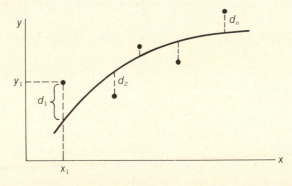

Figure 4–4   Error between the regression curve and the scatter diagram.

In this figure, the *difference* between the regression curve and the corresponding value of y at any x is designated as d_i, $i = 1, 2, \ldots, n$. The criterion of goodness of fit that is employed here is that

$$d_1^2 + d_2^2 + \ldots + d_n^2 = \text{a minimum} \qquad (4\text{--}21)$$

Such a criterion leads to a *least-squares regression curve* and is the criterion that is most often employed. Note that the least-squares criterion weights errors on either side of the regression curve equally and also weights large errors more than small errors.

Having decided upon a criterion to use, the next step is to select the type of the equation that is to be fitted to the data. This choice is based largely on the nature of the data, but most often a polynomial of the form

$$y = a + bx + cx^2 + \ldots + kx^j$$

is used. Although it is possible to fit an $(n - 1)$-degree polynomial to n data points, one would never want to do this because it would provide no smoothing of the data. That is, the resulting polynomial would go through each data point and the resulting least-squares error would be zero. Since the data is random, one is more interested in a regression curve that approximates the mean value of the data. Thus, in most cases, a first or second degree polynomial is employed. Our discussion in this section is limited to using a first degree polynomial in order to perserve simplicity while conveying the essential aspects of the method. This technique is referred to as *linear regression*.

The linear regression equation becomes

$$y = a + bx \qquad (4\text{--}22)$$

in which it is necessary to determine the values of a and b that satisfy (4–21). These are determined by writing

$$\sum_{i=1}^{n} [y_i - (a + bx_i)]^2 = \text{a minimum}$$

In order to minimize this expression, one would differentiate partially with respect to a and b and set the derivatives equal to zero. This leads to two equations that may be solved simultaneously for the values of a and b. The equations are

$$\sum_{i=1}^{n} y_i = an + b \sum_{i=1}^{n} x_i$$

and

$$\sum_{i=1}^{n} x_i y_i = a \sum_{i=1}^{n} x_i + b \sum_{i=1}^{n} x_i^2$$

Table 4–3.   Data for Breakdown Voltage *vs* Temperature.

i	1	2	3	4	5	6	7	8	9	10
°C, x_i	10	20	30	40	50	60	70	80	90	100
V_B, y_i	420	410	360	360	340	290	300	270	210	200

The resulting values of a and b are

$$b = \frac{n \sum_{i=1}^{n} x_i y_i - \sum_{i=1}^{n} x_i \sum_{i=1}^{n} y_i}{n \sum_{i=1}^{n} x_i^2 - \left(\sum_{i=1}^{n} x_i\right)^2} \tag{4–23}$$

and

$$a = \frac{\sum_{i=1}^{n} y_i \sum_{i=1}^{n} x_i^2 - \sum_{i=1}^{n} x_i \sum_{i=1}^{n} x_i y_i}{n \sum_{i=1}^{n} x_i^2 - \left(\sum_{i=1}^{n} x_i\right)^2} = \frac{\sum_{i=1}^{n} y_i - b \sum_{i=1}^{n} x_i}{n} \tag{4–24}$$

Although these are fairly complicated expressions, they can be evaluated readily by computer or programmable calculator.

An example serves to illustrate the technique. A manufacturer of capacitors wishes to determine the relationship between the breakdown voltage and the ambient temperature in which the capacitor operates. He tests 10 capacitors at different temperatures and obtains the data displayed in Table 4–3. From equations (4–23) and 4–24) the values of a and b become $a = 451.33$ and $b = -2.406$. Thus, the desired mathematical relationship is

$$\text{(Breakdown Voltage)} = 451.33 - 2.406 \text{ (Temperature)}$$

The corresponding scatter diagram and linear regression curve are shown in Figure 4–5.

Similar techniques can be used to fit higher degree polynomials to experimental data. Obviously, the difficulty in determining the best values for the polynomial coefficients increases as the degree of the polynomial increases. However, there are very effective matrix formulations of the problem that lend themselves readily to computational methods.

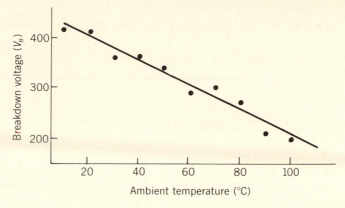

Figure 4–5 Linear regression curve for capacitor breakdown voltage *vs* ambient temperature.

Exercise 4–6.1

Four light bulbs are tested to establish a relationship between lifetime and operating voltage. The resulting data are shown in the following table:

i	1	2	3	4
V, x_i	105	110	115	120
Hrs., y_i	1200	1000	920	750

Find the coefficients of the linear regression curve and plot it and the scatter diagram.

Answers: 4185, −28.6

Exercise 4–6.2

Assume that the linear regression curve determined in Exercise 4–6.1 holds for all values of voltage. Find the expected lifetime of a light bulb operating at a voltage of

a) 95 volts

b) 125 volts

c)   117 volts.

   Answers: 838.8, 610, 1468

PROBLEMS

4–2.1   A calculator with a random number generator produces the following sequence of random numbers: 0.276, 0.123, 0.072, 0.324, 0.815, 0.312,
0.432, 0.283, 0.717,

a)   Find the sample mean.

b)   If the calculator produces three digit random numbers that are uniformly distributed between 0.000 and 0.999, find the variance of the
sample mean.

c)   How large should the sample size be in order to obtain a sample
mean whose standard deviation is no greater than 0.01?

4–2.2   A political poll is assessing the relative strengths of two presidential can-
didates. A value of $+1$ is assigned to every person who states a preference for candidate A and a value of -1 is assigned to anyone who indicates a preference for candidate B.

a)   Find the sample mean if 60 percent of those polled indicate a preference for candidate A.

b)   Write an expression for the sample mean as a function of the sample
size and the percentage of those polled that are in favor of candidate A.

c)   Find the sample size necessary to estimate the percentage of persons
in favor of candidate A with a standard deviation no greater than 0.1
percent.

4–2.3   In a class of 50 students, the result of a particular examination is a true
mean of 70 and a true variance of 12. It is desired to estimate the mean
by sampling, without replacement, a subset of the scores.

a)   Find the standard deviation of the sample mean if only 10 scores are
used.

b)   How large should the sample size be for the standard deviation of the sample mean to be one percentage point (out of 100)?

c)   How large should the sample size be for the standard deviation of the sample mean to be one percent of the true mean?

4–2.4   The HYGAYN Transistor Company produces a line of bipolar transistors that have an average current gain of 120 with a standard deviation of 10. Another company, ACE Electronics, produces a similar line of transistors with the same average current gain but with a standard deviation of 5. Ed Engineer purchases 20 transistors from each company and mixes them together.

a)   If Ed selects a random sample of 5 transistors with replacement, find the variance of the sample mean.

b)   If Ed selects a random sample of 5 transistors without replacement, find the variance of the sample mean.

c)   How large a sample size should Ed use, without replacement, in order to obtain a standard deviation of the sample mean of 2?

4–2.5   For the transistors of Problem 4–2.4, assume that the current gains are independent Gaussian random variables.

a)   If Ed selects a random sample of 10 transistors with replacement, find the probability that the sample mean is within 2 percent of the true mean.

b)   Repeat part (a) if the sampling is without replacement.

4–3.1   a)   For the random numbers given in Problem 4–2.1, find the sample variance if an unbiased estimator is used.

b)   Find the variance of this estimate of the population variance.

4–3.2   A zero-mean Gaussian random time function is sampled so as to obtain independent sample values. How many sample values are required to obtain an unbiased estimate of the variance of the time function with a standard deviation that is two percent of the true variance?

4–3.3   It is desired to estimate the variance of a random phase angle that is uniformly distributed over a range of 2π. Find the number of independent samples that are required to estimate this variance with a standard deviation that is five percent of the true variance if an unbiased estimate is used.

4–4.1　a)　Calculate the value of the Student's t probability density function for $t = 2$ and for 6 degrees of freedom.

　　　　b)　Repeat (a) for 12 degrees of freedom.

4–4.2　A very large population of bipolar transistors has a current gain with a mean value of 120 and a standard deviation of 10. The values of current gain may be assumed to be independent Gaussian random variables.

　　　　a)　Find the confidence limits for a confidence level of 90 percent on the sample mean if it is computed from a sample size of 150.

　　　　b)　Repeat part (a) if the sample size is 21.

4–4.3　Repeat Problem 4–4.2 if a one-sided confidence interval is considered. That is, find the value of current gain above which 90 percent of the sample means would lie.

4–5.1　The resistance of coils manufactured by a certain company is claimed to have a mean value of resistance of 100 Ω. A sample of 9 coils is taken and it is found that the sample mean is 115 Ω and the sample standard deviation is 20 Ω.

　　　　a)　Is the claim justified if a 95 percent confidence level is used?

　　　　b)　Is the claim justified if a 90 percent confidence level is used?

4–5.2　Repeat Problem 4–5.1 if the sample size is 50 coils, the sample mean is still 115 Ω, and the sample standard deviation is 10 Ω.

4–5.3　A manufacturer of traveling wave tubes claims the mean lifetime is at least 4 years. Twenty of these tubes are installed in a communication satellite and a record kept of their performance. It is found that the mean lifetime of this sample is 3.7 years and the standard deviation of the sample is 1 year.

　　　　a)　For what confidence level would the company's claim be valid?

　　　　b)　What must the mean lifetime of the tubes have been in order for the claim to be valid at a confidence level of 90 percent?

4–5.4 A manufacturer of capacitors claims the breakdown voltage has a mean value of at least 100 V. A test of nine capacitors yielded breakdown voltages of 97, 104, 95, 98, 106, 92, 110, 103, and 93 V.

a) Find the sample mean.

b) Find the sample variance using an unbiased estimate.

c) Is the manufacturer's claim valid if a confidence level of 95 percent is employed?

4–6.1 Data is taken for a random variable Y as a function of another variable X. The x-values are 1, 3, 4, 6, 8, 9, 11, 14 and the corresponding y-values are 1, 2, 4, 4, 5, 7, 8, 9.

a) Plot the scatter diagram for this data.

b) Find the linear regression curve that best fits this data.

4–6.2 A test is made of the breakdown voltage of capacitors as a function of the capacitance. For capacitance values of 0.0001, 0.001, 0.01, 0.1, 1, 10 microfarads the corresponding breakdown voltages are 310, 290, 285, 270, 260, and 225 V.

a) Plot the scatter diagram for this data on a semi-log coordinate system.

b) Find the linear regression curve that best fits this data on a semi-log coordinate system.

References

Mosteller, F., S. E. Feinberg, and R. E. K. Rourke, *Beginning Statistics with Data Analysis*. Reading, Mass.: Addison-Wesley, 1983.

A recent undergraduate text that emphasizes the data analysis aspects of statistics. The mathematical level of the text is somewhat low for engineering students, but the numerous excellent examples illustrate the concepts very well.

Spiegel, M. R., *Theory and Problems of Probability and Statistics*. Schaum's Outline Series in Mathematics. New York: McGraw-Hill, Inc., 1975.

The chapters on statistics in this outline give concise and well-organized definitions of many of the concepts presented in this chapter. In addition, there are many excellent problems for which answers are provided.

CHAPTER 5

Random Processes

5—1 Introduction

It was noted in Chapter 2 that a *random process* is a collection of time functions and an associated probability description. The probability description may consist of the marginal and joint probability density functions of all random variables that are point functions of the process at specified time instants. This type of probability description is the only one that will be considered here.

The entire collection of time functions is an *ensemble* and will be designated as $\{x(t)\}$, where any particular member of the ensemble, $x(t)$, is a *sample function* of the ensemble. In general, only one sample function of a random process can ever be observed; the other sample functions represent all of the other possible realizations that might have occurred but did not. An arbitrary sample function is denoted $X(t)$. The values of $X(t)$ at any time t_1 define a random variable denoted as $X(t_1)$ or simply X_1.

The extension of the concepts of random variables to those of random processes is quite simple as far as the mechanics are concerned; in fact, all of the essential ideas have already been considered. A more difficult step, however, is the conceptual one of relating the mathematical representations for random variables to the physical properties of the process. Hence, the purpose of this chapter is to help clarify this relationship by means of a number of illustrative examples.

Many different classes of random processes arise in engineering problems. Since methods of representing these processes most efficiently do depend upon the nature of the process under consideration, it is necessary to classify random processes in a manner that assists in determining an appropriate type of representation. Furthermore, it is important to develop a terminology that enables us to specify the class of process under consideration in a concise, but complete, manner so that there is no uncertainty as to which process is being discussed.

Therefore, one of the first steps in discussing random processes is that of developing a terminology that can be used as a "short-cut" in the description of the characteristics of any given process. A convenient way of doing this is to use a set of descriptors, arranged in pairs, and to select one name from each pair to describe the process. Those pairs of descriptors that are appropriate in the present discussion are:

1. Continuous; discrete
2. Deterministic; nondeterministic
3. Stationary; nonstationary
4. Ergodic; nonergodic

Exercise 5–1.1

a) If it is assumed that any random process can be described by picking one descriptor from each pair of descriptors shown above, how many classes of random processes can be described?

b) It is also possible to consider mixed processes in which two or more random processes of the type described in (a) above are combined to form a single random process. If two random processes of the type described in (a) are combined, what is the total number of classes of random processes that can be described now by the above list of descriptors.?

   Answers: 16, 256

Exercise 5–1.2

a) A time function is generated by flipping two coins once every second. A value of +1 is assigned to each head and a value of −1 is assigned to each tail. The time function has a constant value equal to that obtained from the sum of the two coins for one second and then changes to the new value determined by the outcome on the next flip of the coins. Sketch a typical sample function of the random process defined in this way. Let the sample function be eight seconds long and let it exhibit all possible states with the correct probabilities.

b) How many possible sample functions, each eight seconds long, does the entire ensemble of sample functions for this random process have?

   Answer: 6561

5–2   Continuous and Discrete Random Processes

These terms normally apply to the possible values of the random variables. A *continuous random process* is one in which random variables such as $X(t_1)$, $X(t_2)$, and so on, can assume *any* value within a specified range of possible values. This range may be finite, infinite, or semi-infinite. Such things as thermal agitation noise in conductors, shot noise in electron tubes or transistors, and wind velocity are examples of continuous random processes. A sketch of a typical sample function and the corresponding probability density function is shown in Figure 5–1. In this example, the range of possible values is semi-infinite.

A more precise definition for continuous random processes would be that the probability distribution function is continuous. This would also imply that the density function has no δ functions in it.

A *discrete random process* is one in which the random variables can assume only certain isolated values (possibly infinite in number) and no other values. For example, a voltage that is either 0 or 100 because of random opening and closing of a switch would be a sample function from a discrete random process. This is

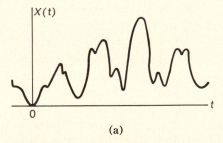

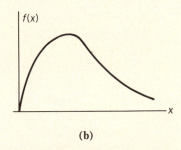

(a)   (b)

Figure 5–1   A continuous random process: (a) typical sample function and (b) probability density function.

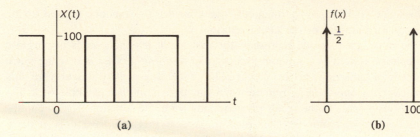

Figure 5–2 A discrete random process: (a) typical sample function and (b) probability density function.

illustrated in Figure 5–2. Note that the probability density function contains *only* δ functions.

It is also possible to have *mixed* processes, which have both continuous and discrete components. For example, the current flowing in an ideal rectifier may be zero for one-half the time, as shown in Figure 5–3. The corresponding probability density has both a continuous part and a δ function.

Some other examples of random processes will serve to further illustrate the concept of continuous and discrete random processes. Thermal noise in an electronic circuit is a typical example of a continuous random process since its amplitude can take on any positive or negative value. The probability density function of thermal noise is a continuous function from minus infinity to plus infinity. Quantizing error associated with analog-to-digital conversion, as discussed in Section 2–7, is another example of a continuous random process since this error may have any value within a finite range of values determined by the size of the increment between quantization levels. The probability density function for the quantizing error is usually assumed to be uniformly distributed over the range of possible errors. This case represents a minor departure from the strict mathematical definition for a continuous probability density function since the uniform density function is not continuous at the end points. Nevertheless, since the density func-

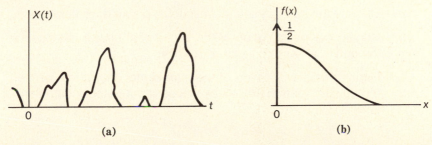

Figure 5–3 A mixed random process: (a) typical sample function and (b) probability density function.

tion does not contain any δ functions, we consider the random process to be continuous for purposes of our classification.

On the other hand, if one represents the number of telephone calls in progress in a telephone system as a random process, the resulting process is discrete since the number of calls must be an integer. The probability density function for this process contains *only* a large number of δ functions. Another example of a discrete random process is the result of quantizing a sample function from a continuous random process into another random process that can have only a finite number of possible values. For example, an 8-bit analog-to-digital converter takes an input signal that may have a continuous probability density function and converts it into one that has a discrete probability density function with 256 δ functions.

Finally we consider some mixed processes that have both a continuous component and a discrete component. One such example is the rectified time function as noted above. Another example might be a system containing a limiter such that when the output magnitude is less than the limiting value, it has the same value as the input. However, the output magnitude can never exceed the limiting value regardless of how large the input becomes. Thus, a sample function from a continuous random process on the input will produce a sample function from a mixed random process on the output and the probability density function of the output will have both a continuous part and a pair of δ functions.

In all of the cases just mentioned, the sample functions are continuous in *time;* that is, a random variable may be defined for any time. Situations in which the random variables exist for particular time instants only (referred to as *point processes* or *time series*) are not discussed in this chapter.

Exercise 5–2.1

Gaussian noise having zero mean and a variance of 0.1 V^2 is added to a random binary signal having values of ± 1 V.

a) Classify the sum signal as continuous, discrete, or mixed.

b) Repeat the classification of the sum signal after it has passed through a limiter that limits at ± 1.1 V.

c) Repeat the classification of the sum signal after it has passed through an ideal hard limiter whose input-output characteristic is

$$V_{out} = \text{sgn}\,(V_{in})$$

Answers: Mixed, continuous, discrete

Exercise 5–2.2

A random time function has a mean value of 10 and an amplitude that is Rayleigh distributed. This random time function is multiplied by a sinusoid having a maximum value of 10 and a random phase that is uniformly distributed between 0 and 2π.

a) Classify the product as continuous, discrete, or mixed.

b) Classify the product after it has been passed through an ideal half-wave rectifier.

c) Suppose the sinusoid is passed through an ideal half-wave rectifier before it multiplies the Rayleigh distributed time function. Classify the product.

Answers: Continuous, mixed, mixed

5–3 Deterministic and Nondeterministic Random Processes

In most of the discussion so far, it has been implied that each sample function is a random function of time and, as such, its future values cannot be exactly predicted from the observed past values. Such a random process is said to be *nondeterministic*. Almost all natural random processes are nondeterministic, because the basic mechanism that generates them is either unobservable or extremely complex. All the examples presented in Section 5–2 are nondeterministic.

It is possible, however, to define random processes for which the future values of any sample function can be exactly predicted from a knowledge of the past values. Such a process is said to be *deterministic*. As an example, consider a random process for which each sample function of the process is of the form

$$X(t) = A \cos (\omega t + \theta) \tag{5–1}$$

where A and ω are constants and θ is a random variable with a specified probability distribution. That is, for any one sample function, θ has the same value for all t but different values for the other members of the ensemble. In this case, the only random variation is over the ensemble—not with respect to time. It is still possible to define random variables $X(t_1)$, $X(t_2)$, and so on, and to determine probability density functions for them.

As a second example of a deterministic process, consider a periodic random process having sample functions of the form

$$X(t) = \sum_{n=0}^{\infty} [A_n \cos (2\pi n f_o t) + B_n \sin (2\pi n f_o t)] \qquad (5\text{-}2)$$

in which the A_n and the B_n are independent random variables that are fixed for any one sample function but are different from sample function to sample function. Given the past history of any sample function, one can determine these coefficients and predict exactly all future values of $X(t)$.

It is not necessary that deterministic processes be periodic, although this is probably the most common situation that arises in practical applications. For example, a deterministic random process might have sample functions of the form

$$X(t) = A \exp (-\beta t) \qquad t \geq 0 \qquad (5\text{-}3)$$

in which A and β are random variables that are fixed for any one sample function but vary from sample function to sample function.

Although the concept of deterministic random processes may seem a little artificial, it often is convenient to obtain a probability model for signals that are known except for one or two parameters. The process described by (5–1), for example, may be suitable to represent a radio signal in which the magnitude and frequency are known, but the phase is not because the precise distance (within a fraction of a wavelength) between transmitter and receiver is not.

Exercise 5–3.1

A sample function of the random process defined by Equation (5–1) is observed at three different time instants and found to have the following values:

$$X(0) = 0 \qquad X(1) = 10 \qquad X(2) = 0$$

There are no zeros between $t = 0$ and $t = 2$.

a) Find the values of A, ω, and θ.

b) Find the value of $X(2.5)$.

Answers: 1.57, 10, −7.07, −1.57

Exercise 5–3.2

A random process has sample functions of the form

$$X(t) = \sum_{n=-\infty}^{\infty} A_n f(t - nt_1)$$

where the A_n are independent random variables that are uniformly distributed from 0 to 10, and

$$f(t) = 1 \qquad 0 \le t \le (1/2)t_1$$
$$= 0 \qquad \text{elsewhere}$$

a)   Is this process deterministic or nondeterministic? Why?

b)   Is this process continuous, discrete, or mixed? Why?

Answers: Nondeterministic, mixed

5—4   Stationary and Nonstationary Random Processes

It has been noted that one can define a probability density function for random variables of the form $X(t_1)$ but so far no mention has been made of the dependence of this density function on the value of time t_1. If all marginal and joint density functions of the process do not depend upon the choice of time origin, the process is said to be *stationary*. In this case, all of the mean values and moments discussed previously are constants that do not depend upon the absolute value of time.

If any of the probability density functions do change with the choice of time origin, the process is *nonstationary*. In this case, one or more of the mean values or moments will also depend on time. Since the analysis of systems responding to nonstationary random inputs is more involved than in the stationary case, all future discussions are limited to the stationary case unless it is specifically stated to the contrary.

In a rigorous sense, there are no stationary random processes that actually exist physically, since any process must have started at some finite time in the past and must presumably stop at some finite time in the future. However, there are many physical situations in which the process does not change appreciably during the time it is being observed. In these cases the stationary assumption leads to a convenient mathematical model, which closely approximates reality.

Determining whether or not the stationary assumption is reasonable for any given situation may not be easy. For nondeterministic processes, it depends upon the mechanism of generating the process and upon the time duration over which the process is observed. As a rule of thumb, it is customary to assume stationarity unless there is some obvious change in the source or unless common sense dictates otherwise. For example, the thermal noise generated by the random motion of electrons in a resistor might reasonably be considered stationary under normal conditions. However, if this resistor were being intermittently heated by a current through it, the stationary assumption is obviously false. As another example, it might be reasonable to assume that random wind velocity comes from a stationary source over a period of one hour, say, but common sense indicates that applying this same assumption to a period of one week might be unreasonable.

Deterministic processes are usually stationary only under certain very special conditions. It is customary to assume that these conditions exist, but one must be aware that this is a deliberate choice and not necessarily a natural occurrence. For example, in the case of the random process defined by (5–1), the reader may easily show (by calculating the mean value) that the process may be (and, in fact, is) stationary when θ is uniformly distributed over a range from 0 to 2π, but that it is definitely not stationary when θ is uniformly distributed over a range from 0 to π. The random process defined by (5–2) can be shown to be stationary if the A_n and the B_n are independent, zero mean, Gaussian random variables, with coefficients of the same index having equal variances. Under most other situations, however, this random process will be nonstationary. The random process defined by (5–3) is nonstationary under all circumstances.

The requirement that all marginal and joint density functions be independent of the choice of time origin is frequently more stringent than is necessary for systems analysis. A more relaxed requirement, which is often adequate, is that the mean value of any random variable, $X(t_1)$, is independent of the choice of t_1 *and* that the correlation of two random variables, $\overline{X(t_1)X(t_2)}$ depends only upon the time difference, $t_2 - t_1$. Processes that satisfy these two conditions are said to be *stationary in the wide sense*. This wide-sense stationarity is adequate to guarantee that the mean value, mean-square value, variance, and correlation coefficient of any pair of random variables are constants independent of the choice of time origin.

In subsequent discussions of the response of systems to random inputs it will be found that the evaluation of this response is made much easier when the processes may be assumed either strictly stationary or stationary in the wide sense. Since the results are identical for either type of stationarity, it is not necessary to distinguish between the two in any future discussion.

Exercise 5–4.1

a) For the random process described in Exercise 5–3.2, find the mean value of the random variable $X(t_1/4)$.

b) Find the mean value of the random variable $X(3t_1/4)$.

c)   Is the process stationary? Why?

   Answers: No, 5, 0

Exercise 5–4.2

A random process has sample functions of the form

$$X(t) = A \cos (\omega t + \theta)$$

in which A and ω are constants and θ is a random variable.

a)   Prove that this process is stationary in the wide sense if θ is uniformly distributed between 0 and 2π.

b)   Prove that this process cannot be stationary if θ is not uniformly distributed over a range of 2π.

5–5   Ergodic and Nonergodic Random Processes

Some stationary random processes possess the property that almost every member[1] of the ensemble exhibits the same statistical behavior as the whole ensemble. Thus, it is possible to determine this statistical behavior by examining only one typical sample function. Such processes are said to be *ergodic*.

For ergodic processes, the mean values and moments can be determined by time averages as well as by ensemble averages. Thus, for example, the *n*th moment is given by

$$\overline{X^n} = \int_{-\infty}^{\infty} x_n f(x) \, dx = \lim_{T \to \infty} \frac{1}{2T} \int_{-T}^{T} X^n(t) \, dt \qquad (5-4)$$

It should be emphasized, however, that this condition cannot exist unless the process is stationary. Thus, ergodic processes are also stationary processes.

A process that does not possess the property of (5–4) is *nonergodic*. All non-

---

[1]The term "almost every member" implies that a set of sample functions having total probability of zero may not exhibit the same behavior as the rest of the ensemble. But having zero probability does *not* mean that such a sample function is impossible.

stationary processes are nonergodic, but it is also possible for stationary processes to be nonergodic. For example, consider sample functions of the form

$$X(t) = Y \cos (\omega t + \theta) \qquad (5-5)$$

where ω is a constant, Y is a random variable (with respect to the ensemble), and θ is a random variable that is uniformly distributed over 0 to 2π, with θ and Y being statistically independent. This process can be shown to be stationary but nonergodic, since Y is a constant in any one sample function but is different for different sample functions.

It is generally difficult, if not impossible, to prove that ergodicity is a reasonable assumption for any physical process, since only one sample function of the process can be observed. Nevertheless, it is customary to assume ergodicity unless there are compelling physical reasons for not doing so.

Exercise 5–5.1

A random process has sample functions of the form

$$X(t) = A$$

where A is a zero-mean, Gaussian random variable having a variance of 4.

    a)   Is this process stationary in the wide sense?

    b)   Is the process ergodic? Why?

        Answers: No, yes

Exercise 5–5.2

A random process has sample functions of the form

$$X(t) = \sum_{n=-\infty}^{\infty} Af(t - nT - t_o)$$

where A and T are constants and t_o is a random variable that is uniformly distributed between 0 and T. The function $f(t)$ is defined by

$$f(t) = 1 \qquad 0 \leq t \leq T/2$$

and is zero elsewhere.

a) Find $\overline{X}$ and $\overline{X^2}$.

b) Find $< x >$ and $< x^2 >$, where $< \cdot >$ implies a time average.

c) Can this process be stationary?

d) Can this process be ergodic?

Answers: A/2, A²/2, yes, yes

5–6   Measurement of Process Parameters

The statistical parameters of a random process are the sets of statistical parameters (such as mean, mean-square, and variance) associated with the $X(t)$ random variables at various times t. In the case of a stationary process these parameters are the same for all such random variables, and, hence, it is customary to consider only one set of parameters.

A problem of considerable practical importance is that of estimating the process parameters from the observations of a single sample function (since one sample function of finite length is all that is ever available). Because there is only one sample function it is not possible to make an ensemble average in order to obtain estimates of the parameters. The only alternative, therefore, is to make a time average. If the process is ergodic, this is a reasonable approach because a time average (over infinite time) is equivalent to an ensemble average, as indicated by (5–4). Of course, in most practical situations, we cannot prove that the process is ergodic and it is usually necessary to *assume* that it is ergodic unless there is some clear physical reason why it should not be. Furthermore, it is not possible to take a time average over an infinite time interval, and a time average over a finite time interval will always be just an approximation to the true value. The following discussion is aimed at determining how good this approximation is, and upon what aspects of the measurement the goodness of the approximation depends.

Consider first the problem of estimating the mean value of an ergodic random process $\{x(t)\}$. This estimate will be designated as $\hat{X}$ and will be computed from a finite time average. Thus, for an arbitrary member of the ensemble, let

$$\hat{X} = \frac{1}{T} \int_0^T X(t) \, dt \tag{5–6}$$

It should be noted that although $\hat{X}$ is a single number in any one experiment, it is also a random variable, since a different number would be obtained if a different time interval were used or if a different sample function had been observed. Thus,

$\hat{\overline{X}}$ will not be identically equal to the true mean value $\overline{X}$, but if the measurement is to be useful it should be close to this value. Just how close it is likely to be is discussed below.

Since $\hat{\overline{X}}$ is a random variable, it has a mean value and a variance. If $\hat{\overline{X}}$ is to be a good estimate of $\overline{X}$, then the mean value of $\hat{\overline{X}}$ should be equal to $\overline{X}$ and the variance should be small. From (5–6) the mean value of $\hat{\overline{X}}$ is

$$E[\hat{\overline{X}}] = E\left[\frac{1}{T}\int_0^T X(t)\,dt\right] = \frac{1}{T}\int_0^T E[X(t)]\,dt$$

$$= \frac{1}{T}\int_0^T \overline{X}\,dt = \frac{1}{T}\left[\overline{X}t\bigg|_0^T\right] = \overline{X}$$

(5–7)

The interchange of expectation and integration is permissible in this case and represents a common type of operation. The conditions where such interchanges are possible will be discussed in more detail in Chapter 8. It is clear from (5–7) that $\hat{\overline{X}}$ has the proper mean value. The evaluation of the variance of $\hat{\overline{X}}$ is considerably more involved and requires a knowledge of autocorrelation functions, a topic that is considered in the next chapter. However, the variance of such estimates is considered for the following discrete time case. It is sufficient to note here that the variance turns out to be proportional to $1/T$. Thus, a better estimate of the mean is found by averaging the sample function over a longer time interval. As T approaches infinity, the variance approaches zero and the estimate becomes equal with probability one to the true mean, as it must for an ergodic process.

As a practical matter, the integration required by (5–6) can seldom be carried out analytically because $X(t)$ cannot be expressed in an explicit mathematical form. The alternative is to perform numerical integration upon samples of $X(t)$ observed at equally spaced time instants. Thus, if $X_1 = X(\Delta t)$, $X_2 = X(2\Delta t)$, . . . , $X_N = X(N\Delta t)$, then the estimate of $\overline{X}$ may be expressed as

$$\hat{\overline{X}} = \frac{1}{N}\sum_{i=1}^N X_i$$

(5–8)

This is the discrete time counterpart of (5–6).

The estimate $\hat{\overline{X}}$ is still a random variable and has an expected value of

$$E[\hat{\overline{X}}] = E\left[\frac{1}{N}\sum_{i=1}^N X_i\right] = \frac{1}{N}\sum_{i=1}^N E[X_i]$$

$$= \frac{1}{N}\sum_{i=1}^N \overline{X} = \overline{X}$$

(5–9)

Hence, the estimate still has the proper mean value.

In order to evaluate the variance of $\hat{\overline{X}}$ it is assumed that the observed samples are spaced far enough apart in time so that they are statistically independent. This assumption is made for convenience at this point; a more general derivation can be made after considering the material in Chapter 6. The mean-square value of $\hat{\overline{X}}$ can be expressed as

$$E[(\hat{\overline{X}})^2] = E\left[\frac{1}{N^2} \sum_{i=1}^{N} \sum_{j=1}^{N} X_i X_j\right] = \frac{1}{N^2} \sum_{i=1}^{N} \sum_{j=1}^{N} E[X_i X_j] \qquad (5\text{--}10)$$

where the double summation comes from the product of two summations. Since the sample values have been assumed to be statistically independent, it follows that

$$E[X_i X_j] = \overline{X^2} \qquad i = j$$
$$= (\overline{X})^2 \qquad i \neq j$$

Thus,

$$E[(\hat{\overline{X}})^2] = \frac{1}{N^2} [N\overline{X^2} + (N^2 - N)(\overline{X})^2] \qquad (5\text{--}11)$$

This results from the fact that the double summation of (5–10) contains N^2 terms all together, but only N of these correspond to $i = j$. Equation (5–11) can be written as

$$E[(\hat{\overline{X}})^2] = \frac{1}{N} \overline{X^2} + \left(1 - \frac{1}{N}\right)(\overline{X})^2$$
$$= \frac{1}{N} \sigma_X^2 + (\overline{X})^2 \qquad (5\text{--}12)$$

The variance of $\hat{\overline{X}}$ can now be written as

$$\text{Var}(\hat{\overline{X}}) = E[(\hat{\overline{X}})^2] - \{E[\hat{\overline{X}}]\}^2 = \frac{1}{N} \sigma_X^2 + (\overline{X})^2 - (\overline{X})^2$$
$$= \frac{1}{N} \sigma_X^2 \qquad (5\text{--}13)$$

This result says that the variance of the estimate of the mean value is simply $1/N$ times the variance of the process. Thus, the quality of the estimate can be made better by averaging a larger number of samples.

As an illustration of the above result, suppose it is desired to estimate the *variance* of a zero-mean Gaussian random process by passing it through a square law device and estimating the *mean value* of the output. Suppose it is also desired to find the number of sample values that must be averaged in order to be assured that the standard deviation of the resulting estimate is less than 10 percent of the true mean value.

Let the observed sample function of the zero-mean Gaussian process be $Y(t)$ and have a variance of σ_Y^2. After this sample function is squared, it is designated as $X(t)$. Thus,

$$X(t) = Y^2(t)$$

From (2–27) it follows that

$$\overline{X} = E[Y^2] = \sigma_Y^2$$
$$\overline{X^2} = E[Y^4] = 3\sigma_Y^4$$

Hence,

$$\sigma_X^2 = \overline{X^2} - (\overline{X})^2 = 3\sigma_Y^4 - \sigma_Y^4 = 2\sigma_Y^4$$

It is clear from this, that an estimate of $\overline{X}$ is also an estimate of σ_Y^2. Furthermore, the variance of the estimate of $\overline{X}$ must be $0.01(\overline{X})^2 = 0.01\sigma_Y^4$ to meet the requirement of an error of less than 10 percent. From (5–13)

$$\text{Var } (\hat{\overline{X}}) = \frac{1}{N} \sigma_X^2 = \frac{1}{N} (2\sigma_Y^4) = 0.01\sigma_Y^4$$

Thus, $N = 200$ statistically independent samples are required to achieve the desired accuracy.

The preceding not only illustrates the problems in estimating the mean value of a random process, but also indicates how the variance of a zero-mean process might be estimated. The same general procedures can obviously be extended to estimate the variance of a nonzero-mean random process.

When the process whose variance is to be estimated has an unknown mean value, the procedure for estimating the variance becomes a little more involved. At first thought, it would seem that the logical thing to do is to find the average of the X_i^2 and then subtract out the square of the estimated mean as given by Equation (5–8). It turns out, however, that the resulting estimate of the variance is biased—that is, the mean value of the estimate is not the true variance. This result occurs because the true mean is unknown. It is possible, however, to correct for this lack of knowledge by defining the estimate of the variance as

$$\hat{\sigma}_X^2 = \frac{1}{(N-1)} \sum_{i=1}^{N} X_i^2 + \frac{N}{(N-1)} (\hat{\overline{X}})^2 \qquad (5\text{–}14)$$

It is left as an exercise for the student to show that the mean value of this estimate is indeed the true variance. The student should also compare this result with a similar result shown in Equation (4–8) of the preceding chapter.

Exercise 5–6.1

Ten independent measurements of a voltage from a Gaussian random process have the following values:

207	198
202	197
184	213
204	191
206	201

a) Estimate the mean value of this process.

b) Find the variance of this estimate of the mean.

c) Estimate the variance of this process.

Answer: 4.9; 69.34, 200.3

Exercise 5–6.2

Show that the estimate of the variance given by Equation (5–14) is an unbiased estimate. That is,

$$E[\hat{\sigma}_X^2] = \sigma_X^2$$

PROBLEMS

5–1.1 A sample function from a random process is generated by rolling a die five times. During the interval from i-1 to i the value of the sample function is equal to the outcome of the ith roll of the die.

a) Sketch the resulting sample function if the outcomes of the five rolls are 5, 2, 6, 4, 1.

b) How many different sample functions does the ensemble of this random process contain?

c) What is the probability that the particular sample function observed in part (a) will occur?

d) What is the probability that the sample function consisting entirely of threes will occur?

5–1.2 The random number generator in a computer generates three digit numbers that are uniformly distributed between 0.000 and 0.999 at a rate of one random number per second starting at $t = 0$. A sample function from a random process is generated by summing the ten most recent random numbers and assigning this sum as the value of the sample function during each one second time interval. The sample functions are denoted as $X(t)$ for $t \geq 0$.

a) Find the mean value of the random variable X (4.5).

b) Find the mean value of the random variable X (9.5).

c) Find the mean value of the random variable X (20.5).

5–2.1 Classify each of the following random processes as *continuous*, *discrete*, or *mixed*.

a) A random process in which the random variable is the number of cars per minute passing a given traffic counter.

b) The thermal noise voltage generated by a resistor.

c) The random process defined in Problem 5–1.2.

d) The random process that results when a Gaussian random process is passed through an ideal half-wave rectifier.

e) The random process that results when a Gaussian random process is passed through an ideal full-wave rectifier.

f) A random process having sample functions of the form

$$X(t) = A \cos (Bt + \theta)$$

where A is a constant, B is a random variable that is exponentially distributed from 0 to ∞, and θ is a random variable that is uniformly distributed between 0 and 2π.

5–2.2 A Gaussian random process having a mean value of 2 and a variance of 4 is passed through an ideal half-wave rectifier.

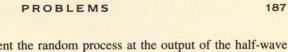

a) Let $X_p(t)$ represent the random process at the output of the half-wave rectifier if the positive portions of the input appear in the output. Determine the probability density function of $X_p(t)$.

b) Let $X_n(t)$ represent the random process at the output of the half-wave rectifier if the negative portions of the input appear in the output. Determine the probability density function of $X_n(t)$.

c) Determine the probability density function of $X_p(t)X_n(t)$.

5–3.1 State whether each of the random processes described in Problem 5–2.1 is deterministic or nondeterministic.

5–3.2 Sample functions from a deterministic random process are described by

$$X(t) = At + B \qquad t \geq 0$$
$$= 0 \qquad t < 0$$

where A is a Gaussian random variable with zero mean and a variance of 9 and B is a random variable that is uniformly distributed between 0 and 6. A and B are statistically independent.

a) Find the mean value of this process.

b) Find the variance of this process.

c) If a particular sample function is found to have a value of 10 at $t = 2$ and a value of 20 at $t = 4$, find the value of the sample function at $t = 8$.

5–4.1 State whether each of the random processes described in Problem 5–2.1 may reasonably be considered to be stationary or nonstationary. If you describe a process as nonstationary, state the reason for this claim.

5–4.2 a) Is the process described in Problem 5–3.2 stationary or nonstationary? Why?

b) A random process is described by

$$X(t) = A + B \cos (\omega t + \theta)$$

where A is a random variable that is uniformly distributed between -5 and $+5$, B is a Gaussian random variable with zero mean and a variance of 25, ω is a constant, and θ is a random variable that is uniformly distributed from $-\pi/2$ to $+3\pi/2$. A, B, and θ are statistically independent. Calculate the mean and variance of this process. Is this process stationary in the wide sense?

5–5.1 State whether each of the processes described in Problem 5–2.1 is ergodic or nonergodic. If you claim a process is nonergodic, explain why.

5–5.2 State whether each of the processes described in Problem 5–4.2 is ergodic or nonergodic and give reasons for your decision.

5–6.1 A stationary random process is sampled at time instants separated by 0.01 seconds. The resulting sample values are tabulated below.

i	$x(i)$	i	$x(i)$	i	$x(i)$
0	0.19	7	−1.24	14	1.45
1	0.29	8	−1.88	15	−0.82
2	1.44	9	−0.31	16	−0.25
3	0.83	10	1.18	17	0.23
4	−0.01	11	1.70	18	−0.91
5	−1.23	12	0.57	19	−0.19
6	−1.47	13	0.95	20	0.24

a) Estimate the mean value of this process.

b) If the process has a true variance of 1.0, find the variance of your estimate of the mean.

5–6.2 Estimate the variance of the process in Problem 5–6.1.

References

See the References for Chapter 1—particularly, Davenport and Root, Papoulis, and Helstrom.

CHAPTER 6

Correlation Functions

6–1 Introduction

The subject of correlation between two random variables was introduced in Section 3–4. Now that the concept of a random process has also been introduced, it is possible to relate these two subjects to provide a statistical (rather than a probabilistic) description of random processes. Although a probabilistic description is the most complete one, since it incorporates all the knowledge that is available about a random process, there are many engineering situations in which this degree of completeness is neither needed nor possible. If the major interest in a random quantity is in its average power, or the way in which that power is distributed with frequency, then the entire probability model is not needed. If the probability distributions of the random quantities are not known, use of the probability model is not even possible. In either case, a partial statistical description, in terms of certain average values, may provide an acceptable substitute for the probability description.

It was noted in Section 3–4 that the *correlation* between two random variables was the expected value of their product. If the two random variables are defined as samples of a random process at two different time instants, then this expected value depends upon how rapidly the time functions can change. We would expect that the random variables would be highly correlated when the two time instants are very close together, because the time function cannot change rapidly enough to be greatly different. On the other hand, we would expect to find very little correlation between the values of the random variables when the two time instants are widely separated, because almost any change can take place. Because the correlation does depend upon how rapidly the values of the random variable can change with respect to time, we expect that this correlation may also be related to the manner in which the energy content of a random process is distributed with respect to frequency. This is because a time function must have appreciable energy at high frequencies in order to be able to change rapidly with time. This aspect of random processes is discussed in more detail in subsequent chapters.

The previously defined correlation was simply a number since the random variables were not necessarily defined as being associated with time functions. In the following case, however, every pair of random variables can be related by the time separation between them, and the correlation will be a function of this separation. Thus, it becomes appropriate to define a *correlation function* in which the argument is the time separation of the two random variables. If the two random variables come from the same random process, this function will be known as the *autocorrelation function*. If they come from different random processes, it will be called the *crosscorrelation function*. We will consider autocorrelation functions first.

If $X(t)$ is a sample function from a random process, and the random variables are defined to be

$$X_1 = X(t_1)$$

$$X_2 = X(t_2)$$

then the autocorrelation function is defined to be

$$R_X(t_1, t_2) = E[X_1 X_2] = \int_{-\infty}^{\infty} dx_1 \int_{-\infty}^{\infty} x_1 x_2 \, f(x_1, x_2) \, dx_2 \qquad (6\text{--}1)$$

This definition is valid for both stationary and nonstationary random processes. However, our interest is primarily in stationary processes, for which further simplification of (6–1) is possible. It may be recalled from the previous chapter that for a wide-sense stationary process all such ensemble averages are independent of the time origin. Accordingly, for a wide-sense stationary process,

$$R_X(t_1, t_2) = R_X(t_1 + T, t_2 + T)$$

$$= E[X(t_1 + T)X(t_2 + T)]$$

Since this expression is independent of the choice of time origin, we can set $T = -t_1$ to give

$$R_X(t_1, t_2) = R_X(0, t_2 - t_1) = E[X(0)X(t_2 - t_1)]$$

It is seen that this expression depends only on the time difference $t_2 - t_1$. Setting this time difference equal to $\tau = t_2 - t_1$ and suppressing the zero in the argument of $R_X(0, t_2 - t_1)$, we can rewrite (6–1) as

$$R_X(\tau) = E[X(t_1)X(t_1 + \tau)] \tag{6-2}$$

This is the expression for the autocorrelation function of a stationary process and depends only on τ and not on the value of t_1. Because of this lack of dependence on the particular time t_1 at which the ensemble averages are taken, it is common practice to write (6–2) without the subscript; thus,

$$R_X(\tau) = E[X(t)X(t + \tau)]$$

Whenever correlation functions relate to nonstationary processes, since they are dependent on the particular time at which the ensemble average is taken as well as on the time difference between samples, they must be written as $R_X(t_1,t_2)$ or $R_X(t_1,\tau)$. In all cases in this and subsequent chapters, unless specifically stated otherwise, it is assumed that all correlation functions relate to wide-sense stationary random processes.

It is also possible to define a *time autocorrelation* function for a particular sample function as[1]

$$\mathcal{R}_x(\tau) = \lim_{T \to \infty} \frac{1}{2T} \int_{-T}^{T} x(t)x(t + \tau)\, dt = \langle x(t)x(t + \tau)\rangle \tag{6-3}$$

For the special case of an *ergodic process*, $\langle x(t)x(t + \tau)\rangle$ is the same for every $x(t)$ and equal to $R_X(\tau)$. That is,

$$\mathcal{R}_x(\tau) = R_X(\tau) \qquad \text{for an ergodic process} \tag{6-4}$$

The assumption of ergodicity, where it is not obviously invalid, often simplifies the computation of correlation functions.

From (6–2) it is seen readily that for $\tau = 0$, since $R_X(0) = E[X(t_1)X(t_1)]$, the autocorrelation function is equal to the mean-square value of the process. For values of τ other than $\tau = 0$, the autocorrelation function $R_X(\tau)$ can be thought of as a measure of the similarity of the waveform $X(t)$ and the waveform $X(t + \tau)$. In order to illustrate this point further, let $X(t)$ be a sample function from a zero-mean stationary random process and form the new function

$$Y(t) = X(t) - \rho X(t + \tau)$$

By determining the value of ρ that minimizes the mean-square value of $Y(t)$ we will have a measure of how much of the waveform $X(t + \tau)$ is contained in the waveform $X(t)$. The determination of ρ is made by computing the variance of $Y(t)$, setting the derivative of the variance with respect to ρ equal to zero, and solving for ρ. The operations are as follows:

---

[1]The symbol $\langle \ \rangle$ is used to denote time averaging.

$$E\{[Y(t)]^2\} = E\{[X(t) - \rho X(t + \tau)]^2\}$$

$$= E\{X^2(t) - 2\rho X(t)X(t + \tau) + \rho^2 X^2(t + \tau)\}$$

$$\sigma_Y^2 = \sigma_X^2 - 2\rho R_X(\tau) + \rho^2 \sigma_X^2 \qquad (6-5)$$

$$\frac{d\sigma_Y^2}{d\rho} = -2R_X(\tau) + 2\rho\sigma_X^2 = 0$$

$$\rho = \frac{R_X(\tau)}{\sigma_X^2}$$

It is seen from (6–5) that ρ is directly related to $R_X(\tau)$ and is exactly the *correlation coefficient* defined in Section 3–4. The coefficient ρ can be thought of as the fraction of the waveshape of $X(t)$ remaining after τ seconds have elapsed. It must be remembered that ρ was calculated on a statistical basis; and that it is the average retention of waveshape over the ensemble, and not this property in any particular sample function, that is important. As shown previously, the correlation coefficient ρ can vary from $+1$ to -1. For a value of $\rho = 1$, the waveshapes would be identical—that is, completely correlated. For $\rho = 0$, the waveforms would be completely uncorrelated; that is, no part of the waveform $X(t + \tau)$ would be contained in $X(t)$. For $\rho = -1$, the waveshapes would be identical, except for opposite signs; that is, the waveform $X(t + \tau)$ would be the negative of $X(t)$.

For an ergodic process or for nonrandom signals, the foregoing interpretation can be made in terms of average power instead of variance and in terms of the time correlation function instead of the ensemble correlation function.

Since $R_X(\tau)$ is dependent both on the amount of correlation ρ and the variance of the process, σ_X^2, it is not possible to estimate the significance of some particular value of $R_X(\tau)$ without knowing one or the other of these quantities. For example, if the random process has a zero mean and the autocorrelation function has a positive value, the most that can be said is that the random variables $X(t_1)$ and $X(t_1 + \tau)$ probably have the same sign.[2] If the autocorrelation function has a negative value, it is likely that the random variables have opposite signs. If it is nearly zero, the random variables are about as likely to have opposite signs as they are to have the same sign.

Exercise 6–1.1

A random process has sample functions of the form

$$X(t) = A \qquad 0 \le t \le 1$$

$$= 0 \qquad \text{elsewhere}$$

---

[2]This is strictly true only if $f(x_1)$ is symmetrical about the axis $x_1 = 0$.

where A is a random variable that is uniformly distributed from -12 to 12. Using the basic definition of the autocorrelation function as given by Equation (6–1), find the autocorrelation function of this process.

Answer:

$$R_X(t_1, t_2) = 48 \qquad 0 \le t_1, t_2 \le 1$$

$$= 0 \qquad \text{elsewhere}$$

Exercise 6–1.2

Define a random variable $Z(t)$ as

$$Z(t) = X(t) + X(t + \tau_1)$$

where $X(t)$ is a sample function from a stationary random process whose autocorrelation function is

$$R_X(\tau) = \exp(-\tau^2)$$

Write an expression for the autocorrelation function of the random process $Z(t)$.

Answer:

$$R_Z(\tau) = 2 \exp(-\tau^2) + \exp[-(\tau - \tau_1)^2] + \exp[-(\tau + \tau_1)^2]$$

6–2   Example: Autocorrelation Function of a Binary Process

The above ideas may be made somewhat clearer by considering, as a special example, a random process having a very simple autocorrelation function. Figure 6–1 shows a typical sample function from a discrete, stationary, zero-mean random process in which only two values, $\pm A$, are possible. The sample function either can change from one value to the other every t_a seconds or remain the same, with equal probability. The time t_0 is a random variable with respect to the ensemble of possible time functions and is uniformly distributed over an interval of length t_a. This means, as far as the ensemble is concerned, that changes in value can occur at any time with equal probability. It is also assumed that the value of $X(t)$ in any one interval is statistically independent of its value in any other interval.

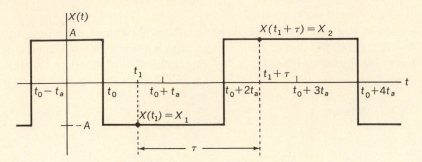

Figure 6–1 A discrete, stationary sample function.

Although the random process described in the above paragraph may seem contrived, it actually represents a very practical situation. In modern digital communication systems, the messages to be conveyed are converted into binary symbols. This is done by first sampling the message at periodic time instants and then quantizing the samples into a finite number of amplitude levels as discussed in Section 2–7 in connection with the uniform probability density function. Each amplitude level is then represented by a block of binary symbols; for example, 256 amplitude levels can each be uniquely represented by a block of 8 binary symbols. The binary symbols can in turn be represented by a voltage level of $+A$ or $-A$. Thus, a sequence of binary symbols becomes a waveform of the type shown in Figure 6–1. Similarly, this waveform is typical of those found in digital computers or in communication links connecting computers together. Hence, the random process being considered here is not only one of the simplest ones to analyze, but is also one of the most practical ones in the real world.

The autocorrelation function of this process will be determined by heuristic arguments rather than by rigorous derivation. In the first place, when $|\tau|$ is larger than t_a, then t_1 and $t_1 + \tau = t_2$ cannot lie in the same interval, and X_1 and X_2 are statistically independent. Since X_1 and X_2 have zero mean, the expected value of their product must be zero, as shown by (3–22); that is,

$$R_X(\tau) = E[X_1 X_2] = \overline{X}_1 \overline{X}_2 = 0 \qquad |\tau| > t_a$$

since $\overline{X}_1 = \overline{X}_2 = 0$. When $|\tau|$ is less than t_a, then t_1 and $t_1 + \tau$ may or may not be in the same interval, depending upon the value of t_0. Since t_0 can be anywhere, with equal probability, the probability that they do lie in the same interval is proportional to the *difference* between t_a and τ. In particular, for $\tau \geq 0$, it is seen that $t_0 \leq t_1 \leq t_1 + \tau < t_0 + t_a$, which yields $t_1 + \tau - t_a < t_0 \leq t_1$. Hence,

Pr (t_1 and $t_1 + \tau$ are in the same interval)

$$= \text{Pr} \left[(t_1 + \tau - t_a < t_0 \leq t_1) \right]$$

$$= \frac{1}{t_a} [t_1 - (t_1 + \tau - t_a)] = \frac{t_a - \tau}{t_a}$$

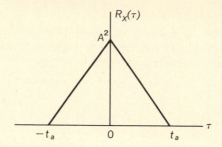

Figure 6–2   Autocorrelation function of the process in Figure 6–1.

since the probability density function for t_0 is just $1/t_a$. When $\tau < 0$, it is seen that $t_0 \leq t_1 + \tau \leq t_1 < t_0 + t_a$, which yields $t_1 - t_a < t_0 \leq t_1 + \tau$. Thus,

Pr $(t_1$ and $t_1 + \tau$ are in the same interval)
$$= \text{Pr}\,[(t_1 - t_a) < t_0 \leq (t_1 + \tau)]$$

$$= \frac{1}{t_a}[t_1 + \tau - (t_1 - t_a)] = \frac{t_a + \tau}{t_a}$$

Hence, in general,

$$\text{Pr}\,(t_1 \text{ and } t_1 + \tau \text{ are in same interval}) = \frac{t_a - |\tau|}{t_a}$$

When they are in the same interval, the product of X_1 and X_2 is always A^2; when they are not, the expected product is zero. Hence,

$$R_X(\tau) = A^2\left[\frac{t_a - |\tau|}{t_a}\right] = A^2\left[1 - \frac{|\tau|}{t_a}\right] \qquad 0 \leq |\tau| \leq t_a$$

$$= 0 \qquad\qquad\qquad\qquad\qquad |\tau| > t_a$$

(6–6)

This function is sketched in Figure 6–2.

It is interesting to consider the physical interpretation of this autocorrelation function in the light of the previous discussion. Note that when $|\tau|$ is small (less than t_a), there is an increased probability that $X(t_1)$ and $X(t_1 + \tau)$ will have the same value, and the autocorrelation function is positive. When $|\tau|$ is greater than t_a, it is equally probable that $X(t_1)$ and $X(t_1 + \tau)$ will have the same value as that they will have opposite values, and the autocorrelation function is zero. For $\tau = 0$ the autocorrelation function yields the mean-square value of A^2.

Exercise 6–2.1

A speech waveform is sampled 8000 times a second and each sample is quantized into 128 amplitude levels. The resulting amplitude levels are represented by a binary voltage having values of ± 2. Assuming that successive binary symbols are statistically independent, write the autocorrelation function of the binary process.

Answer:

$$R_X(\tau) = 4[1 - 56{,}000\,|\tau|] \qquad 0 \le |\tau| \le \frac{1}{56{,}000}$$

$$= 0 \qquad\qquad\qquad \text{elsewhere}$$

Exercise 6–2.2

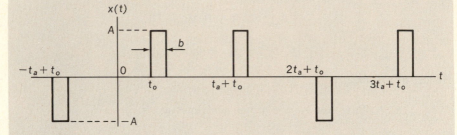

A sample function from a stationary random process is shown above. The quantity t_o is a random variable that is uniformly distributed from 0 to t_a and the pulse amplitudes are $\pm A$ with equal probability and are independent from pulse to pulse. Find the autocorrelation function of this process.

Answer:

$$R_X(\tau) = A^2 \frac{b}{t_a}[1 - |\frac{\tau}{b}|] \qquad |\tau| \le b$$

$$= 0 \qquad\qquad\qquad |\tau| > b$$

6–3   Properties of Autocorrelation Functions

If autocorrelation functions are to play a useful role in representing random processes and in the analysis of systems with random inputs, it is necessary to be able to relate the properties of the autocorrelation function to the properties of the random process it represents. In this section, a number of the properties that are possessed by all autocorrelation functions of stationary and ergodic random processes are summarized. The student should pay particular attention to these properties because they will come up many times in future discussions.

1. $R_X(0) = \overline{X^2}$. Hence, the mean-square value of the random process can always be obtained simply by setting $\tau = 0$.

It should be emphasized that $R_X(0)$ gives the mean-square value whether the process has a nonzero mean value or not. If the process is zero mean, then the mean-square value is equal to the variance of the process.

2. $R_X(\tau) = R_X(-\tau)$. The autocorrelation function is an *even* function of τ.

This is most easily seen, perhaps, by thinking of the time averaged autocorrelation function, which is the same as the ensemble averaged autocorrelation function for an ergodic random process. In this case, the time average is taken over exactly the same product function regardless of which direction one of the time functions is shifted. This symmetry property is extremely useful in deriving the autocorrelation function of a random process because it implies that the derivation needs to be carried out only for positive values of τ and the result for negative τ determined by symmetry. Thus, in the derivation shown in the example in Section 6–2, it would have been necessary to consider only the case for $\tau \geq 0$. For a nonstationary process, the symmetry property does not necessarily apply.

3. $|R_X(\tau)| \leq R_X(0)$. The largest value of the autocorrelation function always occurs at $\tau = 0$. There may be other values of τ for which it is just as big (for example, see the periodic case below), but it cannot be larger. This is shown easily by considering

$$E[(X_1 \pm X_2)^2] = E[X_1^2 + X_2^2 \pm 2X_1X_2] \geq 0$$

$$E[X_1^2 + X_2^2] = 2R_X(0) \geq |E(2X_1X_2)| = |2R_X(\tau)|$$

and thus,

$$R_X(0) \geq |R_X(\tau)| \tag{6–7}$$

4. If $X(t)$ has a dc component or mean value, then $R_X(\tau)$ will have a constant component. For example, if $X(t) = A$, then

$$R_X(\tau) = E[X(t_1)X(t_1 + \tau)] = E[AA] = A^2 \tag{6–8}$$

More generally, if $X(t)$ has a mean value *and* a zero mean component $N(t)$ so that

$$X(t) = \overline{X} + N(t)$$

then

$$
\begin{aligned}
R_X(\tau) &= E\{[\overline{X} + N(t_1)][\overline{X} + N(t_1 + \tau)]\} \\
&= E[(\overline{X})^2 + \overline{X}N(t_1) + \overline{X}N(t_1 + \tau) + N(t_1)N(t_1 + \tau)] \\
&= (\overline{X})^2 + R_N(\tau)
\end{aligned}
\tag{6–9}
$$

since

$$E[N(t_1)] = E[N(t_1 + \tau)] = 0$$

Thus, even in this case, $R_X(\tau)$ contains a constant component.

For ergodic processes the magnitude of the mean value of the process can be determined by looking at the autocorrelation function as τ approaches infinity, provided that any periodic components in the autocorrelation function are ignored in the limit. Since only the *square* of the mean value is obtained from this calculation, it is not possible to determine the sign of the mean value. If the process is stationary, but not ergodic, the value of $R_X(\tau)$ may not yield any information regarding the mean value. For example, a random process having sample functions of the form

$$X(t) = A$$

where A is a random variable with zero mean and variance σ_A^2, has an autocorrelation function of

$$R_X(\tau) = \sigma_A^2$$

for all τ. Thus, the autocorrelation function does not vanish at $\tau = \infty$ even though the process has zero mean. This strange result is a consequence of the process being nonergodic and would not occur for an ergodic process.

5. If $X(t)$ has a periodic component, then $R_X(\tau)$ will also have a periodic component, with the same period. For example, let

$$X(t) = A \cos (\omega t + \theta)$$

where A and ω are constants and θ is a random variable uniformly distributed over a range of 2π. That is,

$$f(\theta) = \frac{1}{2\pi} \qquad 0 \le \theta \le 2\pi$$

$$= 0 \qquad \text{elsewhere}$$

Then

$$
\begin{aligned}
R_X(\tau) &= E[A \cos (\omega t_1 + \theta) A \cos (\omega t_1 + \omega\tau + \theta)] \\
&= E\left[\frac{A^2}{2} \cos (2\omega t_1 + \omega\tau + 2\theta) + \frac{A^2}{2} \cos \omega\tau\right] \\
&= \frac{A^2}{2} \int_0^{2\pi} \frac{1}{2\pi} [\cos (2\omega t_1 + \omega\tau + 2\theta) + \cos \omega\tau] \, d\theta \\
&= \frac{A^2}{2} \cos \omega\tau
\end{aligned}
\tag{6-10}
$$

In the more general case, in which

$$X(t) = A \cos (\omega t + \theta) + N(t)$$

where θ and $N(t_1)$ are statistically independent for all t_1, by the method used in obtaining (5–9), it is easy to show that

$$R_X(\tau) = \frac{A^2}{2} \cos \omega\tau + R_N(\tau) \tag{6–11}$$

Hence, the autocorrelation function still contains a periodic component.

The above property can be extended to consider random processes that contain any number of periodic components. If the random variables associated with the periodic components are statistically independent, then the autocorrelation function of the sum of the periodic components is simply the sum of the periodic autocorrelation functions of each component. This statement is true regardless of whether the periodic components are harmonically related or not.

If every sample function of the random process is periodic and can be represented by a Fourier series, the resulting autocorrelation is also periodic and can also be represented by a Fourier series. However, this Fourier series will include more than just the sum of the autocorrelation functions of the individual terms if the random variables associated with the various components of the sample function are not statistically independent. A common situation in which the random variables are not independent is the case in which there is only one random variable for the process, namely a random delay on each sample function that is uniformly distributed over the fundamental period.

6. If $\{X(t)\}$ is ergodic and zero mean, and has no periodic components, then

$$\lim_{|\tau| \to \infty} R_X(\tau) = 0 \tag{6–12}$$

For large values of τ, since the effect of past values tends to die out as time progresses, the random variables tend to become statistically independent.

7. Autocorrelation functions cannot have an arbitrary shape. One way of specifying shapes that are permissible is in terms of the Fourier transform of the autocorrelation function. That is, if

$$\mathcal{F}[R_X(\tau)] = \int_{-\infty}^{\infty} R_X(\tau)e^{-j\omega\tau}\, d\tau$$

then the restriction is

$$\mathcal{F}[R_X(\tau)] \geq 0 \qquad \text{all } \omega \tag{6–13}$$

The reason for this restriction will become apparent after the discussion of spectral density in Chapter 7. Among other things, this restriction precludes the existence of autocorrelation functions with flat tops, vertical sides, or any discontinuity in amplitude.

There is one further point that should be emphasized in connection with autocorrelation functions. Although a knowledge of the joint probability density functions of the random process is sufficient to obtain a unique autocorrelation function, the converse is not true. There may be many different random processes that can yield the same autocorrelation function. Furthermore, as will be shown later, the effect of linear systems on the autocorrelation function of the input can be computed *without* knowing anything about the probability density functions. Hence, the specification of the correlation function of a random process is not equivalent to the specification of the probability density functions and, in fact, represents a considerably smaller amount of information.

Exercise 6–3.1

a) An ergodic random process has an autocorrelation function of the form

$$R_X(\tau) = 25e^{-4|\tau|} + 16 \cos 20\tau + 36$$

Find the mean-square value, mean value, and variance of this process.

b) An ergodic random process has an autocorrelation function of the form

$$R_X(\tau) = \frac{25\tau^2 + 36}{6.25\tau^2 + 4}$$

Find the mean-square value, mean value, and variance of this process.

Answers: ± 2, 5, ± 6, 9, 41, 77

Exercise 6–3.2

For each of the following functions of τ, determine the largest value of the constant A for which the function could be a valid autocorrelation function:

a) $9 e^{-4|\tau|} - A e^{-6|\tau|}$

b) $10 e^{-4|\tau - A|}$

c) $20 \cos 5\tau + A \sin 5\tau$

Answers: 0, 0, 6

6–4   Measurement of Autocorrelation Functions

Since the autocorrelation function plays an important role in the analysis of linear systems with random inputs, an important practical problem is that of determining these functions for experimentally observed random processes. In general, they cannot be calculated from the joint density functions, since these density functions are seldom known. Nor can an ensemble average be made, because there is usually only one sample function from the ensemble available. Under these circumstances, the only available procedure is to calculate a time autocorrelation function for a finite time interval, under the assumption that the process is ergodic.

In order to illustrate this, assume that a particular voltage or current waveform $x(t)$ has been observed over a time interval from 0 to T seconds. It is then possible to define an *estimated* correlation function as for this particular waveform as

$$\hat{R}_X(\tau) = \frac{1}{T - \tau} \int_0^{T-\tau} x(t)x(t + \tau)\, dt \qquad 0 \leq \tau \ll T \qquad (6\text{--}14)$$

Over the ensemble of sample functions, this estimate is a random variable denoted by $\hat{R}_X(\tau)$. Note that the averaging time is $T - \tau$ rather than T because this is the only portion of the observed data in which both $x(t)$ and $x(t + \tau)$ are available.

In most practical cases it is not possible to carry out the integration called for in (6–14) because a mathematical expression for $x(t)$ is not available. An alternative procedure is to approximate the integral by sampling the continuous time function at discrete instants of time and performing the discrete equivalent to (6–14). Thus, if the samples of a particular sample function are taken at time instants of $0, \Delta t, 2\Delta t, \ldots, N\Delta t$, and if the corresponding values of $x(t)$ are $x_0, x_1, x_2, \ldots, x_N$, the discrete equivalent to (6–14) is

$$\hat{R}_x(n\Delta t) = \frac{1}{N - n + 1} \sum_{k=0}^{N-n} x_k x_{k+n} \qquad n = 0, 1, 2, \ldots, M \qquad (6\text{--}15)$$
$$M \ll N$$

This estimate is also a random variable over the ensemble and, as such, is denoted by $\hat{R}_X(n\Delta t)$. Since N is quite large (on the order of several thousand) this operation is best performed by a digital computer.

In order to evaluate the quality of this estimate it is necessary to determine the mean and the variance of $\hat{R}_x(n\Delta t)$, since it is a random variable whose precise value depends upon the particular sample function being used and the particular set of samples taken. The mean is easy enough to obtain since

$$E[\hat{R}_X(n\Delta t)] = E\left[\frac{1}{N-n+1}\sum_{k=0}^{N-n} X_k X_{k+n}\right]$$

$$= \frac{1}{N-n+1}\sum_{k=0}^{N-n} E[X_k X_{k+n}] = \frac{1}{N-n+1}\sum_{k=0}^{N-n} R_X(n\Delta t)$$

$$= R_X(n\Delta t)$$

Thus, the expected value of the estimate is the true value of the autocorrelation function and this is an *unbiased* estimate of the autocorrelation function.

Although the estimate described by (6–15) is unbiased, it is not necessarily the best estimate in the mean-square error sense and is not the form that is most commonly used. Instead it is customary to use

$$\hat{R}_x(n\Delta t) = \frac{1}{N+1}\sum_{k=0}^{N-n} X_k X_{k+n} \qquad n = 0, 1, 2, \ldots, M \qquad (6\text{–}16)$$

This is a biased estimate, as can be seen readily from the evaluation of $E[\hat{R}_X(n\Delta t)]$ given above for the estimate of (6–15). Since only the factor by which the sum is divided is different in the present case, the expected value of this new estimate is simply

$$E[\hat{R}_X(n\Delta t)] = \left[1 - \frac{n}{N+1}\right]R_X(n\Delta t)$$

Note that if $N \gg n$, the bias is small. Although this estimate is biased, in most cases, the total mean-square error is slightly less than for the estimate of (6–15). Furthermore, (6–16) is slightly easier to calculate. A computer program for carrying out this calculation is given in Appendix G.

It is much more difficult to determine the variance of the estimate, and the details of this are beyond the scope of the present discussion. It is possible to show, however, that the variance of the estimate must be smaller than

$$\text{Var}\,[\hat{R}_X(n\Delta t)] \le \frac{2}{N}\sum_{k=-M}^{M} R_X^2(k\Delta t) \qquad (6\text{–}17)$$

This expression for the variance assumes that the $2M + 1$ estimated values of the autocorrelation function span the region in which the autocorrelation function has a significant amplitude. If the value of $(2M + 1)\Delta t$ is too small, the variance given by (6–17) may be too small. If the mathematical form of the autocorrelation function is known, or can be deduced from the measurements that are made, a more accurate measure of the variance of the estimate is

$$\text{Var}\,[\hat{R}_X(n\Delta t)] \le \frac{2}{T}\int_{-\infty}^{\infty} R_X^2(\tau)\,d\tau \qquad (6\text{–}18)$$

where $T = N\Delta t$ is the length of the observed sample.

As an illustration of what this result means in terms of the number of samples required for a given degree of accuracy, suppose that it is desired to estimate a correlation function of the form shown in Figure 6–2 with 4 points on either side of center ($M = 4$). If an rms error of 5 percent[3] or less is required, then (6–17) implies that (since $t_a = 4\Delta t$)

$$(0.05A^2)^2 \geq \frac{2}{N} \sum_{k=-4}^{4} A^4 \left[1 - \frac{|k|\, \Delta t}{4\Delta t} \right]^2$$

This can be solved for N to obtain

$$N \geq 2200$$

It is clear that long samples of data and extensive calculations are necessary if accurate estimates of correlation functions are to be made.

Exercise 6–4.1

An ergodic random process has an autocorrelation function of the form

$$R_X(\tau) = 10 \left(\frac{\sin \pi\tau}{\pi\tau} \right)^2$$

a) Over what range of τ-values must the autocorrelation function of this process be estimated in order to include the first two zeros of the autocorrelation function?

b) If 21 estimates ($M = 20$) of the autocorrelation function are to be made in the interval specified in (a), what should the sampling interval be?

c) How many sample values of the random process are required so that the rms error of the estimate is less than 5 percent of the true maximum value of the autocorrelation function?

Answers: 0.1, 2, 5331

---

[3]This implies that the standard deviation of the estimate should be no greater than 5 percent of the true mean value of the random variable $\hat{R}_X(n\Delta t)$.

Exercise 6–4.2

Using the variance bounds given by the integral of (6–18), find the number of sample points required for the autocorrelation function estimate of Exercise 6–4.1.

Answer: 5333

6–5 Examples of Autocorrelation Functions

Before going on to consider crosscorrelation functions it is worthwhile to look at some typical autocorrelation functions, suggest the circumstances under which they might arise, and list possible applications. This discussion is not intended to be exhaustive but is intended primarily to introduce some ideas.

The triangular correlation function shown in Figure 6–2 is typical of random binary signals in which the switching must occur at uniformly spaced time intervals. Such a signal arises in many types of communication and control systems in which the continuous signals are sampled at periodic instants of time and the resulting sample amplitudes converted to binary numbers. The correlation function shown in Figure 6–2 assumes that the random process has a mean value of zero, but this is not always the case. If, for example, the random signal could assume values of A and 0 (rather than $-A$) then the process has a mean value of $A/2$ and a mean-square value of $A^2/2$. The resulting autocorrelation function, shown in Figure 6–3, follows from an application of (6–9).

Not all binary time functions have triangular autocorrelation functions, however. For example, another common type of binary signal is one in which the switching occurs at randomly spaced instants of time. If all times are equally probable, then the probability density function associated with the duration of each

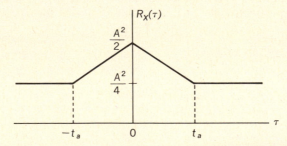

Figure 6–3 Autocorrelation function of a binary process with a non-zero mean value.

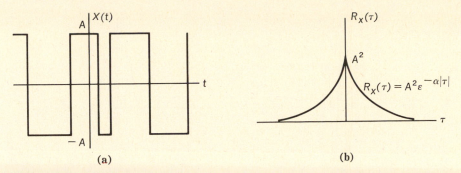

Figure 6–4 (a) A binary signal with randomly spaced switching times and (b) the corresponding autocorrelation function.

interval is exponential, as shown in Section 2–7. The resulting autocorrelation function is also exponential, as shown in Figure 6–4. The usual mathematical representation of such an autocorrelation function is

$$R_X(\tau) = A^2 e^{-\alpha|\tau|} \tag{6–19}$$

where α is the average number of intervals per second.

Binary signals and correlation functions of the type shown in Figure 6–4 frequently arise in connection with radioactive monitoring devices. The randomly occurring pulses at the output of a particle detector are used to trigger a flip-flop circuit that generates the binary signal. This type of signal is a convenient one for measuring either the average time interval between particles or the average rate of occurrence. It is usually referred to in the literature as the *Random Telegraph Wave*.

Nonbinary signals can also have exponential correlation functions. For example, if very wideband noise (having almost any probability density function) is passed through a low pass RC filter, the signal appearing at the output of the filter will have a nearly exponential autocorrelation function. This result is shown in detail in Chapter 8.

Both the triangular autocorrelation function and the exponential autocorrelation function share one feature that is worth noting. That is, in both cases the autocorrelation function has a discontinuous derivative at the origin. Random processes whose autocorrelation functions have this property are said to be *nondifferentiable*. A nondifferentiable process is one whose derivative has an infinite variance. For example, if a random voltage having an exponential autocorrelation function is applied to a capacitor, the resulting current is proportional to the derivative of the voltage, and this current would have an infinite variance. Since this doesn't make sense on a physical basis, the implication is that random processes having truly triangular or truly exponential autocorrelation functions cannot exist in the real world. In spite of this conclusion, which is indeed true, both the triangular and exponential autocorrelation functions provide useful models in many situ-

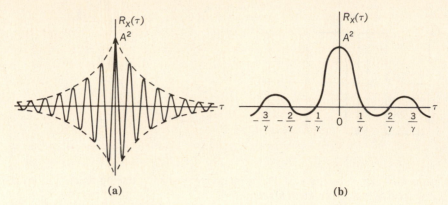

(a)   (b)

Figure 6–5 The autocorrelation functions arising at the outputs of (a) a bandpass filter and (b) an ideal low pass filter.

ations. One must be careful, however, that these models are not used in any situation in which the derivative of the random process is needed because the resulting calculation is almost certain to be wrong.

All of the correlation functions discussed so far have been positive for all values of τ. This is not necessary, however, and two common types of autocorrelation functions that have negative regions are given by

$$R_X(\tau) = A^2 e^{-\alpha|\tau|} \cos \beta\tau \qquad \qquad \textbf{(6–20)}$$

and

$$R_X(\tau) = \frac{A^2 \sin \pi\gamma\tau}{\pi\gamma\tau} \qquad \qquad \textbf{(6–21)}$$

and are illustrated in Figure 6–5. The autocorrelation function of (6–20) arises at the output of the narrow band bandpass filter whose input is very wideband noise, while that of (6–21) is typical of the autocorrelation at the output of an ideal low-pass filter. Both of these results will be derived in Chapters 7 and 8.

Although there are many other types of autocorrelation functions that arise in connection with signal and system analysis, the few discussed here are the ones most commonly encountered. The student should refer to the properties of auto-correlation functions discussed in Section 6–3 and verify that all these correlation functions possess those properties.

Exercise 6–5.1

a) Determine whether each of the random processes described by the autocorrelation functions of (6–20) and (6–21) are differentiable.

b) Indicate whether the following statement is true or false: The product of a function that is differentiable at the origin and a function that is nondifferentiable at the origin is always differentiable. Test your conclusion on the autocorrelation function of (6–20).

Answers: Yes, yes, true

Exercise 6–5.2

Which of the following functions of τ cannot be valid mathematical models for autocorrelation functions? Explain why.

a) $e^{-\tau^2}$

b) $|\tau| e^{-|\tau|}$

c) $10 e^{-(\tau+2)}$

d) $\left[\dfrac{\sin \pi\tau}{\pi\tau} \right]^2$

e) $\dfrac{\tau^2 + 4}{\tau^2 + 8}$

Answers: b, c, e are not valid models.

6–6 Crosscorrelation Functions

It is also possible to consider the correlation between two random variables from different random processes. This situation arises when there is more than one random signal being applied to a system or when one wishes to compare random voltages or currents occurring at different points in the system. If the random processes are jointly stationary in the wide sense, and if sample functions from these processes are designated as $X(t)$ and $Y(t)$, then for two random variables

$$X_1 = X(t_1)$$

$$Y_2 = Y(t_1 + \tau)$$

it is possible to define the *crosscorrelation function*

$$R_{XY}(\tau) = E[X_1 Y_2] = \int_{-\infty}^{\infty} dx_1 \int_{-\infty}^{\infty} x_1 y_2 f(x_1, y_2)\, dy_2 \qquad \text{(6–22)}$$

The order of subscripts is significant; the second subscript refers to the random variable taken at $(t_1 + \tau)$.[4]

There is also another crosscorrelation function that can be defined for the same two time instants. Thus, let

$$Y_1 = Y(t_1)$$

$$X_2 = X(t_1 + \tau)$$

and define

$$R_{YX}(\tau) = E[Y_1 X_2] = \int_{-\infty}^{\infty} dy_1 \int_{-\infty}^{\infty} y_1 x_2 f(y_1, x_2) \, dx_2 \qquad \text{(6–23)}$$

Note that because both random processes are assumed to be *jointly* stationary, these crosscorrelation functions depend only upon the time difference τ.

It is important that the processes be jointly stationary and not just individually stationary. It is quite possible to have two individually stationary random processes that are not jointly stationary. In such a case, the crosscorrelation function depends upon time, as well as the time difference τ.

The *time crosscorrelation functions* may be defined as before for a particular pair of sample functions as

$$\mathcal{R}_{xy}(\tau) = \lim_{T \to \infty} \frac{1}{2T} \int_{-T}^{T} x(t) y(t + \tau) \, dt \qquad \text{(6–24)}$$

and

$$\mathcal{R}_{yx}(\tau) = \lim_{T \to \infty} \frac{1}{2T} \int_{-T}^{T} y(t) x(t + \tau) \, dt \qquad \text{(6–25)}$$

If the random processes are jointly ergodic, then (6–24) and (6–25) yield the same value for every pair of sample functions. Hence, for ergodic processes,

$$\mathcal{R}_{xy}(\tau) = R_{XY}(\tau) \qquad \text{(6–26)}$$

$$\mathcal{R}_{yx}(\tau) = R_{YX}(\tau) \qquad \text{(6–27)}$$

In general, the physical interpretation of crosscorrelation functions is no more concrete than that of autocorrelation functions. It is simply a measure of how much these two random variables depend upon one another. In the later study of system analysis, however, the specific crosscorrelation function between system input and output will take on a very definite and important physical significance.

---

[4]This is an arbitrary convention, which is by no means universal with all authors. The definitions should be checked in every case.

Exercise 6–6.1

Two jointly stationary random processes have sample functions of the form

$$X(t) = 5 \cos(10t + \theta)$$

and

$$Y(t) = 20 \sin(10t + \theta)$$

where θ is a random variable that is uniformly distributed from 0 to 2π. Find the crosscorrelation function $R_{XY}(\tau)$ for these two processes.

Answer: $50 \sin 10\tau$

Exercise 6–6.2

Two sample functions from two random processes have the form

$$x(t) = 5 \cos 10t$$

and

$$y(t) = 20 \sin 10t$$

Find the time crosscorrelation function for $x(t)$ and $y(t + \tau)$.

Answer: $50 \sin 10\tau$

6–7 · Properties of Crosscorrelation Functions

The general properties of all crosscorrelation functions are quite different from those of autocorrelation functions. They may be summarized as follows:

1. The quantities $R_{XY}(0)$ and $R_{YX}(0)$ have *no* particular physical significance and do *not* represent mean-square values. It is true, however, that $R_{XY}(0) = R_{YX}(0)$.

2. Crosscorrelation functions are not generally even functions of τ. There is a type of symmetry, however, as indicated by the relations

$$R_{YX}(\tau) = R_{XY}(-\tau) \tag{6–28}$$

This result follows from the fact that a shift of $Y(t)$ in one direction (in time) is equivalent to a shift of $X(t)$ in the other direction.

3. The crosscorrelation function does not necessarily have its maximum value at $\tau = 0$. It can be shown, however, that

$$|R_{XY}(\tau)| \le [R_X(0)R_Y(0)]^{1/2} \tag{6–29}$$

with a similar relationship for $R_{YX}(\tau)$. The maximum of the crosscorrelation function can occur anywhere, but it cannot exceed the above value. Furthermore, it may not achieve this value anywhere.

4. If the two random processes are statistically independent, then

$$R_{XY}(\tau) = E[X_1, Y_2] = E[X_1]E[Y_2] = \overline{X}\,\overline{Y} \tag{6–30}$$
$$= R_{YX}(\tau)$$

If, in addition, *either* process has zero mean, then the crosscorrelation function vanishes for all τ. The converse of this is not necessarily true, however. The fact that the crosscorrelation function is zero and that one process has zero mean does *not* imply that the random processes are statistically independent, except for jointly Gaussian random variables.

5. If $X(t)$ is a stationary random process and $\dot{X}(t)$ is its derivative with respect to time, the crosscorrelation function of $X(t)$ and $\dot{X}(t)$ is given by

$$R_{X\dot{X}}(\tau) = \frac{dR_X(\tau)}{d\tau} \tag{6–31}$$

in which the right side of (6–31) is the derivative of the autocorrelation function with respect to τ. This is easily shown by employing the fundamental definition of a derivative

$$\dot{X}(t) = \lim_{e \to 0} \frac{X(t + e) - X(t)}{e}$$

Hence,

$$R_{X\dot{X}}(\tau) = E[X(t)\dot{X}(t + \tau)]$$

$$= E\left\{ \lim_{e \to 0} \frac{X(t)X(t + \tau + e) - X(t)X(t + \tau)}{e} \right\}$$

$$= \lim_{e \to 0} \frac{R_X(\tau + e) - R_X(\tau)}{e} = \frac{dR_X(\tau)}{d(\tau)}$$

The interchange of the limit operation and the expectation is permissible whenever $\dot{X}(t)$ exists. If the above process is repeated, it is also possible to show that the autocorrelation function of $\dot{X}(t)$ is

$$R_{\dot{X}}(\tau) = R_{\dot{X}\dot{X}}(\tau) = -\frac{d^2R_X(\tau)}{d\tau^2} \tag{6–32}$$

where the right side is the second derivative of the basic autocorrelation function with respect to τ.

It is worth noting that the requirements for the existence of crosscorrelation functions are more relaxed than those for the existence of autocorrelation functions. Crosscorrelation functions are generally not even functions of τ, their Fourier transforms do not have to be positive for all values of ω, and it is not even necessary that the Fourier transforms be real. These latter two points are discussed in more detail in the next chapter.

Exercise 6–7.1

Prove the inequality shown in Equation (6–29). This is most easily done by evaluating the expected value of the quantity

$$\left[\frac{X_1}{\sqrt{R_X(0)}} \pm \frac{Y_2}{\sqrt{R_Y(0)}}\right]^2$$

Exercise 6–7.2

Two random processes have sample functions of the form

$$X(t) = A\cos(\omega_0 t + \theta) \text{ and } Y(t) = B\sin(\omega_0 t + \theta)$$

where θ is a random variable that is uniformly distributed between 0 and 2π and A and B are constants.

a) Find the crosscorrelation functions $R_{XY}(\tau)$ and $R_{YX}(\tau)$.

b) What is the significance of the values of these crosscorrelation functions at $\tau = 0$?

Answer: $\left(\dfrac{1}{2}\right)\sin\omega_0\tau$

6–8 Examples and Applications of Crosscorrelation Functions

It is noted previously that one of the applications of crosscorrelation functions is in connection with systems with two or more random inputs. In order to explore this in more detail, consider a random process whose sample functions are of the form

$$Z(t) = X(t) \pm Y(t)$$

in which $X(t)$ and $Y(t)$ are also sample functions of random processes. Then defining the random variables as

$$Z_1 = X_1 \pm Y_1 = X(t_1) \pm Y(t_1)$$
$$Z_2 = X_2 \pm Y_2 = X(t_1 + \tau) \pm Y(t_1 + \tau)$$

the autocorrelation function of $Z(t)$ is

$$R_Z(\tau) = E[Z_1 Z_2] = E[(X_1 \pm Y_1)(X_2 \pm Y_2)]$$

$$= E[X_1 X_2 + Y_1 Y_2 \pm X_1 Y_2 \pm Y_1 X_2] \qquad \text{(6–33)}$$

$$= R_X(\tau) + R_Y(\tau) \pm R_{XY}(\tau) \pm R_{YX}(\tau)$$

This result is easily extended to the sum of any number of random variables. In general, the autocorrelation function of such a sum will be the sum of *all* the autocorrelation functions plus the sum of *all* the crosscorrelation functions.

If the two random processes being considered are statistically independent and one of them has zero mean, then both of the crosscorrelation functions in (6–33) vanish and the autocorrelation function of the sum is just the sum of the autocorrelation functions. An example of the importance of this result arises in connection with the extraction of periodic signals from random noise. Let $X(t)$ be a desired signal sample function of the form

$$X(t) = A \cos (\omega t + \theta) \qquad \text{(6–34)}$$

where θ is a random variable. It is shown previously that the autocorrelation function of this process is

$$R_X(\tau) = \frac{1}{2} A^2 \cos \omega \tau$$

Next, let $Y(t)$ be a sample function of zero-mean random noise that is statistically independent of the signal and specify that it has an autocorrelation function of the form

$$R_Y(\tau) = B^2 e^{-\alpha|\tau|}$$

The observed quantity is $Z(t)$, which from (6–33) has an autocorrelation function of

$$R_Z(\tau) = R_X(\tau) + R_Y(\tau) \qquad \text{(6–35)}$$

$$= \frac{1}{2} A^2 \cos \omega \tau + B^2 e^{-\alpha|\tau|}$$

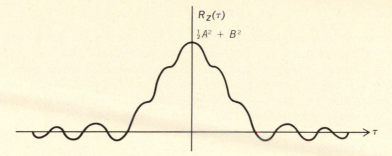

Figure 6–6 Autocorrelation function of sinusoidal signal plus noise.

This function is sketched in Figure 6–6 for a case in which the average noise power, Y^2, is much larger than the average signal power, $\frac{1}{2}A^2$. It is clear from the sketch that for large values of τ, the autocorrelation function depends mostly upon the signal, since the noise autocorrelation function tends to zero as τ tends to infinity. Thus, it should be possible to extract tiny amounts of sinusoidal signal from large amounts of noise by using an appropriate method for measuring the autocorrelation function of the received signal plus noise.

Another method of extracting a small known signal from a combination of signal and noise is to perform a crosscorrelation operation. A typical example of this might be a radar system that is transmitting a signal $X(t)$. The signal that is returned from any target is a very much smaller version of $X(t)$ and has been delayed in time by the propagation time to the target and back. Since noise is always present at the input to the radar receiver, the total received signal $Y(t)$ may be represented as

$$Y(t) = aX(t - \tau_1) + N(t) \tag{6–36}$$

where a is a number very much smaller than 1, τ_1 is the round-trip delay time of the signal, and $N(t)$ is the receiver noise. In a typical situation the average power of the returned signal, $aX(t - \tau_1)$, is very much smaller than the average power of the noise, $N(t)$.

The crosscorrelation function of the transmitted signal and the total receiver input is

$$R_{XY}(\tau) = E[X(t)Y(t + \tau)]$$

$$= E[aX(t)X(t + \tau - \tau_1) + X(t)N(t + \tau)] \tag{6–37}$$

$$= aR_X(\tau - \tau_1) + R_{XN}(\tau)$$

Since the signal and noise are statistically independent and have zero mean (because they are RF bandpass signals), the crosscorrelation function between $X(t)$ and $N(t)$ is zero for all values of τ. Thus, (6–37) becomes

$$R_{XY}(\tau) = aR_X(\tau - \tau_1) \qquad \text{(6–38)}$$

Remembering that autocorrelation functions have their maximum values at the origin, it is clear that if τ is adjusted so that the measured value of $R_{XY}(\tau)$ is a maximum, then $\tau = \tau_1$ and this value indicates the distance to the target.

In some situations involving two random processes it is possible to observe both the sum and the difference of the two processes but not each one individually. In this case, one may be interested in the crosscorrelation between the sum and difference as a means of learning something about them. Suppose, for example, that we have available two processes described by

$$U(t) = X(t) + Y(t) \qquad \text{(6–39)}$$
$$V(t) = X(t) - Y(t) \qquad \text{(6–40)}$$

in which $X(t)$ and $Y(t)$ are not necessarily zero mean nor statistically independent. The crosscorrelation function between $U(t)$ and $V(t)$ is

$$
\begin{aligned}
R_{UV}(\tau) &= E[U(t)V(t + \tau)] \\
&= E[X(t) + Y(t)][X(t + \tau) - Y(t + \tau)] \\
&= E[X(t)X(t + \tau) + Y(t)X(t + \tau) - X(t)Y(t + \tau) - Y(t)Y(t + \tau)]
\end{aligned}
\qquad \text{(6–41)}
$$

Each of the expected values in (6–41) may be identified as an autocorrelation function or a crosscorrelation function. Thus,

$$R_{UV}(\tau) = R_X(\tau) + R_{YX}(\tau) - R_{XY}(\tau) - R_Y(\tau) \qquad \text{(6–42)}$$

In a similar way, the reader may verify easily that the other crosscorrelation function is

$$R_{VU}(\tau) = R_X(\tau) - R_{YX}(\tau) + R_{XY}(\tau) - R_Y(\tau) \qquad \text{(6–43)}$$

If both X and Y are zero mean and statistically independent, both crosscorrelation functions reduce to the same function, namely

$$R_{UV}(\tau) = R_{VU}(\tau) = R_X(\tau) - R_Y(\tau) \qquad \text{(6–44)}$$

The actual measurement of crosscorrelation functions can be carried out in much the same way as that suggested for measuring autocorrelation functions in Section 6–4. This type of measurement is still unbiased when crosscorrelation functions are being considered, but the result given in (6–17) for the variance of the estimate is no longer strictly true—particularly if one of the signals contains additive uncorrelated noise, as in the radar example just discussed. Generally speaking, the number of samples required to obtain a given variance in the estimate of a crosscorrelation function is much greater than that required for an autocorrelation function.

Exercise 6–8.1

A random process has sample functions of the form $X(t) = A$ in which A is a random variable that has a mean value of 10 and a variance of 25. Sample functions from this process can be observed only in the presence of independent noise having an autocorrelation function of

$$R_N(\tau) = 100 \exp\left(-10|\tau|\right)$$

a) Find the autocorrelation function of the sum of these two processes.

b) If the autocorrelation function of the sum is observed, find the value of τ at which this autocorrelation function is within 1% of its value at $\tau = \infty$.

Answers: 0.439, $125 + 100 \exp(-10|\tau|)$

Exercise 6–8.2

A random binary process such as that described in Section 6–2 has sample functions with amplitudes of ± 10 and $t_a = 0.01$. It is applied to the half–wave rectifier circuit shown below.

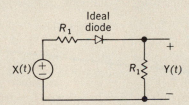

a) Find the autocorrelation function of the output, $R_Y(\tau)$.

b) Find the crosscorrelation function $R_{XY}(\tau)$.

c) Find the crosscorrelation function $R_{YX}(\tau)$.

Answers: $25 + 25\left(1 - \dfrac{|\tau|}{0.01}\right)$, $50[1 - |\tau|]$

6—9   Correlation Matrices for Sampled Functions

The discussion of correlation thus far has concentrated on only two random variables. Thus, for stationary processes the correlation functions can be expressed as a function of the single variable τ. There are many practical situations, however, in which there may be many random variables and it is necessary to develop some convenient method for representing the many autocorrelations and crosscorrelations that arise. The use of vector notation provides a convenient way of representing a set of random variables, and the product of vectors that is necessary to obtain correlations results in a matrix. It is important, therefore, to discuss some situations in which the vector representation is useful and to describe some of the properties of the resulting correlation matrices. A situation in which vector notation is useful in representing a signal arises in the case of a single time function that is sampled at periodic time instants. If only a finite number of such samples are to be considered, say N, then each sample value can become a component of an $(N \times 1)$ vector. Thus, if the sampling times are $t_1, t_2, \ldots, t_N$, the vector representing the time function $X(t)$ may be expressed as

$$X = \begin{bmatrix} X(t_1) \\ X(t_2) \\ \cdot \\ \cdot \\ \cdot \\ X(t_N) \end{bmatrix}$$

If $X(t)$ is a sample function from a random process, then each of the components of the vector X is a random variable.

It is now possible to define a correlation matrix that is $(N \times N)$ and gives the correlation between every pair of random variables. Thus,

$$R_X = E[XX^T] = E \begin{bmatrix} X(t_1)X(t_1) & X(t_1)X(t_2) & \cdots & X(t_1)X(t_N) \\ X(t_2)X(t_1) & X(t_2)X(t_2) & & \\ \cdot & & & \\ \cdot & & & \\ \cdot & & & \\ X(t_N)X(t_1) & & \cdots & X(t_N)X(t_N) \end{bmatrix}$$

where X^T is the transpose of X. When the expected value of each element of the matrix is taken, that element becomes a particular value of the autocorrelation function of the random process from which $X(t)$ came. Thus,

$$\boldsymbol{R}_X = \begin{bmatrix} R_X(t_1, t_1) & R_X(t_1, t_2) & \cdots & R_X(t_1, t_N) \\ R_X(t_2, t_1) & R_X(t_2, t_2) & & \\ & \cdot & & \\ & \cdot & & \\ & \cdot & & \\ R_X(t_N, t_1) & & \cdots & R_X(t_N, t_N) \end{bmatrix} \qquad (6\text{--}45)$$

When the random process from which $X(t)$ came is wide-sense stationary, then all the components of $\boldsymbol{R}_X$ become functions of time difference only. If the interval between sample values is Δt, then

$$t_2 = t_1 + \Delta t$$
$$t_3 = t_1 + 2\Delta t$$

$$\cdot$$

$$\cdot$$

$$\cdot$$

$$t_N = t_1 + (N - 1)\,\Delta t$$

and

$$\boldsymbol{R}_X = \begin{bmatrix} R_X[0] & R_X[\Delta t] & \cdots & R_X[(N-1)\,\Delta t] \\ R_X[\Delta t] & R_X[0] & & \\ & \cdot & & \\ & \cdot & & \\ & \cdot & & \\ R_X[(N-1)\,\Delta t] & & \cdots & R_X[0] \end{bmatrix} \qquad (6\text{--}46)$$

where use has been made of the symmetry of the autocorrelation function; that is, $R_X[i\,\Delta t] = R_X[-i\,\Delta t]$. Note that as a consequence of the symmetry, $\boldsymbol{R}_X$ is a symmetric matrix (even in the nonstationary case), and that as a consequence of stationarity, the major diagonal (and all diagonals parallel to it) have identical elements.

Although the $\boldsymbol{R}_X$ just defined is a logical consequence of previous definitions, it is not the most customary way of designating the correlation matrix of a random vector consisting of sample values. A more common procedure is to define a *covariance matrix*, which contains the variances and covariances of the random variables. The general covariance between two random variables is defined as

$$E\{[X(t_i) - \overline{X}(t_i)][X(t_j) - \overline{X}(t_j)]\} = \sigma_i \sigma_j \rho_{ij} \qquad (6\text{--}47)$$

where $\overline{X}(t_i) = $ mean value of $X(t_i)$
$\overline{X}(t_j) = $ mean value of $X(t_j)$

$$\sigma_i^2 = \text{variance of } X(t_i)$$

$$\sigma_j^2 = \text{variance of } X(t_j)$$

$$\rho_{ij} = \text{normalized covariance coefficient of } X(t_i) \text{ and } X(t_j)$$

$$= 1, \text{ when } i = j$$

The covariance matrix is defined as

$$\boldsymbol{\Lambda}_X = E[X - \overline{X})(X^T - \overline{X}^T)] \tag{6–48}$$

where $\overline{X}$ is the mean value of X. Using the covariance definitions leads immediately to

$$
\boldsymbol{\Lambda}_X = \begin{bmatrix}
\sigma_1^2 & \sigma_1\sigma_2\rho_{12} & \cdots & \sigma_1\sigma_N\rho_{1N} \\
\sigma_2\sigma_1\rho_{21} & \sigma_2^2 & & \\
& \cdot & & \\
& \cdot & & \\
& \cdot & & \\
\sigma_N\sigma_1\rho_{N1} & & \cdots & \sigma_N^2
\end{bmatrix} \tag{6–49}
$$

since $\rho_{ii} = 1$, for $i = 1, 2, \ldots, N$. By expanding (6–49) it is easy to show that $\boldsymbol{\Lambda}_X$ is related to $\boldsymbol{R}_X$ by

$$\boldsymbol{\Lambda}_X = \boldsymbol{R}_X - \overline{X}\,\overline{X}^T \tag{6–50}$$

If the random process has a zero mean, then $\boldsymbol{\Lambda}_X = \boldsymbol{R}_X$.

The above representation for the covariance matrix is valid for both stationary and nonstationary processes. In the case of a wide-sense stationary process, however, all the variances are the same and the correlation coefficients in a given diagonal are the same. Thus,

$$\sigma_i^2 = \sigma_j^2 = \sigma^2 \qquad i, j = 1, 2, \ldots, N$$

$$\rho_{ij} = \rho_{|i-j|} \qquad i, j = 1, 2, \ldots, N$$

and

$$
\boldsymbol{\Lambda}_X = \sigma^2 \begin{bmatrix}
1 & \rho_1 & \rho_2 & \cdots & & & \rho_{N-1} \\
\rho_1 & 1 & \rho_1 & \cdots & & & \rho_{N-2} \\
\rho_2 & \rho_1 & 1 & \rho_1 & & & \\
& & \cdot & \cdot & \cdot & & \\
\cdot & & & \cdot & \cdot & \cdot & \\
\cdot & & & & \cdot & \cdot & \cdot \\
\cdot & & & & & 1 & \rho_1 \\
\rho_{N-1} & & & & & \rho_1 & 1
\end{bmatrix} \tag{6–51}
$$

Such a matrix is said to be *Toeplitz*.

As an illustration of some of the above concepts, suppose we have a stationary random process whose autocorrelation function is given by

$$R_X(\tau) = 10e^{-|\tau|} + 9 \qquad (6\text{--}52)$$

To keep the example simple, assume that three random variables separated by one second are to be considered. Thus, $N = 3$ and $\Delta t = 1$. Evaluating (6–52) for $\tau = 0, 1, 2$ yields the values that are needed for the correlation matrix. Thus, the correlation matrix becomes

$$\boldsymbol{R}_X = \begin{bmatrix} 19 & 12.68 & 10.35 \\ 12.68 & 19 & 12.68 \\ 10.35 & 12.68 & 19 \end{bmatrix}$$

Since the variance of this process is 10 and its mean value is ± 3, the covariance matrix is

$$\boldsymbol{\Lambda}_X = 10 \begin{bmatrix} 1 & 0.368 & 0.135 \\ 0.368 & 1 & 0.368 \\ 0.135 & 0.368 & 1 \end{bmatrix}$$

Another situation in which the use of vector notation is convenient arises when the random variables come from different random processes. In this case, the vector representing all the random variables might be written as

$$X(t) = \begin{bmatrix} X_1(t) \\ X_2(t) \\ \cdot \\ \cdot \\ \cdot \\ X_N(t) \end{bmatrix}$$

The correlation matrix is now defined as

$$\boldsymbol{R}_X(\tau) = E[X(t)X^T(t + \tau)] \qquad (6\text{--}53)$$
$$= \begin{bmatrix} R_1(\tau) & R_{12}(\tau) & \cdots & R_{1N}(\tau) \\ R_{21}(\tau) & R_2(\tau) & \cdots & \\ & \cdot & & \\ & \cdot & & \\ & \cdot & & \\ R_{N1}(\tau) & & \cdots & R_N(\tau) \end{bmatrix}$$

in which

$$R_i(\tau) = E[X_i(t)X_i(t + \tau)]$$
$$R_{ij}(\tau) = E[X_i(t)X_j(t + \tau)]$$

Note that in this case, the elements of the correlation matrix are functions of τ rather than numbers as they were in the case of the correlation matrix associated with samples taken from a single random process. Situations in which such a correlation matrix might occur arise in connection with antenna arrays or arrays of seismic detectors. In such systems, the noise signals at each antenna element, or each seismic detector, may be from different, but correlated, random processes.

Before we leave the subject of covariance matrices, it is worth noting the important role that these matrices play in connection with the joint probability density function for N random variables from a Gaussian process. It was noted earlier that the Gaussian process was one of the few for which it is possible to write a joint probability density function for any number of random variables. The derivation of this joint density function is beyond the scope of this discussion, but it can be shown that it becomes

$$f(\boldsymbol{x}) = f[x(t_1), x(t_2), \ldots, x(t_N)]$$

$$= \frac{1}{(2\pi)^{N/2}|\boldsymbol{\Lambda_X}|^{1/2}} \exp\left[-\frac{1}{2}(\boldsymbol{x}^T - \bar{\boldsymbol{x}}^T)\boldsymbol{\Lambda_x}^{-1}(\boldsymbol{x} - \bar{\boldsymbol{x}})\right] \qquad (6\text{--}54)$$

where $|\boldsymbol{\Lambda_X}|$ is the determinant of $\boldsymbol{\Lambda_X}$ and $\boldsymbol{\Lambda_X}^{-1}$ is its inverse.

Exercise 6–9.1

A random process has an autocorrelation function of the form

$$R_X(\tau) = 10e^{-|\tau|}\cos 2\tau$$

Write the correlation matrix associated with four random variables defined for time instants separated by 0.5 seconds.

Answers: Elements in the first row include 3.677, 2.228, 10.0, 6.064

Exercise 6–9.2

A covariance matrix for a stationary random process has the form

$$\begin{bmatrix} 1 & 0.6 & 0.4 & - \\ - & 1 & 0.6 & - \\ 0.4 & 0.6 & - & 0.6 \\ 0.2 & - & - & 1 \end{bmatrix}$$

Fill in the blank spaces in this matrix.

Answers: 1, 0.6, 0.2, 0.4

PROBLEMS

6–1.1 A stationary random process having sample functions of $X(t)$ has an autocorrelation function of

$$R_X(\tau) = 5e^{-5|\tau|}$$

Another random process has sample functions of

$$Y(t) = X(t) + bX(t - 0.1)$$

    a) Find the value of b that minimizes the mean-square value of $Y(t)$.

    b) Find the value of the minimum mean-square value of $Y(t)$.

    c) If $|b| \le 1$, find the maximum mean-square value of $Y(t)$.

6–1.2 For each of the autocorrelation functions given below, state whether the process if represents might be wide-sense stationary or cannot be wide-sense stationary.

    a) $R_X(t_1, t_2) = e^{t_1}e^{-t_2}$

    b) $R_X(t_1, t_2) = \cos t_1 \cos t_2 + \sin t_1 \sin t_2$

    c) $R_X(t_1, t_2) = e^{(t_1^2 - t_2^2)}$

    d)
$$R_X(t_1, t_2) = \frac{\sin t_1 \cos t_2 - \cos t_1 \sin t_2}{t_1 - t_2}$$

6–2.1 Consider a stationary random process having sample functions of the form shown below:

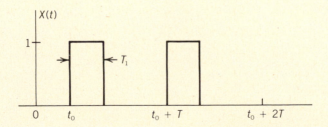

At periodic time instants $t_0 \pm nT$, a rectangular pulse of unit height and width T_1 may appear, or not appear, with equal probability and independently from interval to interval. The time t_0 is a random variable that is uniformly distributed over the period T and $T_1 \leq T/2$.

a)   Find the mean value and the mean-square value of this process.

b)   Find the autocorrelation function of this process.

6–2.2   Find the time autocorrelation function of the sample function in Problem 6–2.1.

6–2.3   Consider a stationary random process having sample functions of the form

$$X(t) = \sum_{n=-\infty}^{\infty} A_n g(t - t_0 - nT)$$

in which the A_n are independent random variables that are $+1$ or -1 with equal probability and t_0 is a random variable that is uniformly distributed over the period T. Define a function

$$G(\tau) = \int_{-\infty}^{\infty} g(t)g(t + \tau)\, dt$$

and express the autocorrelation function of the process in terms of $G(\tau)$.

6–3.1   Which of the functions shown below cannot be valid autocorrelation functions? For each case explain why it is not an autocorrelation function.

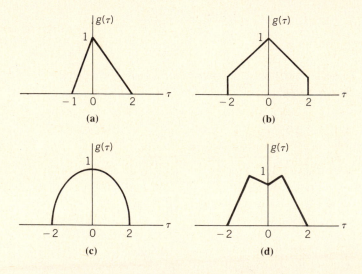

(a)

(b)

(c)

(d)

6–3.2   A random process has sample functions of the form

$$X(t) = Y \cos(\omega_0 t + \theta)$$

in which Y, ω_0, and θ are statistically independent random variables. Assume the Y has a mean value of 3 and a variance of 9, that θ is uniformly distributed from $-\pi$ to π, and that ω_0 is uniformly distributed from -6 to $+6$.

a) Is this process stationary? Is it ergodic?

b) Find the mean and mean-square value of the process.

c) Find the autocorrelation function of the process.

6–3.3 A stationary random process has an autocorrelation function of the form

$$R_X(\tau) = 100e^{-\tau^2} \cos 2\pi\tau + 10 \cos 6\pi\tau + 36$$

a) Find the mean value, mean-square value, and the variance of this process.

b) What discrete frequency components are present?

c) Find the smallest value of τ for which the random variables $X(t)$ and $X(t + \tau)$ are uncorrelated.

6–3.4 Consider a function of τ of the form

$$V(\tau) = \left[1 - \frac{|\tau|}{2} \right] \qquad |\tau| \le T$$

$$= 0 \qquad\qquad |\tau| > T$$

Take the Fourier transform of this function and show that it is a valid autocorrelation function *only* for $T = 2$.

6–4.1 A stationary random process is sampled at time instants separated by 0.01 seconds. The sample values are:

k	x_k	k	x_k	k	x_k
0	0.19	7	-1.24	14	1.45
1	0.29	8	-1.88	15	-0.82
2	1.44	9	-0.31	16	-0.25
3	0.83	10	1.18	17	0.23
4	-0.01	11	1.70	18	-0.91
5	-1.23	12	0.57	19	-0.19
6	-1.47	13	0.95	20	0.24

a) Find the sample mean.

b) Find the estimated autocorrelation function $\hat{R}(0.01\ n)$ for $n = 0, 1, 2, 3$ using Equation (6–15).

c) Repeat (b) using Equation (6–16).

6–4.2 a) For the data of Problem 6–4.1, find an upper bound on the variance of the estimated autocorrelation function using the estimated values of part (b).

b) Repeat (a) using the estimated values of part (c).

6–4.3 Assume that the true autocorrelation function of the random process from which the data of Problem 6–4.1 comes has the form

$$R(\tau) = A\left[1 - \frac{|\tau|}{T}\right] \qquad |\tau| \le T$$

and is zero elsewhere.

a) Find the values of A and T that provide the best fit to the estimated autocorrelation function values of Problem 6–4.1(b) in the least-mean-square sense. (See Sec. 4–6.)

b) Using the results of part (a) and Equation (6–18), find another upper bound on the variance of the estimate of the autocorrelation function. Compare with the result of Problem 6–4.2(a).

6–4.4 A random process has an autocorrelation function of the form

$$R_X(\tau) = 10e^{-5|\tau|}\cos20\tau$$

If this process is sampled every 0.01 seconds, find the number of samples required to estimate the autocorrelation function with a standard deviation that is no more than 1% of the variance of the process.

6–5.1 Consider a random process having sample functions of the form shown in Figure 6–4(a) and assume that the time intervals between switching times are independent, exponentially distributed random variables. (see Sec. 2–7.) Show that the autocorrelation function of this process is a two-sided exponential as shown in Figure 6–4(b).

6–5.2 Suppose that each sample function of the random process in Problem 6–

5.1 is switching between 0 and $2A$ instead of between $\pm A$. Find the autocorrelation function of the process now.

6–5.3 Determine the mean value and the variance of each of the random processes having the following autocorrelation functions:

a) $10e^{-\tau^2}$

b) $10e^{-\tau^2} \cos2\pi\tau^2$

c) $10 \dfrac{\tau^2 + 8}{\tau^2 + 4}$

6–5.4 Consider a random process having an autocorrelation function of

$$R_X(\tau) = 10e^{-2|\tau|} - 5e^{-4|\tau|}$$

a) Find the mean and variance of this process.

b) Is this process differentiable? Why?

6–7.1 Two independent stationary random processes having sample functions of $X(t)$ and $Y(t)$ have autocorrelation functions of

$$R_X(\tau) = 25e^{-10|\tau|} \cos100\pi\tau$$

and

$$R_Y(\tau) = 16 \frac{\sin50\pi\tau}{50\pi\tau}$$

a) Find the autocorrelation function of $X(t) + Y(t)$.

b) Find the autocorrelation function of $X(t) - Y(t)$.

c) Find both crosscorrelation functions of the two processes defined by (a) and (b).

d) Find the autocorrelation function of $X(t)Y(t)$.

6–7.2 For the two processes of Problem 6–7.1(c) find the maximum value that the crosscorrelation functions can have using the bound of Equation (6–29). Compare this bound with the actual maximum values that these crosscorrelation functions have.

6–7.3 A stationary random process has an autocorrelation function of

$$R_X(\tau) = \frac{\sin\tau}{\tau}$$

a) Find $R_{\dot{X}X}(\tau)$.

b) Find $R_{\dot{X}}(\tau)$.

6–7.4 Two stationary random processes have a crosscorrelation function of

$$R_{XY}(\tau) = 16e^{-(\tau-1)^2}$$

Find the crosscorrelation function of the derivative of $X(t)$ and $Y(t)$. That is, find $R_{\dot{X}Y}(\tau)$.

6–8.1 A sinusoidal signal has the form

$$X(t) = 0.01 \sin(100t + \theta)$$

in which θ is a random variable that is uniformly distributed between $-\pi$ and π. This signal is observed in the presence of independent noise whose autocorrelation function is

$$R_N(\tau) = 10e^{-100|\tau|}$$

a) Find the value of the autocorrelation function of the sum of signal and noise at $\tau = 0$.

b) Find the smallest value of τ for which the peak value of the autocorrelation function of the signal is ten times larger than the autocorrelation function of the noise.

6–8.2 One way of detecting a sinusoidal signal in noise is to use a correlator. In this device, the incoming signal plus noise is multiplied by a locally generated reference signal having the same form as the signal to be detected and the average value of the product is extracted with a low pass filter. Suppose the signal and noise of Problem 6–8.1 are multiplied by a reference signal of the form

$$r(t) = 10 \cos(100t + \phi)$$

The product is

$$Z(t) = r(t)X(t) + r(t)N(t)$$

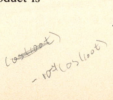

a) Find the expected value of $Z(t)$ where the expectation is taken with respect to the noise and ϕ is assumed to be a fixed, but unknown, value.

b) For what value of ϕ is the expected value of $Z(t)$ the greatest?

6–8.3 Vibration sensors are mounted on the front and rear axles of a moving vehicle to pick up the random vibrations due to the roughness of the road surface. The signal from the front sensor may be modeled as

$$f(t) = s(t) + n_1(t)$$

where the signal $s(t)$ and the noise $n_1(t)$ are from independent random processes. The signal from the rear sensor is modeled as

$$r(t) = s(t - \tau_1) + n_2(t)$$

where $n_2(t)$ is noise that is independent of both $s(t)$ and $n_1(t)$. All processes have zero mean. The delay τ_1 depends upon the spacing of the sensors and the speed of the vehicle.

a) If the sensors are placed 5 meters apart, derive a relationship between τ_1 and the vehicle speed v.

b) Sketch a block diagram of a system that can be used to measure vehicle speed over a range of 5 meters per second to 50 meters per second. Specify the maximum and minimum delay values that are required if an analog correlator is used.

c) Why is there a minimum speed that can be measured this way?

d) If a digital correlator is used, and the signals are each sampled at a rate of 12 samples per second, what is the maximum vehicle speed that can be measured?

6–8.4 The angle to distant stars can be measured by crosscorrelating the outputs of two widely separated antennas and measuring the delay required to maximize the crosscorrelation function. The geometry to be considered is shown on p. 228. In this system, the distance between antennas is nominally 500 meters, but has a standard deviation of 0.01 meters. It is desired to measure the angle θ with a standard deviation of no more than 1 milliradian for any θ between 0 and 1.4 radians. Find an upper bound on the standard deviation of the delay measurement in order to accomplish this. Hint: Use the total differential to linearize the relation.

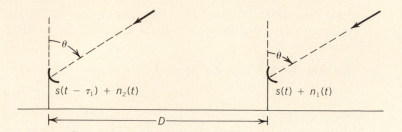

6–9.1   A stationary random process having an autocorrelation function of

$$R_X(\tau) = 36e^{-2|\tau|}\cos\pi\tau$$

is sampled at periodic time instants separated by 0.5 seconds. Write the covariance matrix for four consecutive samples taken from this process.

6–9.2   A Gaussian random vector

$$X = \begin{bmatrix} X_1 \\ X_2 \\ X_3 \end{bmatrix}$$

has a covariance matrix of

$$\Lambda = \begin{bmatrix} 1 & 0.5 & 0 \\ 0.5 & 1 & 0.5 \\ 0 & 0.5 & 1 \end{bmatrix}$$

Find the expected value, $E[X^T\Lambda^{-1}X]$.

6–9.3   A transversal filter is a tapped delay line with the outputs from the various taps weighted and summed as shown below.

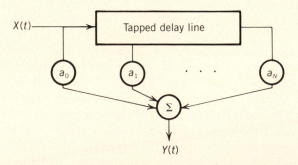

If the delay between taps is Δt the outputs from the taps can be expressed as a vector by

$$X(t) = \begin{bmatrix} X(t) \\ X(t - \Delta t) \\ \cdot \\ \cdot \\ \cdot \\ X(t - N\Delta t) \end{bmatrix}$$

Likewise, the weighting factors on the various taps can be written as a vector

$$\mathbf{a} = \begin{bmatrix} a_0 \\ a_1 \\ \cdot \\ \cdot \\ \cdot \\ a_N \end{bmatrix}$$

a)   Write an expression for the output of the transversal filter, $Y(t)$, in terms of the vectors $X(t)$ and $\mathbf{a}$.

b)   If $X(t)$ is from a stationary random process with an autocorrelation function of $R_X(\tau)$, write an expression for the autocorrelation function $R_Y(\tau)$.

6–9.4   Let the input to the transversal filter of Problem 6–9.3 have an autocorrelation function of

$$R_X(\tau) = 1 - \frac{|\tau|}{\Delta t} \qquad |\tau| \leq \Delta t$$

and is zero elsewhere.

a)   If the transversal filter has 4 taps (i.e., $N = 3$) and the weighting factor for each tap is $a_i = 1$ for all i, determine and sketch the autocorrelation function of the output.

b)   Repeat part (a) if the weighting factors are $a_i = 4 - i$, $i = 0$, 1, 2, 3.

References

See the References for Chapter 1. Of particular interest for the material of this chapter are the books by Davenport and Root, Helstrom, and Papoulis.

CHAPTER 7

Spectral Density

7–1   Introduction

The use of Fourier transforms and Laplace transforms in the analysis of linear systems is widespread and frequently leads to much saving in labor. The principal reason for this simplification is that the convolution integral of time-domain methods is replaced by simple multiplication when frequency-domain methods are used.

In view of this widespread use of frequency-domain methods, it is natural to ask if such methods are still useful when the inputs to the system are random. The answer to this question is that they *are* still useful but that some modifications are required and that a little more care is necessary in order to avoid pitfalls. However, when properly used, frequency-domain methods offer essentially the same advantages in dealing with random signals as they do with nonrandom signals.

Before beginning this discussion, it is desirable to review briefly the frequency-domain representation of a nonrandom time function. The most natural representation of this sort is the Fourier transform, which leads to the concept of frequency spectrum. Thus, the Fourier transform of some nonrandom time function, $f(t)$, is defined to be

$$F(\omega) = \int_{-\infty}^{\infty} f(t) e^{-j\omega t} \, dt \tag{7–1}$$

If $f(t)$ is a voltage, say, then $F(\omega)$ has the units of volts per rads/s and represents the *relative* magnitude and phase of steady-state sinusoids (of frequency ω) that

230

can be summed to produce the original $f(t)$. Thus, the magnitude of the Fourier transform has the physical significance of being the *amplitude density* as a function of frequency and as such gives a clear indication of how the energy of $f(t)$ is distributed with respect to frequency.

It might seem reasonable to use exactly the same procedure in dealing with *random* signals—that is, the use the Fourier transform of any particular sample function $x(t)$, defined by

$$F_x(\omega) = \int_{-\infty}^{\infty} x(t)e^{-j\omega t} \, dt$$

as the frequency-domain representation of the random process. This is not possible, however, for at least two reasons. In the first place, the Fourier transform will be a random variable over the ensemble (for any fixed ω) since it will have a different value for each member of the ensemble of possible sample functions. Hence, it cannot be a frequency representation of the process, but only of one member of the process. However, it might still be possible to use this function by finding its expected value (or mean) over the emsemble if it were not for the second reason. The second, and more basic, reason for not using the $F_x(\omega)$ just defined is that—for stationary processes, at least—it almost never exists! It may be recalled that one of the conditions for a time function to be Fourier transformable is that it be absolutely integrable; that is,

$$\int_{-\infty}^{\infty} |x(t)| dt < \infty \tag{7-2}$$

This condition can never be satisfied by any nonzero sample function from a wide-sense stationary random process. The Fourier transform in the ordinary sense will never exist in this case, although it may occasionally exist in the sense of generalized functions, including impulses, and so forth.

Now that the usual Fourier transform has been ruled out as a means of obtaining a frequency-domain representation for a random process, the next thought is to use the Laplace transform, since this contains a built-in convergence factor. Of course, the usual one-sided transform, which considers $f(t)$ for $t \geq 0$ only, is not applicable for a wide-sense stationary process; however, this is no real difficulty since the two-sided Laplace transform is good for negative values of time as well as positive. Once this is done, the Laplace transform for almost any sample function from a stationary random process will exist.

It turns out, however, that this approach is not so promising as it looks since it merely transfers the existence problems from the transform to the inverse transform. A study of these problems requires a knowledge of complex variable theory that is beyond the scope of the present discussion. Hence, it appears that the simplest mathematically-acceptable approach is to return to the Fourier transform and employ an artifice that will insure existence. Even in this case it will not be possible to justify rigorously all the steps, and a certain amount of the procedure will have to be accepted on faith.

7–2   Relation of Spectral Density to the Fourier Transform

In order to use the Fourier transform technique it is necessary to modify the sample functions of a stationary random process in such a way that the transform of each sample function exists. There are many ways in which this might be done, but the simplest one is to define a new sample function having finite duration. Thus, let

$$X_T(t) = X(t) \qquad |t| \leq T < \infty$$
$$ = 0 \qquad\quad |t| > T$$
(7–3)

and note that the truncated time function $X_T(t)$ will satisfy the condition of (7–2), as long as T remains finite, provided that the stationary process from which it is taken has a finite mean-square value. Hence, $X_T(t)$ will be Fourier transformable. In fact, $X_T(t)$ will satisfy the more stringent requirement for integrable square functions; that is

$$\int_{-\infty}^{\infty} |X_T(t)|^2 \, dt \quad < \infty$$
(7–4)

This condition will be needed in the subsequent development.

Since $X_T(t)$ is Fourier transformable, its transform may be written as

$$F_X(\omega) = \int_{-\infty}^{\infty} X_T(t) e^{-j\omega t} \, dt \qquad T < \infty$$
(7–5)

Eventually, it will be necessary to let T increase without limit; the purpose of the following discussion is to show that the *expected value* of $|F_X(\omega)|^2$ does exist in the limit even though the $F_X(\omega)$ for any one sample function does not. The first step in demonstrating this is to apply Parseval's theorem to $X_T(t)$ and $F_X(\omega)$.[1] Thus, since $x_T(t) = 0$ for $|t| > T$,

$$\int_{-T}^{T} X_T^2(t) \, dt = \frac{1}{2\pi} \int_{-\infty}^{\infty} |F_X(\omega)|^2 \, d\omega$$
(7–6)

---

[1] Parseval's theorem states that if $f(t)$ and $g(t)$ are transformable time functions with transforms of $F(\omega)$ and $G(\omega)$ respectively, then

$$\int_{-\infty}^{\infty} f(t)g(t) \, dt = \frac{1}{2\pi} \int_{-\infty}^{\infty} F(\omega)G(-\omega) \, d\omega$$

Note that $|F_X(\omega)|^2 = F_X(\omega)F_X(-\omega)$ since $F_X(-\omega)$ is the complex conjugate of $F_X(\omega)$ when $X_T(t)$ is a real time function.

Since the quantity being sought is the distribution of average power as a function of frequency, the next step is to average both sides of (7–6) over the total time, $2T$. Hence, dividing both sides by $2T$ gives

$$\frac{1}{2T}\int_{-T}^{T} X_T^2(t)\,dt = \frac{1}{4\pi T}\int_{-\infty}^{\infty} |F_X(\omega)|^2\,d\omega \tag{7–7}$$

The left side of (7–7) is seen to be proportional to the average power of the sample function in the time interval from $-T$ to T. More exactly, it is the square of the *effective* value of $X_T(t)$. Furthermore, for an ergodic process, this quantity would approach the mean-square value of the process as T approached infinity.

However, it is not possible at this stage to let T approach infinity, since $F_X(\omega)$ simply does not exist in the limit. It should be remembered, though, that $F_X(\omega)$ is a random variable with respect to the ensemble of sample functions from which $X(t)$ was taken. It is reasonable to suppose (and can be rigorously proved) that the limit of the *expected value* of $(1/T)|F_X(\omega)|^2$ does exist since the integral of this "always positive" quantity certainly does exist, as shown by (7–4). Hence, taking the expectation of both sides of (7–7), interchanging the expectation and integration and then taking the limit as $T \to \infty$ we obtain

$$E\left\{\frac{1}{2T}\int_{-T}^{T} X_T^2(t)\,dt\right\} = E\left\{\frac{1}{4\pi T}\int_{-\infty}^{\infty} |F_X(\omega)|^2\,d\omega\right\}$$

$$\lim_{T\to\infty}\frac{1}{2T}\int_{-T}^{T} \overline{X^2}\,dt = \lim_{T\to\infty}\frac{1}{4\pi T}\int_{-\infty}^{\infty} E\{|F_X(\omega)|^2\}d\omega \tag{7–8}$$

$$\langle\overline{X^2}\rangle = \frac{1}{2\pi}\int_{-\infty}^{\infty} \lim_{T\to\infty}\frac{E\{|F_X(\omega)|^2\}}{2T}\,d\omega$$

For a stationary process, the time average of the mean-square value is equal to the mean-square value and (7–8) can be written as

$$\overline{X^2} = \frac{1}{2\pi}\int_{-\infty}^{\infty} \lim_{T\to\infty}\frac{E\{|F_X(\omega)|^2\}}{2T}\,d\omega \tag{7–9}$$

The integrand of the right side of (7–9), which will be designated by the symbol $S_X(\omega)$, is called the *spectral density* of the random process. Thus,

$$S_X(\omega) = \lim_{T\to\infty}\frac{E[|F_X(\omega)|^2]}{2T} \tag{7–10}$$

and it must be remembered that it is not possible to let $T \to \infty$ *before* taking the expectation. If $X(t)$ is a voltage, say, then $S_X(\omega)$ has the units of volts^2 per hertz, and its integral, as shown by (7–9), leads to the mean-square value; that is,

$$\overline{X^2} = \frac{1}{2\pi} \int_{-\infty}^{\infty} S_X(\omega)\, d\omega \qquad\qquad (7\text{–}11)$$

The physical interpretation of spectral density can be made somewhat clearer by thinking in terms of average power, although this is a fairly specialized way of looking at it. If $X(t)$ is a voltage or current associated with a 1 Ω resistance, then $\overline{X^2}$ is just the average power dissipated in that resistance. The spectral density, $S_X(\omega)$, can then be interpreted as the average power associated with a bandwidth of 1 Hz centered at $\omega/2\pi$ Hz. [Note that the unit of bandwidth is the hertz (or cycle per second) and not the radian per second, because of the factor of $1/(2\pi)$ in the integral of (7–11).] Because the relationship of the spectral density to the average power of the random process is often the one of interest, the spectral density is frequently designated as the "power density spectrum."

The spectral density defined above is sometimes referred to as the "two-sided spectral density" since it exists for both positive and negative values of ω. Some authors prefer to define a "one-sided spectral density," which is usually expressed as a function of $f = \omega/2\pi$ and exists only for positive values of f. If this one-sided spectral density is designated by $G_X(f)$, then the mean-square value of the random process is given by

$$\overline{X^2} = \int_0^{\infty} G_X(f)\, df \qquad\qquad (7\text{–}12)$$

Since the one-sided spectral density is defined for positive frequencies only, it may be related to the two-sided spectral density by

$$
\begin{aligned}
G_X(f) &= 2S_X(2\pi f) \qquad f \geq 0 \\
&= 0 \qquad\qquad\quad f < 0
\end{aligned}
\qquad\qquad (7\text{–}13)
$$

Although both the one-sided spectral density and the two-sided spectral density are commonly used in the engineering literature, in the interest of consistency this text will use only the two-sided spectral density. However, the reader is cautioned that other references may use either and it is essential to be aware of the definition being employed.

The foregoing analysis of spectral density has been carried out in somewhat more detail than is customary in an introductory discussion. The reason for this is an attempt to avoid some of the mathematical pitfalls that a more superficial approach might gloss over. There is no doubt that this method makes the initial study of spectral density more difficult for the reader, but it is felt that the additional rigor is well worth the effort. Furthermore, even if all of the implications of the discussion are not fully understood, it should serve to make the reader aware of the existence of some of the less obvious difficulties of frequency-domain methods.

Another approach to spectral density, which treats it as a defined quantity based on the autocorrelation function, is given in Section 7–6. From the standpoint of application, such a definition is probably more useful than the more basic approach given here and is also easier to understand. It does not, however, make the physical interpretation as apparent as the basic derivation does.

Before turning to a more detailed discussion of the properties of spectral densities, it may be noted that in system analysis the spectral density of the input random process will play the same role as does the transform of the input in the case of nonrandom signals. The major difference is that spectral density represents a *power* density rather than a *voltage* density. Thus, it will be necessary to define a *power transfer function* for the system rather than a *voltage transfer function*.

Exercise 7–2.1

A stationary random process has two-sided spectral density given by

$$S_X(\omega) = 16\pi \qquad a \le |\omega| \le b$$

$$= 0 \qquad \text{elsewhere}$$

a) Find the mean-square value of this process if $a = 0$ and $b = 2$.

b) Find the mean square value of this process if $a = 2$ and $b = 3$.

Answers: 16, 32

Exercise 7–2.2

A stationary random process has a two-sided spectral density given by

$$S_X(\omega) = \frac{32}{\omega^2 + 16}$$

a) Find the average power (on a 1-Ω basis) of this random process.

b) Find the average power (on a 1-Ω basis) associated with a range of ω–values from -4 to $+4$.

Answers: 2, 4

7–3   Properties of Spectral Density

Most of the important properties of spectral density are summarized by the simple statement that it is a *real, positive, even* function of ω. It is known from the study of Fourier transforms that their *magnitude* is certainly real and positive. Hence, the expected value will also possess the same properties.

A special class of spectral densities, which is more commonly used than any other, is said to be *rational,* since it is composed of a ratio of polynomials. Since the spectral density is an even function of ω, these polynomials involve only even powers of ω. Thus, it is represented by

$$S_X(\omega) = \frac{S_0(\omega^{2n} + a_{2n-2}\omega^{2n-2} + \cdots + a_2\omega^2 + a_0)}{\omega^{2m} + b_{2m-2}\omega^{2m-2} + \cdots + b_2\omega^2 + b_0} \tag{7-14}$$

If the mean-square value of the random process is finite, then the area under $S_X(\omega)$ must also be finite, from (7–11). In this case, it is necessary that $m > n$. This condition will always be assumed here except for a very special case of *white noise.* White noise is a term applied to a random process for which the spectral density is constant for all ω; that is, $S_X(\omega) = S_0$. Although such a process cannot exist physically (since it has infinite mean-square value), it is a convenient mathematical fiction, which greatly simplifies many computations that would otherwise be very difficult. The justification and illustration of the use of this concept are discussed in more detail later.

As an example of a rational spectral density consider the function

$$S_X(\omega) = \frac{16(\omega^4 + 12\omega^2 + 32)}{\omega^6 + 18\omega^4 + 92\omega^2 + 120}$$

$$= \frac{16(\omega^2 + 4)(\omega^2 + 8)}{(\omega^2 + 2)(\omega^2 + 6)(\omega^2 + 10)}$$

Note that this function satisfies all of the requirements that spectral densities be real, positive, and even functions of ω. In addition, the denominator is of higher degree than the numerator so that the spectral density vanishes at $\omega = \infty$. Thus, the process described by this spectral density will have a finite mean-square value. The factored form of the spectral density is often useful in evaluating the integral required to obtain the mean-square value of the process. This operation is discussed in more detail in a subsequent section.

It is also possible to have spectral densities that are not rational. A typical example of this is the spectral density

$$S_X(\omega) = \left(\frac{\sin 5\omega}{5\omega}\right)^2$$

As is seen later, this is the spectral density of a random binary signal.

Spectral densities of this type are continuous and, as such, cannot represent random processes having dc or periodic components. The reason is not difficult to understand when spectral density is interpreted as average power per unit bandwidth. Any dc component in a random process represents a finite average power in *zero* bandwidth, since this component has a discrete frequency spectrum. Finite power in zero bandwidth is equivalent to an infinite power density. Hence, we would expect the spectral density in this case to be infinite at zero frequency but finite elsewhere; that is, it would contain a δ function at $\omega = 0$. A similar argument for periodic components would justify the existence of δ functions at these discrete frequencies. A rigorous derivation of these results will serve to make the argument more precise and, at the same time, illustrate the use of the defining equation, (7–10), in the calculation of spectral densities.

In order to carry out the desired derivation, consider a stationary random process having sample functions of the form

$$X(t) = A + B \cos(\omega_1 t + \theta) \qquad (7\text{–}15)$$

where A, B, and ω_1 are constants and θ is a random variable uniformly distributed from 0 to 2π; that is,

$$f(\theta) = \frac{1}{2\pi} \qquad 0 \le \theta \le 2\pi$$

$$= 0 \qquad \text{elsewhere}$$

The Fourier transform of the truncated sample function, $X_T(t)$, is

$$F_X(\omega) = \int_{-T}^{T} [A + B \cos(\omega_1 t + \theta)]e^{-j\omega t}\, dt$$

$$= A \left. \frac{e^{-j\omega t}}{-j\omega} \right|_{-T}^{T} + \left. \frac{B}{2} \frac{e^{j[(\omega_1 - \omega)t + \theta]}}{j(\omega_1 - \omega)} \right|_{-T}^{T} + \left. \frac{B}{2} \frac{e^{-j[(\omega_2 + \omega)t + \theta]}}{-j(\omega_1 + \omega)} \right|_{-T}^{T}$$

Substituting in the limits and simplifying leads immediately to

$$F_X(\omega) = \frac{2A \sin \omega T}{\omega} + B\left[\frac{e^{j\theta} \sin(\omega - \omega_1)T}{(\omega - \omega_1)} + \frac{e^{-j\theta} \sin(\omega + \omega_1)T}{(\omega + \omega_1)}\right] \qquad (7\text{–}16)$$

The square of the magnitude of $F_X(\omega)$ will have nine terms, some of which are independent of the random variable θ and the rest of which involve either $e^{\pm j\theta}$ or $e^{\pm j2\theta}$. In anticipation of the result that the expectation of all terms involving θ will vanish, it is convenient to write the squared magnitude in symbolic form without bothering to determine all the coefficients. Thus,

$$|F_X(\omega)|^2 = \frac{4A^2 \sin^2 \omega T}{\omega^2} + B^2\left[\frac{\sin^2(\omega - \omega_1)T}{(\omega - \omega_1)^2} + \frac{\sin^2(\omega + \omega_1)T}{(\omega + \omega_1)^2}\right]$$

$$+ C(\omega)e^{j\theta} + C(-\omega)e^{-j\theta} + D(\omega)e^{j2\theta} + D(-\omega)e^{-j2\theta} \qquad (7\text{–}17)$$

Now consider the expected value of any term involving θ. These are all of the form $G(\omega)e^{jn\theta}$, and the expected value is

$$E[G(\omega)e^{jn\theta}] = G(\omega) \int_0^{2\pi} \frac{1}{2\pi} e^{jn\theta}\,d\theta = \frac{G(\omega)}{2\pi} \frac{e^{jn\theta}}{jn}\bigg|_0^{2\pi}$$

$$= 0 \qquad n = \pm 1, \pm 2, \ldots$$

(7–18)

Thus, the last four terms of (7–17) will vanish and the expected value will become

$$E[|F_X(\omega)|^2] = 4A^2 \left[\frac{\sin^2 \omega T}{\omega^2}\right] + B^2 \left[\frac{\sin^2 (\omega - \omega_1)T}{(\omega - \omega_1)^2} + \frac{\sin^2 (\omega + \omega_1)T}{(\omega + \omega_1)^2}\right]$$

(7–19)

From (7–10), the spectral density is

$$S_X(\omega) = \lim_{T\to\infty} \left\{ 2A^2 T \left[\frac{\sin \omega T}{\omega T}\right]^2 + \frac{B^2 T}{2}\left[\frac{\sin (\omega - \omega_1)T}{(\omega - \omega_1)T}\right]^2 \right.$$
$$\left. + \frac{B^2 T}{2}\left[\frac{\sin (\omega + \omega_1)T}{(\omega + \omega_1)T}\right]^2 \right\}$$

(7–20)

In order to investigate the limit, consider the essential part of the first term; that is,

$$\lim_{T\to\infty} T \left(\frac{\sin \omega T}{\omega T}\right)^2 = ?$$

This limit is clearly zero when ω is *not* zero since $\sin^2 \omega T$ cannot exceed 1 and the denominator increases as T. When $\omega = 0$, however,

$$\frac{\sin \omega T}{\omega T}\bigg|_{\omega=0} = 1$$

and the limit is infinite. Hence, one can write

$$\lim_{T\to\infty} T \left(\frac{\sin \omega T}{\omega T}\right)^2 = K\delta(\omega)$$

(7–21)

where K represents the area of the δ function and has not yet been evaluated. The value of K can be found by equating the areas of both sides of (7–21); that is,

$$\lim_{T\to\infty} \int_{-\infty}^{\infty} T \left(\frac{\sin \omega T}{\omega T}\right)^2 d\omega = \int_{-\infty}^{\infty} K\delta(\omega)\,d\omega$$

(7–22)

The integral on the left is tabulated and has a value of π for *all* values of $T > 0$. Thus, the limiting operation becomes trivial, and (7–22) leads to

$$\pi = K$$

An exactly similar procedure can be used for the other terms in (7–20). It is left as an exercise for the reader to show that the final result becomes

$$S_X(\omega) = 2\pi A^2 \delta(\omega) + \frac{\pi}{2} B^2 \delta(\omega - \omega_1) + \frac{\pi}{2} B^2 \delta(\omega + \omega_1) \qquad (7\text{-}23)$$

This spectral density is shown in Figure 7–1.

It is of interest to determine the area of the spectral density in order to verify that (7–23) does, in fact, lead to the proper mean-square value. Thus, according to (7–11)

$$\overline{X^2} = \frac{1}{2\pi} \int_{-\infty}^{\infty} \left[2\pi A^2 \delta(\omega) + \frac{\pi}{2} B^2 \delta(\omega - \omega_1) + \frac{\pi}{2} B^2 \delta(\omega + \omega_1) \right] d\omega \qquad (7\text{-}24)$$

$$= \frac{1}{2\pi} \left[2\pi A^2 + \frac{\pi}{2} B^2 + \frac{\pi}{2} B^2 \right] = A^2 + \frac{1}{2} B^2$$

The reader can verify easily that this same result would be obtained from the ensemble average of $X^2(t)$.

A numerical example serves to illustrate discrete spectral densities. Suppose we have a stationary random process having sample functions of the form

$$X(t) = 5 + 10 \sin (6t + \theta_1) + 8 \cos (12t + \theta_2)$$

in which θ_1 and θ_2 are independent random variables and both are uniformly distributed between 0 and 2π. Note that because the phases are uniformly distributed over 2π radians, there is no difference between sine terms and cosine terms and both can be handled with the results just discussed. This would not be true if the distribution of phases were not uniform over this range. Using (7–23), the spectral density of this process can be written immediately as

$$S_X(\omega) = 2\pi(5)^2 \delta(\omega) + \frac{\pi}{2}(10)^2 \delta(\omega - 6) + \frac{\pi}{2}(10)^2 \delta(\omega + 6)$$

$$+ \frac{\pi}{2}(8)^2 \delta(\omega - 12) + \frac{\pi}{2}(8)^2 \delta(\omega + 12)$$

$$= \pi[50\delta(\omega) + 50\delta(\omega - 6) + 50\delta(\omega + 6)$$

$$+ 32\delta(\omega - 12) + 32\delta(\omega + 12)]$$

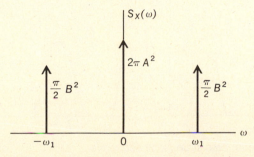

Figure 7–1   Spectral density of dc and sinusoidal components.

The mean-square value of the process can be obtained from (7–24) as

$$\overline{X^2} = (5)^2 + \left(\frac{1}{2}\right)(10)^2 = \left(\frac{1}{2}\right)(8)^2 = 107$$

It is apparent from this example that finding the spectral density and mean-square value of random discrete frequency components is quite simple and straightforward.

It is also possible to have spectral densities with both a continuous component and discrete components. An example of this sort that arises frequently in connection with communication systems or sampled data control systems is the random amplitude pulse sequence shown in Figure 7–2. It is assumed here that all of the pulses have the same shape but their amplitudes are random variables that are statistically independent from pulse to pulse. However, all the amplitude variables have the same mean, $\overline{Y}$, and the same variance, σ_Y^2. The repetition period for the pulses is t_1, a constant, and the reference time for any sample function is t_0, which is a random variable uniformly distributed over an interval of t_1.

The complete derivation of the spectral density is too lengthy to be included here, but the final result indicates some interesting points. This result may be expressed in terms of the Fourier transform $F(\omega)$ of the basic pulse shape $f;(t)$, and is

$$S_X(\omega) = |F(\omega)|^2 \left[\frac{\sigma_Y^2}{t_1} + \frac{2\pi(\overline{Y})^2}{t_1^2} \sum_{n=-\infty}^{\infty} \delta\left(\omega - \frac{2\pi n}{t_1}\right) \right] \qquad \textbf{(7–25)}$$

If the basic pulse shape is rectangular, with a width of t_2, the corresponding spectral density will be as shown in Figure 7–3. From (7–25) the following general conclusions are possible:

1. Both the continuous spectrum amplitude and the areas of the δ functions are proportional to the squared magnitude of the Fourier transform of the basic pulse shape.

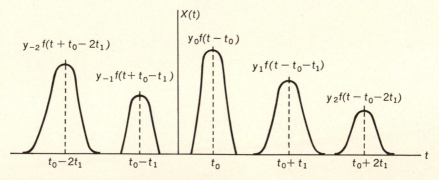

Figure 7–2 Random amplitude pulse sequence.

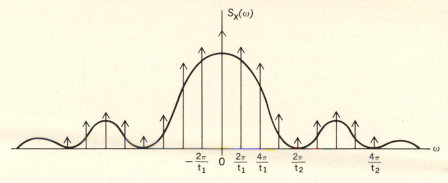

Figure 7–3 Spectral density for rectangular pulse sequence with random amplitudes.

2. If the mean value of the pulse amplitude is zero, there will be *no discrete spectrum* even though the pulses occur periodically.
3. If the variance of the pulse amplitude is zero, there will be *no* continuous spectrum.

The above result is illustrated by considering a sequence of rectangular pulses having random amplitudes. Let each pulse have the form

$$f(t) = 1 \qquad -0.01 \le t \le 0.01$$
$$= 0 \qquad \text{elsewhere}$$

and assume that these are repeated periodically every 0.1 second and have independent random amplitudes that are uniformly distributed between 0 and 12. The first step is to find the Fourier transform of the pulse shape. This is

$$F(\omega) = \int_{-0.01}^{0.01} (1)e^{-j\omega t}\, dt = 0.02\, \frac{\sin 0.01\omega}{0.01\omega}$$

Next we need to find the mean and variance of the random amplitudes. Since the amplitudes are uniformly distributed the mean value is

$$\bar{Y} = \left(\frac{1}{2}\right)(0 + 12) = 6$$

and the variance is

$$\sigma_Y^2 = \left(\frac{1}{12}\right)(12 - 0)^2 = 12$$

The spectral density may now be obtained from (7–25) as

$$S_X(\omega) = \left[0.02 \frac{\sin 0.01\omega}{0.01\omega}\right]^2 \left[\frac{12}{0.01} + \frac{2\pi(6)^2}{0.01^2} \sum_{n=-\infty}^{\infty} \delta\left(\omega - \frac{2\pi n}{0.01}\right)\right]$$

$$= \left[\frac{\sin 0.01\omega}{0.01\omega}\right]^2 \left[0.48 + 288 \sum_{n=-\infty}^{\infty} \delta(\omega - 200\pi n)\right]$$

Again it may be seen that there is a continuous part to the spectral density as well as an infinite number of discrete frequency components.

Another property of spectral densities concerns the derivative of the random process. Suppose that $\dot{X}(t) = dX(t)/dt$ and that $X(t)$ has a spectral density of $S_X(\omega)$, which was defined as

$$S_X(\omega) = \lim_{T\to\infty} \frac{E[|F_X(\omega)|^2]}{2T}$$

The truncated version of the derivative, $\dot{X}_T(t)$, will have a Fourier transform of $j\omega F_X(\omega)$, with the possible addition of two constant terms (arising from the discontinuities at $\pm T$) that will vanish in the limit. Hence, the spectral density of the derivative becomes

$$S_{\dot{X}}(\omega) = \lim_{T\to\infty} \frac{E[|j\omega F_X(\omega)(-j\omega)F_X(-\omega)|]}{2T}$$

$$= \omega^2 \lim_{T\to\infty} \frac{E[|F_X(\omega)|^2]}{2T} = \omega^2 S_X(\omega)$$

(7–26)

It is seen, therefore, that differentiation creates a new process whose spectral density is simply ω^2 times the spectral density of the original process. In this connection, it should be noted that if $S_X(\omega)$ is finite at $\omega = 0$, then $S_{\dot{X}}(\omega)$ will be zero at $\omega = 0$. Furthermore, if $S_X(\omega)$ does not drop off more rapidly than $1/\omega^2$ as $\omega \to \infty$, then $S_{\dot{X}}(\omega)$ will approach a constant at large ω and the mean-square value for the derivative will be infinite. This corresponds to the case of nondifferentiable random processes.

Exercise 7–3.1

A stationary random process has a spectral density of the form

$$S_X(\omega) = 8\pi\delta(\omega) + 36\pi\delta(\omega - 16) + 36\pi\delta(\omega + 16)$$

a) List all of the discrete frequencies present in this process.

b) Find the mean value of the process.

c) Find the variance of the process.

Answers: 0, ±2, ±16, 36

Exercise 7–3.2

A random process consists of a sequence of rectangular pulses having a duration of 1 millisecond and occurring every 5 milliseconds. The
pulse amplitudes are independent random variables that are uniformly
distributed between A and B. For each of the following sets of values
for A and B, determine if the spectral density has a continuous component, discrete components, both, or neither.

a) $A = -5, B = 5$

b) $A = 5, B = 15$

c) $A = 8, B = 8$

d) $A = 0, B = 8$

Answers: Both, neither, discrete only, continuous only

7–4 Spectral Density and the Complex Frequency Plane

In the discussion so far, the spectral density has been expressed as a function of
the real angular frequency ω. However, for applications to system analysis, it is
very convenient to express it in terms of the complex frequency s, since system
transfer functions are more convenient in this form. This change can be made very
simply by replacing $j\omega$ with s. Hence, along the $j\omega$-axis of the complex frequency
plane, the spectral density will be the same as that already discussed.

The formal conversion to complex frequency representation is accomplished by
replacing ω by $-js$ or ω^2 by $-s^2$. The resulting spectral density should properly
be designated as $S_X(-js)$, but this notation is somewhat clumsy. Therefore, spectral density in the *s-plane* will be designated simply as $S_X(s)$. It is evident that
$S_X(s)$ and $S_X(\omega)$ are somewhat different functions of their respective arguments,
so that the notation is symbolic rather than precise.

For the special case of rational spectral densities, in which only even powers of

ω occur, this substitution is equivalent to replacing ω^2 by $-s^2$. For example, consider the rational spectrum

$$S_X(\omega) = \frac{10(\omega^2 + 5)}{\omega^4 + 10\omega^2 + 24}$$

When expressed as a function of s, this becomes

$$S_X(s) = S_X(-js) = \frac{10(-s^2 + 5)}{s^4 - 10s^2 + 24} \tag{7-27}$$

Any spectral density can also be represented (except for a constant of proportionality) in terms of its pole-zero configuration in the complex frequency plane. Such a representation is often convenient in carrying out certain calculations, which will be discussed in the following sections. For purposes of illustration, consider the spectral density of (7–27). This may be factored as

$$S_X(s) = \frac{-10(s + \sqrt{5})(s - \sqrt{5})}{(s + 2)(s - 2)(s + \sqrt{6})(s - \sqrt{6})}$$

and the pole-zero configuration plotted as shown in Figure 7–4. This plot also illustrates the important point that such configurations are always symmetrical about the $j\omega$ axis. When the spectral density is not rational, the substitution is the same but may not be quite as straightforward. For example, the spectral density given by (7–25) could be expressed in the complex frequency plane as

$$S_X(s) = F(s)F(-s)\left[\frac{\sigma_Y^2}{t_1} + \frac{2\pi(\bar{Y})^2}{t_1^2}\sum_{n=-\infty}^{\infty}\delta\left(s - j\frac{2\pi n}{t_1}\right)\right] \tag{7-28}$$

where $F(s)$ is the *Laplace* transform of the basic pulse shape $f(t)$.

In addition to making spectral densities more convenient for system analysis, the use of the complex frequency s also makes it more convenient to evaluate mean-square values. This application is discussed in the following section.

Exercise 7–4.1

A stationary random process has a spectral density of the form

$$S_X(\omega) = \frac{25(\omega^2 + 16)}{\omega^4 + 34\omega^2 + 225}$$

Find the pole and zero locations for this spectral density in the complex frequency plane.

Answers: $\pm 3, \pm 4, \pm 5$

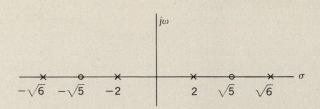

Figure 7–4 Pole-zero configuration for a spectral density.

Exercise 7–4.2

A stationary random process has a spectral density of the form

$$S_X(\omega) = \frac{\omega^2 (\omega^2 + 25)}{\omega^6 - 33\omega^4 + 463\omega^2 + 7569}$$

a) Verify that this spectral density is positive for all values of ω.

b) Find the pole and zero locations for this spectral density in the complex frequency plane.

Answers: 0, ±3, ±5, ±2, ±j5

7–5   Mean-Square Values from Spectral Density

It was shown in the course of defining the spectral density that the mean-square value of the random process was given by

$$\overline{X^2} = \frac{1}{2\pi} \int_{-\infty}^{\infty} S_X(\omega) \, d\omega \tag{7–11}$$

Hence, the mean-square value is proportional to the *area* of the spectral density.

The evaluation of an integral such as (7–11) may be very difficult if the spectral density has a complicated form or if it involves high powers of ω. A classical way of carrying out such integration is to convert the variable of integration to a complex variable (by substituting s for $j\omega$) and then to utilize some powerful theorems concerning integration around closed paths in the complex plane. This is probably the easiest and most satisfactory way of obtaining mean-square values but, unfortunately, requires a knowledge of complex variables that the reader may not possess. The *mechanics* of the procedure is discussed at the end of this section, however, for those interested in this method.

An alternative method, which will be discussed first, is to utilize some tabulated results for spectral densities that are rational. These have been tabulated in general form for polynomials of various degrees and their use is simply a matter of substituting in the appropriate numbers. The existence of such general forms is primarily a consequence of the symmetry of the spectral density. As a result of this symmetry, it is always possible to factor rational spectral densities into the form

$$S_X(s) = \frac{c(s)\, c(-s)}{d(s)\, d(-s)} \qquad (7\text{--}29)$$

where $c(s)$ contains the left-half-plane (lhp) zeros, $c(-s)$ the right-half-plane (rhp) zeros, $d(s)$ the lhp poles, and $d(-s)$ the rhp poles.

When the real integration of (7–11) is expressed in terms of the complex variable s, the mean-square value becomes

$$\overline{X^2} = \frac{1}{2\pi j} \int_{-j\infty}^{j\infty} S_X(s)\, ds = \frac{1}{2\pi j} \int_{-j\infty}^{j\infty} \frac{c(s)\, c(-s)}{d(s)\, d(-s)}\, ds \qquad (7\text{--}30)$$

For the special case of rational spectral densities, $c(s)$ and $d(s)$ are polynomials in s and may be written as

$$c(s) = c_{n-1}s^{n-1} + c_{n-2}s^{n-2} + \cdots + c_0$$

$$d(s) = d_n s^n + d_{n-1}s^{n-1} + \cdots + d_0$$

Some of the coefficients of $c(s)$ may be zero, but $d(s)$ must be of higher degree than $c(s)$ and must not have any coefficients missing.

Integrals of the form in (7–30) have been tabulated for values of n up to 10, although beyond $n = 3$ or 4 the general results are so complicated as to be of doubtful value. An abbreviated table is given in Table 7–1.

Table 7–1.   Table of Integrals.

$$I_n = \frac{1}{2\pi j} \int_{-j\infty}^{j\infty} \frac{c(s)\, c(-s)}{d(s)\, d(-s)}\, ds$$

$$c(s) = c_{n-1}s^{n-1} + c_{n-2}s^{n-2} + \cdots + c_0$$

$$d(s) = d_n s^n + d_{n-1}s^{n-1} + \cdots + d_0$$

$$I_1 = \frac{c_0^2}{2d_0 d_1}$$

$$I_2 = \frac{c_1^2 d_0 + c_0^2 d_2}{2d_0 d_1 d_2}$$

$$I_3 = \frac{c_2^2 d_0 d_1 + (c_1^2 - 2c_0 c_2)d_0 d_3 + c_0^2 d_2 d_3}{2d_0 d_3(d_1 d_2 - d_0 d_3)}$$

As an example of this calculation, consider the spectral density

$$S_X(\omega) = \frac{\omega^2 + 4}{\omega^4 + 10\omega^2 + 9}$$

When ω is replaced by $-js$, this becomes

$$S_X(s) = \frac{-(s^2 - 4)}{s^4 - 10s^2 + 9} = \frac{-(s^2 - 4)}{(s^2 - 1)(s^2 - 9)} \qquad (7\text{--}31)$$

This can be factored into

$$S_X(s) = \frac{(s + 2)(-s + 2)}{(s + 1)(s + 3)(-s + 1)(-s + 3)} \qquad (7\text{--}32)$$

from which it is seen that

$$c(s) = s + 2$$
$$d(s) = (s + 1)(s + 3) = s^2 + 4s + 3$$

This is a case in which $n = 2$ and

$c_1 = 1$
$c_0 = 2$
$d_2 = 1$
$d_1 = 4$
$d_0 = 3$

From Table 7–1, I_2 is given by

$$I_2 = \frac{c_1^2 d_0 + c_0^2 d_2}{2d_0 d_1 d_2} = \frac{(1)^2(3) + (2)^2(1)}{2(3)(4)(1)} = \frac{3 + 4}{24} = \frac{7}{24}$$

However, $\overline{X^2} = I_2$, so that

$$\overline{X^2} = \frac{7}{24}$$

The procedure just presented is a mechanical one and in order to be a useful tool does not require any deep understanding of the theory. Some precautions are necessary, however. In the first place, as noted above, it is necessary that $c(s)$ be of lower degree than $d(s)$. Second, it is necessary that $c(s)$ and $d(s)$ have roots *only* in the left half plane. Finally, it is necessary that $d(s)$ have *no* roots *on* the $j\omega$-axis.

In the example given above the spectral density is rational and, hence, does not contain any δ functions. Thus, the random process that it represents has a mean value of zero and the mean-square value that was calculated is also the variance of the process. There may be situations, however, in which the continuous part of the spectral density is rational but there are also discrete components resulting from a nonzero mean value or from periodic components. In cases such as this, it is necessary to treat the continuous portion of the spectral density and the discrete portions of the spectral density separately when finding the mean-square value. An example will serve to illustrate the technique. Consider a spectral density of the form

$$S_X(\omega) = 8\pi\delta(\omega) + 36\pi\delta(\omega - 16) + 36\pi\delta(\omega + 16) + \frac{25(\omega^2 + 16)}{\omega^4 + 34\omega^2 + 225}$$

From the discussion in Section 7–3 and Equation (7–24), it is clear that the contribution to the mean-square value from the discrete components is simply

$$\overline{X^2}_{discrete} = \left(\frac{1}{2\pi}\right)(8\pi + 36\pi + 36\pi) = 40$$

Note that this includes a mean value of ± 2. The continuous portion of the spectral density may now be written as a function of s as

$$S_{X_c}(s) = \frac{25(-s^2 + 16)}{s^4 - 34s^2 + 225}$$

which, in factored form becomes

$$S_{X_c}(s) = \frac{[5(s + 4)][5(-s + 4)]}{[(s + 3)(s + 5)][(-s + 3)(-s + 5)]}$$

It is now clear that

$$c(s) = 5(s + 4) = 5s + 20$$

from which $c_0 = 20$ and $c_1 = 5$. Also

$$d(s) = (s + 3)(s + 5) = s^2 + 8s + 15$$

from which $d_0 = 15$, $d_1 = 8$, and $d_2 = 1$. Using the expression for I_2 in Table 7–1 yields

$$\overline{X^2}_{cont.} = \frac{(5)^2(15) + (20)^2(1)}{2(15)(8)(1)} = 3.229$$

Hence, the total mean-square value of this process is

$$\overline{X^2} = 40 + 3.229 = 43.229$$

Since the mean value of the process is ± 2, the variance of the process becomes $\sigma_X^2 = 43.229 - (2)^2 = 39.229$.

It was noted previously that the use of complex integration provides a very general, and very powerful, method of evaluating integrals of the form given in

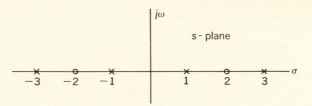

Figure 7–5 Pole-zero configuration for a spectral density.

(7–30). A brief summary of the theory of such integration is given in Appendix I, and these ideas will be utilized here to demonstrate another method of evaluating mean-square values from spectral density. As a means of acquainting the student with the potential usefulness of this general procedure, only the mechanics of this method are discussed. The student should be aware, however, that there are many pitfalls associated with using mathematical tools without having a thorough grasp of their theory. All students are encouraged to acquire the proper theoretical understanding as soon as possible.

The method considered here is based on the evaluation of residues, in much the same way as is done in connection with finding inverse Laplace transforms. Consider, for example, the spectral density given above in (7–31) and (7–32). This spectral density may be represented by the pole-zero configuration shown in Figure 7–5. The path of integration called for by (7–30) is along the $j\omega$-axis, but the methods of complex integration discussed in Appendix I require a closed path. Such a closed path can be obtained by adding a semicircle at infinity that encloses either the left half plane or the right half plane. Less difficulty with the algebraic signs is encountered if the left half plane is used, so the path shown in Figure 7–6 will be assumed from now on. In order for the integral around this closed path to the same as the integral along the $j\omega$–axis, it is necessary for the contribution due to the semicircle to vanish as $R \to \infty$. For rational spectral densities this will be true whenever the denominator polynomial is of higher degree than the numerator polynomial (since only even powers are present).

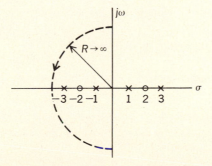

Figure 7–6 Path of integration for evaluating mean-square value.

A basic result of complex variable theory states that the value of an integral around a simple closed contour in the complex plane is equal to $2\pi j$ times the sum of the residues at the poles contained *within* that contour (see (I–3), Appendix I). Since the expression for the mean-square value has a factor of $1/(2\pi j)$, and since the chosen contour completely encloses the left half plane, it follows that the mean-square value can be expressed in general as

$$\overline{X^2} = \Sigma \text{ (residues at lhp poles)} \tag{7–33}$$

For the example being considered, the only lhp poles are at -1 and -3. The residues can be evaluated easily by multiplying $S_X(s)$ by the factor containing the pole in question and letting s assume the value of the pole. Thus,

$$K_{-1} = [(s + 1)S_X(s)]_{s=-1} = \left[\frac{-(s + 2)(s - 2)}{(s - 1)(s + 3)(s - 3)}\right]_{s=-1} = \frac{3}{16}$$

$$K_{-3} = [(s + 3)S_X(s)]_{s=-3} = \left[\frac{-(s + 2)(s - 2)}{(s + 1)(s - 1)(s - 3)}\right]_{s=-3} = \frac{5}{48}$$

From (7–33) it follows that

$$\overline{X^2} = \frac{3}{16} + \frac{5}{48} = \frac{7}{24}$$

which is the same value obtained above.

If the poles are not simple, the more general procedures discussed in Appendix I may be employed for evaluating the residues. However, the mean-square value is still obtained from (7–33).

Exercise 7–5.1

A stationary random process has a spectral density of

$$S_X(\omega) = \frac{16}{\omega^4 + 13\omega^2 + 36}$$

a) Find the mean-square value of this process using the results of Table 7–1.

b) Repeat using contour integration.

Answer: 4/15

Exercise 7–5.2

In discussing the use of Table 7–1, the statement is made that the polynomial $d(s)$ must not have any missing coefficients.

    a) Explain why this condition is necessary.

    b) Check your conclusion by first verifying that the following is a valid spectral density and then attempting to use the table to evaluate the mean-square value.

$$S_X(\omega) = \frac{\omega^2 + 1}{\omega^4 - 4\omega^2 + 4}$$

7–6 Relation of Spectral Density to the Autocorrelation Function

The autocorrelation function is shown in Chapter 6 to be the expected value of the product of time functions. In this chapter, it has been shown that the spectral density is related to the expected value of the product of Fourier transforms. It would appear, therefore, that there should be some direct relationship between these two expected values. Almost intuitively one would expect the spectral density to be the Fourier (or Laplace) transform of the autocorrelation function, and this turns out to be the case.

We consider first the case of a nonstationary random process and then specialize the result to a stationary process. In (7–10) the spectral density was defined as

$$S_X(\omega) = \lim_{T \to \infty} \frac{E[|F_X(\omega)|^2]}{2T} \tag{7–10}$$

where $F_X(\omega)$ is the Fourier transform of the truncated sample function. Thus,

$$F_X(\omega) = \int_{-T}^{T} X_T(t)e^{-j\omega t}\, dt \qquad T < \infty \tag{7–34}$$

Substituting (7–34) into (7–10) yields

$$S_X(\omega) = \lim_{T \to \infty} \frac{1}{2T} E\left[\int_{-T}^{T} X_T(t_1)e^{+j\omega t_1}\, dt_1 \int_{-T}^{T} X_T(t_2)e^{-j\omega t_2}\, dt_2\right] \tag{7–35}$$

since $|F_X(\omega)|^2 = F_X(\omega)F_X(-\omega)$. The subscripts on t_1 and t_2 have been introduced so that we can distinguish the variables of integration when the *product* of integrals is rewritten as an *iterated double integral*. Thus, write (7–35) as

$$S_X(\omega) = \lim_{T\to\infty} \frac{1}{2T} E\left[\int_{-T}^{T} dt_2 \int_{-T}^{T} e^{-j\omega(t_2-t_1)} X_T(t_1) X_T(t_2)\, dt_1\right]$$

$$= \lim_{T\to\infty} \frac{1}{2T} \int_{-T}^{T} dt_2 \int_{-T}^{T} e^{-j\omega(t_2-t_1)} E[X_T(t_1) X_T(t_2)]\, dt_1$$

(7–36)

Moving the expectation operation inside the double integral can be shown to be valid in this case, but the details are not discussed here.

The expectation in the integrand above is recognized as the autocorrelation function of the truncated process. Thus,

$$E[X_T(t_1) X_T(t_2)] = R_X(t_1, t_2) \qquad |t_1|, |t_2| \le T$$

$$= 0 \qquad\qquad \text{elsewhere}$$

(7–37)

Making the substitution

$$t_2 - t_1 = \tau$$

$$dt_2 = d\tau$$

we can write (7–37) as

$$S_X(\omega) = \lim_{T\to\infty} \frac{1}{2T} \int_{-T-t_1}^{T-t_1} d\tau \int_{-T}^{T} e^{-j\omega\tau} R_X(t_1, t_1 + \tau)\, dt_1$$

when the limits on t_1 are imposed by (7–37). Interchanging the order of integration and moving the limit inside the τ-integral gives

$$S_X(\omega) = \int_{-\infty}^{\infty} \left\{ \lim_{T\to\infty} \frac{1}{2T} \int_{-T}^{T} R_X(t_1, t_1 + \tau)\, dt_1 \right\} e^{-j\omega\tau}\, d\tau$$

(7–38)

From (7–38) it is apparent that the spectral density is the Fourier transform of the time average of the autocorrelation function. This may be expressed in shorter notation as follows:

$$S_X(\omega) = \mathscr{F}\{\langle R_X(t, t + \tau)\rangle\}$$

(7–39)

The relationship given in (7–39) is valid for nonstationary processes also.

If the process in question is a stationary random process, the autocorrelation function is independent of time; therefore,

$$\langle R_X(t_1, t_1 + \tau)\rangle = R_X(\tau)$$

Accordingly, the spectral density of a wide-sense stationary random process is just the Fourier transform of the autocorrelation function; that is,

$$S_X(\omega) = \int_{-\infty}^{\infty} R_X(\tau)e^{-j\omega\tau}\,d\tau$$

$$= \mathscr{F}\{R_X(\tau)\}$$

(7–40)

The relationship in (7–40), which is known as the Wiener-Khinchine relation, is of fundamental importance in analyzing random signals because it provides the link between the time domain (correlation function) and the frequency domain (spectral density). Because of the uniqueness of the Fourier transform it follows that the autocorrelation function of a wide-sense stationary random process is the inverse transform of the spectral density. In the case of a nonstationary process, the autocorrelation function cannot be recovered from the spectral density—only the time average of the correlation function, as seen from (7–39). In subsequent discussions, we will deal only with wide-sense stationary random processes for which (7–40) is valid.

As a simple example of this result, consider an autocorrelation function of the form

$$R_X(\tau) = Ae^{-\beta|\tau|} \qquad A > 0, \beta > 0$$

The absolute value sign on τ is required by the symmetry of the autocorrelation function. This function is shown in Figure 7–7 (a) and is seen to have a discontinuous derivative at $\tau = 0$. Hence, it is necessary to write (7–40) as the sum of two integrals—one for negative values of τ and one for positive values of τ. Thus,

$$S_X(\omega) = \int_{-\infty}^{0} A\,e^{\beta\tau}e^{-j\omega\tau}\,d\tau + \int_{0}^{\infty} Ae^{-\beta\tau}\,e^{-j\omega\tau}\,d\tau$$

$$= A\,\frac{e^{(\beta-j\omega)\tau}}{\beta - j\omega}\bigg|_{-\infty}^{0} + A\,\frac{e^{-(\beta+j\omega)\tau}}{-(\beta + j\omega)}\bigg|_{0}^{\infty}$$

(7–41)

$$= A\left[\frac{1}{\beta - j\omega} + \frac{1}{\beta + j\omega}\right] = \frac{2A\beta}{\omega^2 + \beta^2}$$

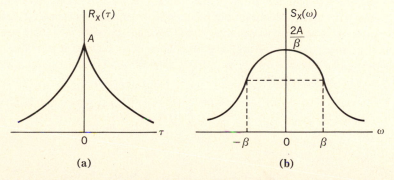

(a)   (b)

Figure 7–7 Relation between (a) autocorrelation function and (b) spectral density.

This spectral density is shown in Figure 7–7 (b).

In the stationary case it is also possible to find the autocorrelation function corresponding to a given spectral density by using the inverse Fourier transform. Thus,

$$R_X(\tau) = \frac{1}{2\pi} \int_{-\infty}^{\infty} S_X(\omega) e^{j\omega\tau} \, d\omega \tag{7–42}$$

An example of the application of this result will be given in the next section.

In obtaining the result in (7–41), the integral was separated into two parts because of the discontinuous slope at the origin. An alternative procedure, which is possible in all cases, is to take advantage of the symmetry of autocorrelation functions. Thus, if (7–40) is written as

$$S_X(\omega) = \int_{-\infty}^{\infty} R_X(\tau)[\cos \omega\tau - j \sin \omega\tau] \, d\tau$$

by expressing the exponential in terms of sines and cosines, it may be noted that $R_X(\tau) \sin \omega\tau$ is an odd function of τ and, hence, will integrate to zero. On the other hand, $R_X(\tau) \cos \omega\tau$ is even, and the integral from $-\infty$ to $+\infty$ is just twice the integral from 0 to $+\infty$. Hence,

$$S_X(\omega) = 2 \int_0^{\infty} R_X(\tau) \cos \omega\tau \, d\tau \tag{7–43}$$

is an alternative form that does not require integrating over the origin. The corresponding inversion formula, for wide-sense stationary processes, is easily shown to be

$$R_X(\tau) = \frac{1}{\pi} \int_0^{\infty} S_X(\omega) \cos \omega\tau \, d\omega \tag{7–44}$$

It was noted earlier that the relationship between spectral density and correlation function can also be expressed in terms of the Laplace transform. However, it should be recalled that the form of the Laplace transform used most often in system analysis requires that the time function being transformed be zero for negative values of time. Autocorrelation functions can never be zero for negative values of τ since they are always even functions of τ. Hence, it is necessary to use the *two-sided Laplace transform* for this application. The corresponding transform pair may be written as

$$S_X(s) = \int_{-\infty}^{\infty} R_X(\tau) e^{-s\tau} \, d\tau \tag{7–45}$$

and

$$R_X(\tau) = \frac{1}{2\pi j} \int_{-j\infty}^{j\infty} S_X(s) e^{s\tau} \, ds \tag{7–46}$$

Since the spectral density of a process having a finite mean-square value can have *no* poles on the $j\omega$-axis, the path of integration in (7–46) can always be on the $j\omega$-axis.

The direct two-sided Laplace transform, which yields the spectral density from the autocorrelation function, is no different from the ordinary one-sided Laplace transform and does not require any special comment. However, the inverse two-sided Laplace transform does require a little more care so that a simple example of this operation is desirable.

Consider the spectral density found in (7–41) and write it as a function of s as

$$S_X(s) = \frac{-2A\beta}{s^2 - \omega^2} = \frac{-2A\beta}{(s + \beta)(s - \beta)}$$

in which there is one pole in the left-half plane and one pole in the right-half plane. Because of the symmetry of spectral densities, there will always be as many rhp poles as there are lhp poles. A partial fraction expansion of the above expression yields

$$S_X(s) = \frac{A}{s + \beta} - \frac{A}{s - \beta}$$

The inverse Laplace transform of the lhp terms in any partial fraction expansion is usually interpreted to represent a time function that exists in positive time only. Hence, in this case we can interpret the inverse transform of the above function to be

$$\frac{A}{s + \beta} \Leftrightarrow Ae^{-\beta\tau} \quad \tau > 0$$

Because we are dealing with an autocorrelation function here it is possible to use the property that such functions are even in order to obtain the value of the autocorrelation function for negative values of τ. However, it is useful to discuss a more general technique that can also be used for crosscorrelation functions, in which this type of symmetry does not exist. Thus, for those factors in the partial fraction expansion that come from rhp poles, it is always possible to (a) replace s by $-s$, (b) find the single-sided inverse Laplace transform of what is now an lhp function, and (c) replace τ by $-\tau$. Using this procedure on the rhp factor above yields

$$\frac{-A}{-s - \beta} = \frac{A}{s + \beta} \Leftrightarrow Ae^{-\beta\tau}$$

Upon replacing τ by $-\tau$ yields

$$\frac{-A}{s + \beta} \Leftrightarrow Ae^{\beta\tau} \quad \tau < 0$$

Thus, the resulting autocorrelation function is

$$R_X(\tau) = Ae^{-\beta|\tau|} \qquad -\infty < \tau < \infty$$

which is exactly the autocorrelation function we started with. The technique illustrated by this example is sufficiently general to handle transformations from spectral densities to autocorrelation functions as well as from cross-spectral densities (which are discussed in a subsequent section) to crosscorrelation functions.

Exercise 7–6.1

A stationary random process has an autocorrelation function of the form

$$R_X(\tau) = 16e^{-2|\tau|} - 8e^{-4|\tau|}$$

Find the spectral density of this process.

Answer: $\dfrac{768}{\omega^4 + 20\omega^2 + 64}$

Exercise 7–6.2

A stationary random process has a spectral density of the form

$$S_X(\omega) = \frac{16}{\omega^4 + 13\omega^2 + 36}$$

Find the autocorrelation function of this process.

Answer: $\dfrac{8}{15}(1.5e^{-2|\tau|} - e^{-3|\tau|})$

7–7   White Noise

The concept of *white noise* was mentioned previously. This term is applied to a spectral density that is constant for all values of ω; that is, $S_X(\omega) = S_0$. It is interesting to determine the correlation function for such a process. This is best done by giving the result and verifying its correctness. Consider an autocorrelation function that is a δ function of the form

$$R_X(\tau) = S_0\delta(\tau)$$

Using this form in (7–40) leads to

$$S_X(\omega) = \int_{-\infty}^{\infty} R_X(\tau)e^{-j\omega\tau}\, d\tau = \int_{-\infty}^{\infty} S_0\delta(\tau)e^{-j\omega\tau}\, d\tau = S_0 \qquad (7\text{–}47)$$

which is the result for white noise. It is clear, therefore, that the autocorrelation function for white noise is just a δ function with *an area equal to the spectral density*.

It was noted previously that the concept of white noise is fictitious because such a process would have an infinite mean-square value, since the area of the spectral density is infinite. This same conclusion is also apparent from the correlation function. It may be recalled that the mean-square value is equal to the value of the autocorrelation function at $\tau = 0$. For a δ function at the origin, this is also infinite. Nevertheless, the white-noise concept is an extremely valuable one in the analysis of linear systems. It frequently turns out that the random signal input to a system has a bandwidth that is much greater than the range of frequencies that the system is capable of passing. Under these circumstances, assuming the input spectral density to be white may greatly simplify the computation of the system response without introducing any significant error. Examples of this sort are discussed in Chapters 8 and 9.

Another concept that is frequently used is that of *bandlimited white noise*. This implies a spectral density that is constant over a finite bandwidth and zero outside this frequency range. For example,

$$\begin{aligned}
S_X(\omega) &= S_0 \qquad |\omega| \leq 2\pi W \\
&= 0 \qquad |\omega| > 2\pi W
\end{aligned} \qquad (7\text{–}48)$$

as shown in Figure 7–8(a). This spectral density is also fictitious even though the mean-square value is finite (in fact, $\overline{X^2} = 2WS_0$). Why? It can be approached arbitrarily closely, however, and is a convenient form for many analysis problems.

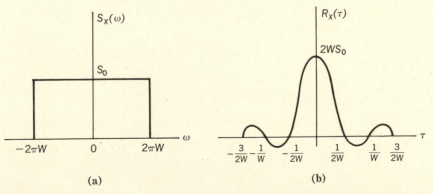

Figure 7–8 Bandlimited white noise: (a) spectral density and (b) autocorrelation function.

The autocorrelation function for such a process is easily obtained from (7–42). Thus,

$$R_X(\tau) = \frac{1}{2\pi} \int_{-\infty}^{\infty} S_X(\omega)e^{j\omega\tau}\,d\omega = \frac{1}{2\pi} \int_{-2\pi W}^{2\pi W} S_0 e^{j\omega\tau}\,d\omega = \frac{S_0}{2\pi} \left.\frac{e^{j\omega\tau}}{j\tau}\right]_{-2\pi W}^{2\pi W}$$

$$= \frac{S_0}{2\pi} \frac{e^{j2\pi W\tau} - e^{-j2\pi W\tau}}{j\tau} = \frac{S_0}{\pi\tau} \sin 2\pi W\tau$$

$$= 2WS_0 \left(\frac{\sin 2\pi W\tau}{2\pi W\tau}\right)$$

This is shown in Figure 7–8(b). Note that in the limit as W approaches infinity this approaches a δ function.

It may be observed from Figure 7–8(b) that random variables from a bandlimited process are uncorrelated if they are separated in time by any multiple of 1/2W seconds. It is known also that bandlimited functions can be represented exactly and uniquely by a set of samples taken at a rate of twice the bandwidth. This is the so-called *sampling theorem*. Hence, if a bandlimited function having a flat spectral density is to be represented by samples, it appears that these samples will be uncorrelated. This lack of correlation among samples may be a significant advantage in carrying out subsequent analysis. In particular, the correlation matrix defined in Section 6–9 for such sampled processes is a diagonal matrix; that is, all terms not on the major diagonal are zero.

Exercise 7–7.1

A stationary random process has a bandlimited white spectral density given by

$$S_X(\omega) = 0.01 \qquad |\omega| \leq 100\pi$$
$$= 0 \qquad |\omega| > 100\pi$$

a) Find the mean-square value of this process.

b) Find the smallest value of τ for which the autocorrelation function is 0.

c) Specify the bandwidth of this process in Hz.

Answers: 0.01, 1, 50

Exercise 7–7.2

It is also possible to have bandlimited white noise processes that are bandpass in nature. Consider such a process having a spectral density of

$$S_X(\omega) = 0.005 \qquad 200\pi \leq |\omega| \leq 250\pi$$

$$= 0 \qquad \text{elsewhere}$$

a) Sketch this spectral density.

b) Specify the bandwidth of this process in Hz.

c) Find the mean-square value of this process.

Answers: 0.25, 25

7–8 Cross-Spectral Density

When two correlated random processes are being considered, such as the input and the output of a linear system, it is possible to define a pair of quantities known as the cross-spectral densities. For purposes of the present discussion, it is sufficient to simply define them and note a few of their properties without undertaking any formal proofs.

If $F_X(\omega)$ is the Fourier transform of a truncated sample function from one process and $F_Y(\omega)$ is a similar transform from the other process, then the two cross-spectral densities may be defined as

$$S_{XY}(\omega) = \lim_{T \to \infty} \frac{E[F_X(-\omega)F_Y(\omega)]}{2T} \qquad \text{(7–49)}$$

$$S_{YX}(\omega) = \lim_{T \to \infty} \frac{E[F_Y(-\omega)F_X(\omega)]}{2T} \qquad \text{(7–50)}$$

Unlike normal spectral densities, cross-spectral densities need not be real, positive, or even functions of ω. They do have the following properties, however:

1. $S_{XY}(\omega) = S_{YX}{}^*(\omega)$ (*implies complex conjugate)

2. Re $[S_{XY}(\omega)]$ is an even function of ω. Also true for $S_{YX}(\omega)$.

3. Im $[S_{XY}(\omega)]$ is an odd function of ω. Also true for $S_{YX}(\omega)$.

Cross-spectral densities can also be related to crosscorrelation functions by the Fourier transform. Thus for jointly stationary processes,

$$S_{XY}(\omega) = \int_{-\infty}^{\infty} R_{XY}(\tau)e^{-j\omega\tau}\,d\tau \qquad\qquad (7\text{-}51)$$

$$R_{XY}(\tau) = \frac{1}{2\pi}\int_{-\infty}^{\infty} S_{XY}(\omega)e^{j\omega\tau}\,d\omega \qquad\qquad (7\text{-}52)$$

$$S_{YX}(\omega) = \int_{-\infty}^{\infty} R_{YX}(\tau)e^{-j\omega\tau}\,d\tau \qquad\qquad (7\text{-}53)$$

$$R_{YX}(\tau) = \frac{1}{2\pi}\int_{-\infty}^{\infty} S_{YX}(\omega)e^{j\omega\tau}\,d\omega \qquad\qquad (7\text{-}54)$$

It is also possible to relate cross-spectral densities and crosscorrelation functions by means of the two-sided Laplace transform, just as was done with the usual spectral density and the autocorrelation function. Thus, for jointly stationary random processes

$$S_{XY}(s) = \int_{-\infty}^{\infty} R_{XY}(\tau)e^{-s\tau}\,d\tau$$

$$R_{XY}(\tau) = \frac{1}{2\pi j}\int_{-j\infty}^{j\infty} S_{XY}(s)e^{s\tau}\,ds$$

$$S_{YX}(s) = \int_{-\infty}^{\infty} R_{YX}(\tau)e^{-s\tau}\,d\tau$$

$$R_{YX}(\tau) = \frac{1}{2\pi j}\int_{-j\infty}^{j\infty} S_{YX}(s)e^{s\tau}\,ds$$

When using the two-sided inverse Laplace transform to find a crosscorrelation function, it is not possible to use symmetry to find the value of the crosscorrelation function for negative values of τ. Instead, the procedure discussed in Section 7–6 must be employed. An example will serve to illustrate this procedure once more. Suppose we have a cross-spectral density given by

$$S_{XY}(\omega) = \frac{96}{\omega^2 - j2\omega + 8}$$

Note that this spectral density is complex for most values of ω. Also, from the properties of cross-spectral densities given above, the other cross-spectral density is simply the conjugate of this one. Thus,

$$S_{YX}(\omega) = \frac{96}{\omega^2 + j2\omega + 8}$$

When $S_{XY}(\omega)$ is expressed as a function of s it becomes

$$S_{XY}(s) = \frac{-96}{s^2 + 2s - 8} = \frac{-96}{(s + 4)(s - 2)}$$

A partial fraction expansion yields

$$S_{XY}(s) = \frac{16}{s + 4} - \frac{16}{s - 2}$$

The lhp pole at $s = -4$ yields the positive τ function

$$\frac{16}{s + 4} \Leftrightarrow 16e^{-4\tau} \qquad \tau > 0$$

In order to handle the rhp pole at $s = 2$, replace s by $-s$ and recognize the inverse transform as

$$\frac{-16}{s - 2} \Leftrightarrow \frac{16}{s + 2} \Leftrightarrow 16e^{-2\tau}$$

If τ is now replaced by $-\tau$ and the two parts combined, the complete crosscorrelation function becomes

$$R_{YX}(\tau) = 16e^{-4\tau} \qquad \tau > 0$$
$$= 16e^{2\tau} \qquad \tau < 0$$

The other crosscorrelation function can be obtained from the relation

$$R_{YX}(\tau) = R_{XY}(-\tau)$$

Thus,

$$R_{XY}(\tau) = 16e^{-2\tau} \qquad \tau > 0$$
$$= 16e^{4\tau} \qquad \tau < 0$$

Exercise 7–8.1

For two jointly stationary random processes, the crosscorrelation function is

$$R_{XY}(\tau) = 9e^{-3\tau} \qquad \tau > 0$$
$$= 0 \qquad \tau < 0$$

a) Find the corresponding cross-spectral density.

b) Find the other cross-spectral density.

Answers: $\dfrac{9}{-j\omega + 3}, \dfrac{9}{j\omega + 3}$

Exercise 7–8.2

Two jointly stationary random processes have a cross-spectral density of

$$S_{XY}(\omega) = \frac{1}{-\omega^2 + j2\omega + 1}$$

Find the corresponding crosscorrelation function.

Answer: $\tau e^{-\tau}$ $\tau > 0$

7–9   Measurement of Spectral Density

When random phenomena are encountered in practical situations, it is often necessary to measure certain parameters of the phenomena in order to be able to determine how best to design the signal processing system. The case that is most easily handled, and the one that will be considered here, is that in which it can be assumed legitimately that the random process involved is ergodic. In such cases, it is possible to make estimates of various parameters of the process from appropriate time averages. The problems associated with estimating the mean and the correlation function have been considered previously; it is now desired to consider how one may estimate the distribution of power throughout the frequency range occupied by the signal—that is, the spectral density. This kind of information is invaluable in many engineering applications. For example, knowing the spectral density of an unwanted or interfering signal often gives valuable clues as to the origin of the signal, and may lead to its elimination. In cases where elimination is not possible, knowledge of the power spectrum often permits design of appropriate filters to reduce the effects of such signals.

As an example of a typical problem of this sort, assume that there is available a continuous recording of a signal $x(t)$ extending over the interval $0 \le t \le T$. The signal $x(t)$ is assumed to be a sample function from an ergodic random process. It is desired to make an estimate of the spectral density $S_X(\omega)$ of the process from which the recorded signal came.

It might be thought that a reasonable way to find the spectral density would be to find the Fourier transform of the observed sample function and let the square of its magnitude be an estimate of the spectral density. This procedure does not work, however. Since the Fourier transform of the entire sample function does not even exist, it is not surprising to find that the Fourier transform of a portion of that sample function is a poor estimator of the desired spectral density. This procedure might be possible if one could take an ensemble average of the squared magnitude of the Fourier transform of all (or even some of) the sample functions of the process but since only one sample function is available no such direct approach is possible.

An alternative to the above is to employ the mathematical relationship between the spectral density and the autocorrelation function, as given by (7–40). Since it is possible to estimate autocorrelation functions from a single sample function, as discussed in Section 6–4, the Fourier transform of this estimate will be an estimate of the spectral density. It is this approach that will be discussed here.

It is shown in (6–14) that an estimate of the autocorrelation function of an ergodic process can be obtained from

$$\hat{R}_X(\tau) = \frac{1}{T - \tau} \int_0^{T-\tau} X(t)X(t + \tau)\, dt \qquad 0 \le \tau \ll T \tag{6–14}$$

when $X(t)$ is an arbitrary member of the ensemble. Since τ must be much smaller than T, the length of the record, let the largest permissible value of τ be designated by τ_m. Thus, $\hat{R}_X(\tau)$ has a value given by (6–14) whenever $|\tau| \le \tau_m$ and is *assumed* to be zero whenever $|\tau| > \tau_m$. A more general way of introducing this limitation on the size of τ is to multiply (6–14) by an even function of τ that is zero when $|\tau| > \tau_m$. Thus, define a new estimate of $R_X(\tau)$ as

$$_w\hat{R}_X(\tau) = \frac{w(\tau)}{T - \tau} \int_0^{T-\tau} X(t)X(t + \tau)\, dt$$

$$= w(\tau)_a\hat{R}_X(\tau) \tag{7–55}$$

where $w(\tau) = 0$ when $|\tau| > \tau_m$ and is an even function of τ and $_a\hat{R}_X(\tau)$ is now *assumed* to exist for all τ. The function $w(\tau)$ is often referred to as a "lag window" since it modifies the estimate of $R_X(\tau)$ by an amount that depends upon the "lag" (that is, the time delay τ) and has a finite width of $2\tau_m$. The purpose of introducing $w(\tau)$, and the choice of a suitable form for it, are extremely important aspects of estimating spectral densities that are all too often overlooked by engineers attempting to make such estimates. The following brief discussion of these topics is hardly adequate to provide a complete understanding, but it may serve to introduce the concepts and to indicate their importance in making meaningful estimates.

Since the spectral density is the Fourier transform of the autocorrelation function, an estimate of spectral density can be obtained by transforming (7–55). Thus,

$$_w\hat{S}_X(\omega) = \mathscr{F}[w(\tau)_a\hat{R}_X(\tau)] \tag{7-56}$$

$$= \frac{1}{2\pi} W(\omega)*_a\hat{S}_X(\omega)$$

where $W(\omega)$ is the Fourier transform of $w(\tau)$ and the symbol $*$ implies convolution of transforms. $_a\hat{S}_X(\omega)$ is the spectral density associated with $_a\hat{R}_X(\tau)$, which is now defined for all τ but cannot be estimated for all τ.

In order to discuss the purpose of the window function, it is important to emphasize that there is a particular window function present even if the problem is ignored. Since (6–14) exists only for $|\tau| \leq \tau_m$, it would be equivalent to (7–55) if a rectangular window defined by

$$w_r(\tau) = 1 \qquad |\tau| \leq \tau_m \tag{7-57}$$

$$= 0 \qquad |\tau| > \tau_m$$

were used. Thus, not assigning any window function is really equivalent to using the rectangular window of (7–57). The significance of using a window function of this type can be seen by noting that the corresponding Fourier transform of the rectangular window is

$$\mathscr{F}[w_r(\tau)] = W_r(\omega) = 2\tau_m \left(\frac{\sin \omega\tau_m}{\omega\tau_m}\right) \tag{7-58}$$

and that this transform is negative half the time, as seen in Figure 7–9. Thus, convolving it with $_a\hat{S}_X(\omega)$ can lead to negative values for the estimated spectral density, even though $_a\hat{S}_X(\omega)$ itself is never negative. Also, the fact that $\hat{R}_X(\tau)$ can be estimated for only a limited range of τ-values (namely, $|\tau| \leq \tau_m$) because of

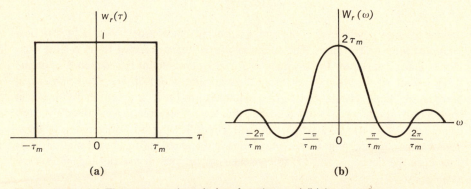

(a)   (b)

Figure 7–9   (a) The rectangular-window function and (b) its transform.

the finite length of the observed data ($\tau_m \ll T$) may lead to completely erroneous estimates of spectral density, regardless of how accurately $\hat{R}_X(\tau)$ is known within its range of values.

The estimate provided by the rectangular window will be designated

$$_r\hat{S}_X(\omega) = \frac{1}{2\pi} W_r(\omega) *_a\hat{S}_X(\omega) \tag{7-59}$$

It should be noted, however, that it is not found by carrying out the convolution indicated (7–59), since $_a\hat{S}_X(\omega)$ cannot be estimated from the limited data available, but instead is just the Fourier transform of $\hat{R}_X(\tau)$ as defined by (6–14). That is,

$$_r\hat{S}_X(\omega) = \mathcal{F}[\hat{R}_X(\tau)] \tag{7-60}$$

where

$$\hat{R}_X(\tau) = \frac{1}{T-\tau} \int_0^{T-\tau} X(t)X(t+\tau)\, dt \qquad 0 \leq \tau \leq \tau_m$$

$$= 0 \qquad\qquad\qquad \tau > \tau_m$$

and

$$\hat{R}_X(\tau) = \hat{R}_X(-\tau) \qquad\qquad \tau < 0$$

Thus, as noted above, $_r\hat{S}_X(\omega)$ is the estimate obtained by ignoring the consequences of the limited range of τ-values. The problem that now arises is how to modify $_r\hat{S}_X(\omega)$ so as to minimize the erroneous results that occur. It is this problem that leads to choosing other shapes for the window function $w(\tau)$.

The source of the difficulty associated with $_r\hat{S}_x(\omega)$ is the sidelobes of $W_r(\omega)$. Clearly, this difficulty could be overcome by selecting a window function that has very small sidelobes in its transform. One such window function that has been used extensively is the so-called "Hamming window," named after the man who suggested it, and given by

$$w_h(\tau) = 0.54 + 0.46 \cos \frac{\pi\tau}{\tau_m} \qquad |\tau| < \tau_m$$

$$= 0 \qquad\qquad\qquad |\tau| > \tau_m \tag{7-61}$$

This window and its transform are shown in Figure 7–10.

The resulting estimate of the spectral density is given formally by

$$_h\hat{S}_X(\omega) = \frac{1}{2\pi} W_h(\omega) *_a\hat{S}_X(\omega) \tag{7-62}$$

but, as before, this convolution cannot be carried out because $_a\hat{S}_X(\omega)$ is not available. However, if it is noted that

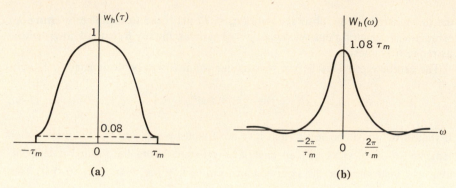

Figure 7–10   (a) The Hamming-window function and (b) its Fourier transform.

$$w_h(\tau) = \left(0.54 + 0.46 \cos \frac{\pi\tau}{\tau_m}\right) w_r(\tau)$$

then it follows that

$$\mathcal{F}[w_h(t)] = W_h(\omega)$$

$$= \left\{0.54\delta(\omega) + 0.23\left[\delta\left(\omega + \frac{\pi}{\tau_m}\right) + \delta\left(\omega - \frac{\pi}{\tau_m}\right)\right]\right\} *W_r(\omega)$$

since the *nontruncated* constant and cosine term of the Hamming window have Fourier transforms that are δ-functions. Substituting this into (7–62), and utilizing (7–59), leads immediately to

$$_h\hat{S}_X(\omega) = 0.54_r\hat{S}_X(\omega) + 0.23\left[_r\hat{S}_X\left(\omega + \frac{\pi}{\tau_m}\right) + _r\hat{S}_X\left(\omega - \frac{\pi}{\tau_m}\right)\right] \qquad \text{(7–63)}$$

Since $_r\hat{S}_X(\omega)$ can be found, by using (7–60), it follows that (7–63) represents the modification of $_r\hat{S}_X(\omega)$ that is needed to insure that the resulting estimate is always positive.

In the discussion of estimating autocorrelation functions in Section 6–4, it was noted that in almost all practical cases the observed record would be sampled at discrete times of 0, Δt, $2\Delta t$, . . . ,$N\Delta t$ and the resulting estimate formed by the summation:

$$\hat{R}_X(n\Delta t) = \frac{1}{N - n + 1}\sum_{k=0}^{N-n} X_k X_{k+n} \qquad n = 0,1, \ldots , M \qquad \text{(7–64)}$$

Since the autocorrelation function is estimated for discrete values of τ only, it is necessary to perform a discrete approximation to the Fourier transform. Although there are techniques for doing this that conserve computer time (the so-called "fast Fourier transform"), it is convenient in this discussion to consider the discrete version of (7–42), which is simply the Fourier cosine transform of the autocorrelation function. Thus, the estimated rectangular-window spectral density is

$$_r\hat{S}_X(q \, \Delta\omega) = \Delta t \left[\hat{R}_X(0) + 2 \sum_{n=1}^{M-1} \hat{R}_X(n \, \Delta t) \cos \frac{qn\pi}{M} + \hat{R}_X(M \, \Delta t) \cos q\pi \right]$$

where $q = 0, 1, 2, \ldots, M$ (7–65)

$$\Delta\omega = \frac{\pi}{M \, \Delta t}$$

The corresponding Hamming-window estimate is

$$_h\hat{S}_X(q \, \Delta\omega) = 0.54\,_r\hat{S}_X(q \, \Delta\omega) + 0.23\{_r\hat{S}_X[(q + 1) \, \Delta\omega] \qquad (7\text{–}66)$$
$$+ \,_r\hat{S}_X[(q - 1) \, \Delta\omega]\}$$

and this represents the final form of the estimate. A computer program for carrying out a Hamming-window estimate of spectral density is given in Appendix G.

In order to illustrate this method of estimating spectral density suppose that we have estimated the autocorrelation function of an ergodic random process, using (7–64) for $M = 5$ with $\Delta t = 0.01$. Let the resulting values of the estimated autocorrelation function be

n	$\hat{R}_X(n \, \Delta t)$
0	10
1	8
2	6
3	4
4	2
5	0

For the specified values of M and Δt, the spacing between spectral estimates becomes

$$\Delta\omega = \frac{\pi}{M \, \Delta t} = \frac{\pi}{(5)(0.01)} = 20\pi \text{ radians/second}$$

Using the estimated values of autocorrelation the rectangular-window estimate of spectral density may be written from (7–65) as

$$_r\hat{S}_X(q \, \Delta\omega) = 0.01 \left[10 + 2 \left(8 \cos\left(q \, \frac{\pi}{5} \right) + 6 \cos\left(2q \, \frac{\pi}{5} \right) \right. \right.$$
$$\left. \left. + 4 \cos\left(3q \, \frac{\pi}{5} \right) + 2 \cos\left(4q \, \frac{\pi}{5} \right) \right) \right]$$

This may be evaluated for values of q ranging from 0 to 5 and the resulting rectangular-window estimate is

q	$_r\hat{S}_X(q\,\Delta\omega)$
0	0.5
1	0.2094
2	0
3	0.0306
4	0
5	0.020

The final Hamming-window estimate is found by using these values in (7–66). The resulting values are shown below.

q	$_h\hat{S}_X(q\,\Delta\omega)$
0	0.3664
1	0.2281
2	0.0552
3	0.0165
4	0.0116
5	0.0108

Although the length of the correlation function sample used is too short to give a very good estimate, this example does illustrate the method of applying a Hamming window and demonstrates the smoothing that such a window achieves.

Many other window functions have been proposed for spectral estimation and some give better results than the Hamming window, although they may not be as easy to use. There is, for example, the Bartlett window, which is simply an isosceles triangle, that can be applied very readily to the autocorrelation estimate, but requires that the actual convolution be carried out when applied to the spectral function. Another well-known window function is the "hanning window," which is a modified form of the Hamming window. Both of these window functions are considered further in the exercises and problems that follow.

Although the problem of evaluating the quality of spectral density estimates is very important, it is also quite difficult. In the first place, Hamming-window estimates are not unbiased, that is, the expected value of the estimate is not the true value of the spectral density. Secondly, it is very difficult to determine the variance of the estimate, although a rough approximation to this variance can be expressed as

$$\text{Var}\,[_h\hat{S}_X(q\Delta\omega)] \simeq \frac{M}{N}\,S_X^2(q\Delta\omega) \tag{7–67}$$

when $2M\Delta t$ is large enough to include substantially all of the autocorrelation function.

When the spectral density being measured is quite nonuniform over the frequency band, the Hamming-window estimate may give rise to serious errors that can be minimized by "whitening"—that is, by modifying the spectrum in a known way to make it more nearly uniform. A particularly severe error of this sort arises when the observed data contains a dc component, since this represents a δ function in the spectral density. In such cases, it is very important to remove the dc component from the data before proceeding with the analysis.

Exercise 7–9.1

A stationary random process having an autocorrelation function of

$$R_X(\tau) = 10\left(\frac{\sin 100\pi\tau}{100\pi\tau}\right)$$

has its autocorrelation function estimated over a range of $|\tau| \leq 0.04$. If a rectangular-window function of this width is used, determine the estimated spectral density at $\omega = 0$ and $\omega = 100\pi$.

Answers: 0.05, 0.1

Exercise 7–9.2

The Bartlett-window function is defined by

$$w_b(\tau) = 1 - \frac{|\tau|}{\tau_m} \qquad |\tau| \leq \tau_m$$

$$= 0 \qquad \tau > \tau_m$$

Find the Fourier transform, $W_b(\omega)$, of this window function.

Answer: $\tau_m\left(\dfrac{\sin\omega\tau_m/2}{\omega\tau_m/2}\right)^2$

7–10   Examples and Applications of Spectral Density

The most important application of spectral density is in connection with the analysis of linear systems having random inputs. However, since this application is

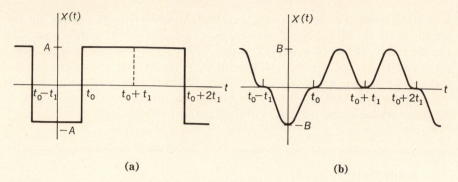

(a)   (b)

Figure 7–11 A binary signal with (a) rectangular pulses and (b) raised cosine pulses.

considered in detail in the next chapter, it will not be discussed here. Instead, some examples are given that emphasize the properties of spectral density and the computational techniques.

The first example considers the signal in a binary communication system—that is, one in which the message is conveyed by the polarities of a sequence of pulses. The obvious form of pulse to use in such a system is the rectangular one shown in Figure 7–11(a). These pulses all have the same amplitude, but the polarities are either plus or minus with equal probability and are independent from pulse to pulse. However, the steep sides of such pulses tend to make this signal occupy more bandwidth than is desirable. An alternative form of pulse is the *raised-cosine* pulse as shown in Figure 7–11(b). The question to be answered concerns the amount by which the bandwidth is reduced by using this type of pulse rather than the rectangular one.

Both of the random processes described above have spectral densities that can be described by the general result in (7–25). In both cases, the mean value of pulse amplitude is zero (since each polarity is equally probable), and the variance of pulse amplitude is A^2 for the rectangular pulse and B^2 for the raised-cosine pulse. (See the discussion of delta distributions in Section 2–7.) Thus, all that is necessary is to find $|F(\omega)|^2$ for each pulse shape.

For the rectangular pulse, the shape function $f(t)$ is

$$f(t) = 1 \qquad |t| \leq \frac{t_1}{2}$$

$$= 0 \qquad |t| > \frac{t_1}{2}$$

Hence, its Fourier transform is

$$F(\omega) = \int_{-t_1/2}^{t_1/2} (1)e^{-j\omega t}\, dt = t_1 \frac{\sin(\omega t_1/2)}{(\omega t_1/2)}$$

and, from (7–25), the spectral density of the binary signal is

$$S_X(\omega) = A^2 t_1 \left[\frac{\sin(\omega t_1/2)}{\omega t_1/2} \right]^2 \tag{7-68}$$

which has a maximum value at $\omega = 0$.

For the raised-cosine pulse the shape function is

$$f(t) = \frac{1}{2} \left(1 + \cos\frac{2\pi t}{t_1} \right) \qquad |t| \leq t_1/2$$

$$= 0 \qquad |t| > t_1/2$$

The Fourier transform of this shape function becomes

$$F(\omega) = \frac{1}{2} \int_{-t_1/2}^{t_1/2} \left(1 + \cos\frac{2\pi t}{t_1} \right) e^{-j\omega t} dt$$

$$= \frac{t_1}{2} \left[\frac{\sin(\omega t_1/2)}{(\omega t_1/2)} \right] \left[\frac{\pi^2}{\pi^2 - (\omega t_1/2)^2} \right]$$

and the corresponding spectral density is

$$S_X(\omega) = \frac{B^2 t_1}{4} \left[\frac{\sin(\omega t_1/2)}{\omega t_1/2} \right]^2 \left[\frac{\pi^2}{\pi^2 - (\omega t_1/2)^2} \right]^2 \tag{7-69}$$

which has a maximum value at $\omega = 0$.

In evaluating the bandwidths of these spectral densities, there are many different criteria that might be employed. However, when one wishes to reduce the interference between two communication systems it is reasonable to consider the bandwidth outside of which the signal spectral density is below some specified fraction (say, 1 percent) of the maximum spectral density. That is, one wishes to find the value of ω_1 such that

$$\frac{S_X(\omega)}{S_X(0)} \leq 0.01 \qquad |\omega| > \omega_1$$

Since $\sin(\omega t_1/2)$ can never be larger than 1, this condition will be assured for (7–68) when

$$\frac{A^2 t_1 \left[\dfrac{1}{(\omega t_1/2)} \right]^2}{A^2 t_1} \leq 0.01$$

from which

$$\omega_1 \simeq \frac{20}{t_1}$$

for the case of rectangular pulses. When raised-cosine pulses are used, this condition becomes

$$\frac{B^2 t_1/4[1/(\omega t_1/2)]^2[\pi^2/(\pi^2 - (\omega t_1/2)^2)]^2}{B^2 t_1/4} \leq 0.01$$

This leads to

$$\omega_1 \simeq \frac{10.68}{t_1}$$

It is clear that the use of raised-cosine pulses, rather than rectangular pulses, has cut the bandwidth almost in half (when bandwidth is specified according to this criterion).

Almost all of the examples of spectral density that have been considered throughout this chapter have been low-pass functions—that is, the spectral density has had its maximum value at $\omega = 0$. However, many practical situations arise in which the maximum value of spectral density occurs at some high frequency, and the second example will illustrate a situation of this sort. Figure 7–12 shows a typical band-pass spectral density and the corresponding pole-zero configuration. The complex frequency representation for this spectral density is obtained easily from the pole-zero plot. Thus,

$$S_X(s) = \frac{S_0(s)(-s)}{(s + \alpha + j\omega_0)(s + \alpha - j\omega_0)(s - \alpha + j\omega_0)(s - \alpha - j\omega_0)} \quad \text{(7–70)}$$

$$= \frac{-S_0 s^2}{[(s + \alpha)^2 + \omega_0^2][(s - \alpha)^2 + \omega_0^2]}$$

where S_0 is a scale factor. Note that this spectral density is zero at zero frequency.

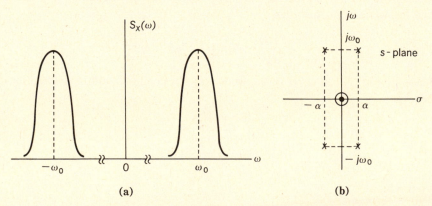

(a)   (b)

Figure 7–12   (a) A band-pass spectral density and (b) the corresponding pole-zero plot.

The mean-square value associated with this spectral density can be obtained by either of the methods discussed in Section 7–5. If Table 7–1 is used, it is seen readily that

$$c(s) = s \qquad c_1 = 1 \qquad c_0 = 0$$

$$d(s) = s^2 + 2\alpha s + \alpha^2 + \omega_0^2 \qquad d_2 = 1 \qquad d_1 = 2\alpha \qquad d_0 = \alpha^2 + \omega_0^2$$

The mean-square value is then related to I_2 by

$$\overline{X^2} = S_0 I_2 = S_0 \frac{c_1^2 d_0 + c_0^2 d_2}{2d_0 d_1 d_2} = S_0 \frac{(1)^2(\alpha^2 + \omega_0^2) + 0}{2(\alpha^2 + \omega_0^2)(2\alpha)(1)}$$

$$= \frac{S_0}{4\alpha}$$

An interesting result of this calculation is that the mean-square value depends only upon the bandwidth parameter α and not upon the center frequency ω_0.

The third example concerns the physical interpretation of spectral density as implied by the relation

$$\overline{X^2} = \frac{1}{2\pi} \int_{-\infty}^{\infty} S_X(\omega) \, d\omega$$

Although this expression only relates the total mean-square value of the process to the total area under the spectral density, there is a further implication that the mean-square value associated with any range of frequencies is similarly related to the partial area under the spectral density within that range of frequencies. That is, if one chooses any pair of frequencies, say ω_1 and ω_2, the mean-square value of that portion of the random process having energy between these two frequencies is

$$\overline{X^2} = \frac{1}{2\pi} \left[\int_{-\omega_2}^{-\omega_1} S_X(\omega) \, d\omega + \int_{\omega_1}^{\omega_2} S_X(\omega) \, d\omega \right] \qquad (7\text{–}71)$$

$$= \frac{1}{\pi} \int_{\omega_1}^{\omega_2} S_X(\omega) \, d\omega$$

The second form in (7–71) is a consequence of $S_X(\omega)$ being an even function of ω.

As an illustration of this concept, consider again the spectral density derived in (7–41). This was

$$S_X(\omega) = \frac{2A\beta}{\omega^2 + \beta^2} \qquad (7\text{–}41)$$

where A is the total mean-square value of the process. Suppose it is desired to find the frequency above which one-half the total mean-square value (or average power) exists. This means that we want to find the ω_1 (with $\omega_2 = \infty$) for which

$$\frac{1}{\pi} \int_{\omega_1}^{\infty} \frac{2A\beta}{\omega^2 + \beta^2} \, d\omega = \frac{1}{2} \left[\frac{1}{\pi} \int_{0}^{\infty} \frac{2A\beta}{\omega^2 + \beta^2} \, d\omega \right] = \frac{1}{2} A$$

Thus,

$$\int_{\omega_1}^{\infty} \frac{d\omega}{\omega^2 + \beta^2} = \frac{\pi}{4\beta}$$

since the A cancels out. The integral becomes

$$\frac{1}{\beta} \tan^{-1} \frac{\omega}{\beta} \bigg|_{\omega_1}^{\infty} = \frac{1}{\beta} \left(\frac{\pi}{2} - \tan^{-1} \frac{\omega_1}{\beta} \right) = \frac{\pi}{4\beta}$$

from which

$$\tan^{-1} \frac{\omega_1}{\beta} = \frac{\pi}{4}$$

and

$$\omega_1 = \beta$$

Thus, one-half of the average power of this process occurs at frequencies above β and one-half below β. Note that in this particular case, β is also the frequency at which the spectral density is one-half of its maximum value at $\omega = 0$. This result is peculiar to this particular spectral density and is not true in general. For example, in the band-limited white-noise case shown in Figure 7–8, the spectral density reaches one-half of its maximum value at $\omega = 2\pi W$, but one-half of the average power occurs at frequencies greater than $\omega = \pi W$. These conclusions are obvious from the sketch.

Exercise 7–10.1

An nth-order Butterworth spectrum is one whose spectral density is given by

$$S_X(\omega) = \frac{1}{1 + (\omega/2\pi W)^{2n}}$$

in which W is the so-called half-power bandwidth.

a)   Find the bandwidth outside of which the spectral density is less than 1 percent of its maximum value.

b) For $n = 1$, find the bandwidth outside of which no more than 1 percent of the average power exists.

Answers: $2\pi W(99)^{1/2n}$, $400W$

Exercise 7–10.2

Suppose that the binary communication system discussed in this section uses triangular pulses instead of rectangular or raised-cosine pulses. Specifically, let

$$f(t) = 1 - \left|\frac{2t}{t_1}\right| \qquad |t| \le t_1/2$$

$$= 0 \qquad\qquad |t| > t_1/2$$

Find the bandwidth of this signal using the same criterion as used in the example.

Answer: $12.56/t_1$

PROBLEMS

7–1.1   A sample function from a random process has the form

$$X(t) = M \qquad |t| \le T$$

and is zero elsewhere. The random variable M is uniformly distributed between -6 and 18.

a) Find the mean value of the random process.

b) Find the Fourier transform of this sample function.

c) Find the expected value of the Fourier transform.

d) What happens to the Fourier transform as T approaches infinity?

7–2.1   a) Use Parseval's theorem to evaluate the following integral:

$$\int_{-\infty}^{\infty} \left(\frac{\sin 4\omega}{4\omega}\right)\left(\frac{\sin 8\omega}{8\omega}\right) d\omega$$

b) Use Parseval's theorem to evaluate

$$\int_{-\infty}^{\infty} \frac{1}{\omega^4 + 5\omega^2 + 4} \, d\omega$$

7–2.2 A stationary random process has a spectral density of

$$S_X(\omega) = 1 - \frac{|\omega|}{8\pi} \qquad |\omega| \leq 8\pi$$

$$= 0 \qquad\qquad \text{elsewhere}$$

Find the mean-square value of this process.

7–2.3 A random process with a spectral density of $S_X(\omega)$ has a mean-square value of 4. Find the mean-square values of random processes having each of the following spectral densities:

a) $4S_X(\omega)$

b) $S_X(4\omega)$

c) $S_X(\omega/4)$

d) $4S_X(4\omega)$

7–3.1 For each of the following functions of ω, state whether it can or cannot be a valid expression for the spectral density of a random process. If it cannot be a spectral density, state why not.

a) $\dfrac{1}{\omega^2 + 3\omega + 1}$

b) $\dfrac{\omega^2 + 16}{\omega^4 + 9\omega^2 + 18}$

c) $10e^{-\omega^2}$

d) $\dfrac{\omega^2 + 4}{\omega^4 - 4\omega^2 + 1}$

e) $\left(\dfrac{1 - \cos \omega}{\omega}\right)^2$

f) $\delta(\omega) + \dfrac{\omega^3}{\omega^4 + 1}$

7–3.2   A stationary random process has sample functions of the form

$$X(t) = M + 5 \cos (10t + \theta_1) + 10 \sin (5t + \theta_2)$$

where M is a random variable that is uniformly distributed between -3 and $+9$, and θ_1 and θ_2 are random variables that are uniformly distributed between 0 and 2π. All three random variables are mutually independent.

   a)   Find the mean value of this process.

   b)   Find the variance of the process. 74.5

   c)   Find the spectral density of the process.

7–3.3   A stationary random process has a spectral density of

$$S_X(\omega) = 32\pi\delta (\omega) + 8\pi\delta (\omega - 6) + 8\pi\delta (\omega + 6)$$
$$+ 32\pi\delta (\omega - 12) + 32\pi\delta (\omega + 12)$$

   a)   Find the mean value of this process.

   b)   Find the variance of this process.

   c)   List all discrete frequency components of the process.

7–3.4

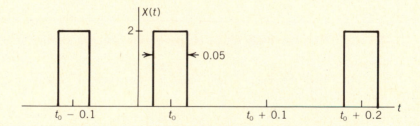

In the random pulse sequence shown above, pulses may occur or not occur with equal probability at periodic time intervals of 0.1 second. The reference time t_0 for any sample function is a random variable uniformly distributed over the interval of 0.1 second.

   a)   Find the mean value of this process.

   b)   Find the variance of this process.

   c)   Find the spectral density of the process.

7–4.1 A stationary random process has a spectral density of

$$S_X(\omega) = \frac{16(\omega^2 + 36)}{\omega^4 + 13\omega^2 + 36}$$

    a)   Write this spectral density as a function of the complex frequency s.

    b)   List all of the pole and zero frequencies.

    c)   Find the value of the spectral density at a frequency of 1 Hz.

    d)   Suppose this spectral density is to be scaled in frequency so that its value at zero frequency is unchanged but its value at 100 Hz is the same as it previously had at 1 Hz. Write the new spectral density as a function of s.

7–4.2 A given spectral density has a value of 10 V²/Hz at zero frequency. Its zeros in the complex frequency plane are at ±5 and its poles are at ±2 $\pm j5$ and ±6 $\pm j3$.

    a)   Write the spectral density as a function of s.

    b)   Write the spectral density as a function of ω.

    c)   Find the value of the spectral density at a frequency of 1 Hz.

7–5.1 a)   Find the mean-square value of the spectral density in Problem 7–3.1(a).

    b)   Find the mean-square value of the spectral density in Problem 7–3.1(d).

7–5.2 a)   Find the mean-square value of the spectral density in Problem 7–3.2 using Table 7–1.

    b)   Repeat part (a) using contour integration in the complex frequency plane.

7–5.3 Find the mean-square value of a stationary random process whose spectral density is

$$S_X(s) = \frac{-s^2}{s^4 - 52s^2 + 576}$$

7–5.4 Find the mean-square value of a stationary random process whose spectral density is

$$S_X(\omega) = \frac{\omega^2 + 10}{\omega^4 + 5\omega^2 + 4} + 8\pi\delta(\omega) + 2\pi\delta(\omega - 3) + 2\pi\delta(\omega + 3)$$

7–6.1 A stationary random process has an autocorrelation function of

$$R_X(\tau) = 10\left[1 - \frac{|\tau|}{0.05}\right] \qquad |\tau| \le 0.05$$
$$= 0 \qquad\qquad \text{elsewhere}$$

a) Find the variance of this process.

b) Find the spectral density of this process.

c) State the relation between the frequencies, in Hz, at which the spectral density is zero and the value of τ at which the autocorrelation function goes to zero.

7–6.2 A stationary random process has an autocorrelation function of

$$R_X(\tau) = 16e^{-5|\tau|} \cos 20\pi\tau + 8 \cos 10\pi\tau$$

a) Find the variance of this process.

b) Find the spectral density of this process.

c) Find the value of the spectral density at zero frequency.

7–6.3 A stationary random process has a spectral density of

$$S_X(\omega) = 5 \qquad 10 \le |\omega| \le 20$$
$$= 0 \qquad \text{elsewhere}$$

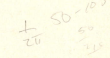

a) Find the mean-square value of this process.

b) Find the autocorrelation function of this process.

c) Find the value of the autocorrelation function at $\tau = 0$.

7–6.4 A *nonstationary* random process has an autocorrelation function of

$$R_X(t, t + \tau) = 8e^{-5|\tau|} (\cos 20\pi t)^2$$

a) Find the spectral density of this process.

b)   Find the autocorrelation function of the *stationary* random process
that has the same spectral density.

7–7.1   A stationary random process has a spectral density of

$$S_X(\omega) = \frac{9}{\omega^2 + 64}$$

a)   Write an expression for the spectral density of *bandlimited white
noise* that has the same value at zero frequency and the same mean-
square value as the above spectral density.

b)   Find the autocorrelation function of the process having the original
spectral density.

c)   Find the autocorrelation function of the bandlimited white noise of
part (a).

d)   Compare the values of these two autocorrelation functions at $\tau =
0$. Compare the areas of these two autocorrelation functions.

7–7.2   A stationary random process has a spectral density of

$$S_X(\omega) = 0.01 \qquad |\omega| \leq 1000\pi$$

$$= 0 \qquad \text{elsewhere}$$

a)   Find the autocorrelation function of this process.

b)   Find the smallest value of τ for which the autocorrelation is zero.

c)   Find the correlation between samples of this process taken at a rate
of 1000 samples/second. Repeat if the sampling rate is 1500 sam-
ples/second.

7–8.1   A stationary random process with sample functions $X(t)$ has a spectral
density of

$$S_X(\omega) = \frac{16}{\omega^2 + 16}$$

and an independent stationary random process with sample functions $Y(t)$
has a spectral density of

$$S_Y(\omega) = \frac{\omega^2}{\omega^2 + 16}$$

A new random variable is formed from $U(t) = X(t) + Y(t)$.

a)   Find the spectral density of $U(t)$.

b)   Find the cross-spectral density $S_{XY}(\omega)$.

c)   Find the cross-spectral density $S_{XU}(\omega)$.

7–8.2   For the two random processes of Problem 7–8.1, a new random process is formed from $V(t) = X(t) - Y(t)$. Find the cross-spectral density $S_{UV}(\omega)$.

7–9.1   Refer to the estimated autocorrelation function from the data given in Problem 6–4.1.

a)   Using the estimated autocorrelation function of part (a), compute the corresponding rectangular-window estimate of the spectral density for $q = 0, 1, 2, 3$.

b)   Using the results of part (a), find the Hamming-window estimate of the spectral density.

c)   Determine the approximate variance of the estimated spectral density for $q = 0$.

7–9.2   One of the earliest window functions employed for smoothing spectra is the so-called "hanning window" defined as

$$w(\tau) = 0.5 + 0.5 \cos \left(\frac{\pi \tau}{\tau_m} \right) \qquad |\tau| \leq \tau_m$$

$$= 0 \qquad\qquad\qquad\qquad \text{elsewhere}$$

a)   Derive an expression for the hanning-window estimate that is similar to (7–66) for the Hamming window.

b)   Compare the sidelobe levels for the Hamming window and the hanning window.

7–9.3   Using the same data as in Problem 7–9.1, find the hanning-window estimate of the spectral density.

7–10.1   Consider a binary communication system using raised-cosine pulses defined by

$$f(t) = \frac{1}{2}(1 + \cos \pi t/t_1) \qquad |t| \leq t_1$$

and zero elsewhere. Note that these pulses are twice as wide as those shown in Figure 7–11(b), but that the message bit duration is still t_1. Thus, the pulses overlap in time, but that at the peak of each pulse, all earlier and later pulses are zero. The objective of this is to reduce the bandwidth still further.

a) Write the spectral density of the resulting sequence of pulses.

b) Find the value of ω_1 such that the spectral density is less than 1 percent of the maximum spectral density for all higher frequencies.

c) What can you conclude about the bandwidth of this communication system as compared to the ones discussed in Section 7–10?

7–10.2 A stationary random process has a spectral density having poles in the complex frequency plane located at $s = \pm 10 \pm j100$.

a) Find the half-power bandwidth in Hz of this spectral density. Half-power bandwidth is simply the frequency increment between frequencies at which the spectral density is one-half of its maximum value.

b) Find the bandwidth between frequencies at which the spectral density is 1 percent of its maximum value.

7–10.3 A binary communication system using rectangular pulses is transmitting messages at a rate of 2400 bits/second. Determine the approximate frequency below which 90% of the average power is contained.

7–10.4 A spectral density having an nth order synchronous shape is of the form

$$S_X(\omega) = \frac{1}{[1 + (\omega/2\pi B_1)^2]^n}$$

a) Express the half-power bandwidth (in Hz) of this spectral density in terms of B_1.

b) Find the value of frequency above which the spectral density is always less than 1 percent of its maximum value.

References

See the References for Chapter 1. Of particular interest for the material in this chapter are the books by Davenport and Root, Helstrom, and Papoulis. The following additional references provide considerable elaboration on the problems of estimating spectral densities.

Blackman, R. B. and J. W. Tukey, *The Measurement of Power Spectra*. New York: Dover Publications, 1958.

This is the classical reference in the field, but is still a valuable source of information and insight into the measurement of spectral densities.

Jenkins, G. M. and D. G. Watts, *Spectral Analysis and Its Applications*. San Francisco: Holden-Day, 1968.

A more advanced treatment of spectral analysis that is still considered to be the most authoritative reference available.

CHAPTER 8

Response of Linear Systems to Random Inputs

8–1 Introduction

The discussion in the preceding chapters has been devoted to finding suitable mathematical representations for random functions of time. The next step is to see how these mathematical representations can be used to determine the response, or output, of a linear system when the input is a random signal rather than a deterministic one.

It is assumed that the student is already familiar with the usual methods of analyzing linear systems in either the time domain or the frequency domain. These methods are restated here in order to clarify the notation, but no attempt is made to review all of the essential concepts. The system itself is represented either in terms of its *impulse response* $h(t)$ or its *system function* $H(\omega)$, which is just the Fourier transform of the impulse response. It is convenient in many cases also to use the *transfer function* $H(s)$, which is the Laplace transform of the impulse response. In most cases the initial conditions are assumed to be zero, for convenience, but any nonzero initial conditions can be taken into account by the usual methods if necessary.

When the input to a linear system is deterministic, either approach will lead to a unique relationship between the input and output. When the input to the system is a sample function from a random process, there is again a unique relationship between the excitation and the response; however, because of its random nature, we do not have an explicit representation of the excitation and, therefore, cannot obtain an explicit expression for the response. In this case we must be content with either a probabilistic or a statistical description of the response, just as we must use this type of description for the random excitation itself.[1] Of these two approaches, statistical and probabilistic, the statistical approach is the most useful. In only a very limited class of problems is it possible to obtain a probabilistic description of the output based on a probabilistic description of the input, whereas in many cases of interest, a statistical model of the output can be obtained readily by performing simple mathematical operations on the statistical model of the input. With the statistical method, such quantities as the mean, correlation function, and spectral density of the output can be determined. Only the statistical approach is considered in the following sections.

8–2 Analysis in the Time Domain

By means of the convolution integral it is possible to determine the response of a linear system to a very general excitation. In the case of time-varying systems or nonstationary random excitations, or both, the details become quite involved; therefore, these cases will not be considered here. To make the analysis more realistic we will further restrict our considerations to physically realizable systems that are bounded-input/bounded-output stable. If the input time function is designated as $x(t)$, the system impulse response as $h(t)$, and the output time function as $y(t)$, as shown in Figure 8–1, then they are all related either by

$$y(t) = \int_0^\infty x(t - \lambda)h(\lambda)\, d\lambda \qquad (8\text{–}1)$$

or by

$$y(t) = \int_{-\infty}^t x(\lambda)h(t - \lambda)\, d\lambda \qquad (8\text{–}2)$$

The physical realizability and stability constraints on the system are given by

$$h(t) = 0 \qquad t < 0 \qquad (8\text{–}3)$$

$$\int_{-\infty}^\infty |h(t)|\, dt < \infty \qquad (8\text{–}4)$$

Starting from these specifications, many important characteristics of the output of a system excited by a stationary random process can be determined.

A simple example of time-domain analysis with a deterministic input signal serves to review the methods and provides the background for extending these

---

[1] By a probabilistic description we mean one in which certain probability functions are specified; by a statistical description we mean one in which certain ensemble averages are specified (for example, mean, variance, correlation).

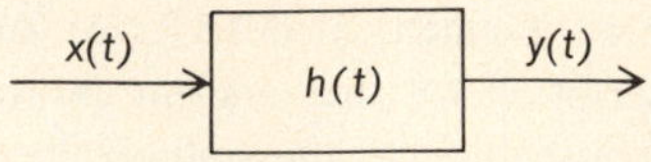

Figure 8–1   Time-domain representation of a linear system.

methods to nondeterministic input signals. Assume that we have a linear system whose impulse response is

$$h(t) = 5e^{-3t} \qquad t \geq 0$$
$$= 0 \qquad t < 0$$

It is clear that this impulse response satisfies the conditions for physical realizability and stability. Let the input be a sample function from a deterministic random process having sample functions of the form

$$X(t) = M + 4 \cos (2t + \theta) \qquad -\infty < t < \infty$$

in which M is a random variable and θ is an independent random variable that is uniformly distributed between 0 and 2π. Note that this process is stationary but not ergodic. Furthermore, since an explicit mathematical form for the input signal is known, an explicit mathematical form for the output signal can be obtained even though the signal comes from a random process. Hence, this situation is quite different from those that form the major concern of this chapter, namely, inputs that come from nondeterministic random processes for which no explicit mathematical representation is possible.

Although either (8–1) or (8–2) may be used to determine the system output, the latter is used here. Thus,

$$Y(t) = \int_{-\infty}^{t} [M + 4 \cos (2\lambda + \theta)]5e^{-3(t-\lambda)} \, d\lambda$$

which may be integrated to yield

$$Y(t) = \frac{5}{3}M + \frac{20}{13}[3 \cos (2t + \theta) + 2 \sin (2t + \theta)]$$

It is clear from this result that the output of the system is still a sample function from a random process and that it contains the same random variables that are associated with the input. Furthermore, if probability density functions for these random variables are specified, it is possible to determine such statistics of the output as the mean and variance. This possibility is illustrated by the Exercises that follow.

Exercise 8–2.1

A linear system has an impulse response of the form

$$h(t) = te^{-5t} \qquad t \geq 0$$
$$= 0 \qquad\qquad t < 0$$

and an input signal that is a sample function from a random process having sample functions of the form

$$X(t) = M \qquad -\infty < t < \infty$$

in which M is a random variable that is uniformly distributed from -6 to $+18$.

a) Write an expression for the output sample function.

b) Find the mean value of the output.

c) Find the variance of the output.

Answers: 48/625, 6/25, M/25

Exercise 8–2.2

A linear system has an impulse response of the form

$$h(t) = 5\delta(t) + 3 \qquad 0 \leq t < 1$$
$$= 0 \qquad\qquad\quad \text{elsewhere}$$

The input is a random sample function of the form

$$X(t) = 4 \sin (2\pi t + \theta) \qquad -\infty < t < \infty$$

where θ is a random variable that is uniformly distributed from 0 to 2π.

a) Write an expression for the output sample function.

b) Find the mean value of the output.

c) Find the variance of the output.

Answers: 0, 200, 20 sin $(2\pi t + \theta)$

8–3 Mean and Mean-Square Value of System Output

The most convenient form of the convolution integral, when the input $X(t)$ is a sample function from a nondeterministic random process, is

$$Y(t) = \int_0^\infty X(t - \lambda)h(\lambda)\, d\lambda \qquad (8\text{–}5)$$

since the limits of integration do not depend on t. Using this form, consider first the mean value of the output $y(t)$. This is given by

$$\bar{Y} = E[Y(t)] = E\left[\int_0^\infty X(t - \lambda)h(\lambda)\, d\lambda\right] \qquad (8\text{–}6)$$

The next logical step is to interchange the sequence in which the time integration and the expectation are performed; that is, to move the expectation operation inside the integral. Before doing this, however, it is necessary to digress a moment and consider the conditions under which such an interchange is justified.

The problem of finding the expected value of an integral whose integrand contains a random variable arises many times. In almost all such cases it is desirable to be able to move the expectation operation inside the integral and thus simplify the integrand. Fortunately, this interchange is possible in almost all cases of practical interest and, hence, is used throughout this book with little or no comment. It is advisable, however, to be aware of the conditions under which this is possible, even though the reasons for these conditions are not fully understood. The conditions may be stated as follows:

If $Z(t)$ is a sample function from a random process (or some function, such as the square, of the sample function) and $f(t)$ is a nonrandom time function, then

$$E\left[\int_{t_1}^{t_2} Z(t)f(t)\, dt\right] = \int_{t_1}^{t_2} E[Z(t)]f(t)\, dt$$

if

1. $\displaystyle \int_{t_1}^{t_2} E[|Z(t)|]|f(t)|\, dt < \infty$

and

2. $Z(t)$ is bounded on the interval t_1 to t_2. Note that t_1 and t_2 may be infinite. (There is *no* requirement that $Z(t)$ be from a stationary process.)

In applying this result to the analysis of linear systems the nonrandom function $f(t)$ is usually the impulse response $h(t)$. For wide-sense stationary input random processes the quantity $E[|Z(t)|]$ is a *constant* not dependent on time t. Hence, the stability condition of (8–4) is sufficient to satisfy condition **1.** The boundedness of $Z(t)$ is always satisfied by physical signals, although there are some mathematical representations that may not be bounded.

Returning to the problem of finding the mean value of the output of a linear system, it follows that

$$\overline{Y} = \int_0^\infty E[X(t - \lambda)]h(\lambda)\,d\lambda = \overline{X} \int_0^\infty h(\lambda)\,d\lambda \qquad (8\text{–}7)$$

when the input process is wide-sense stationary. It should be recalled from earlier work in systems analysis that the *area* of the impulse response is just the *dc gain* of the system—that is, the transfer function of the system evaluated at $\omega = 0$. Hence, (8–7) simply states the obvious fact that the dc component of the output is equal to the dc component of the input times the dc gain of the system. If the input random process has zero mean, the output process will also have zero mean. If the system does not pass direct current, the output process will always have zero mean.

In order to find the mean-square value of the output we must be able to calculate the mean value of the *product* of two integrals. However, such a product may always be written as an iterated double integral if the variables of integration are kept distinct. Thus,

$$\overline{Y^2} = E[Y^2(t)] = E\left[\int_0^\infty X(t - \lambda_1)h(\lambda_1)\,d\lambda_1 \cdot \int_0^\infty X(t - \lambda_2)h(\lambda_2)\,d\lambda_2\right]$$

$$= E\left[\int_0^\infty d\lambda_1 \int_0^\infty X(t - \lambda_1)X(t - \lambda_2)h(\lambda_1)h(\lambda_2)\,d\lambda_2\right] \qquad (8\text{–}8)$$

$$= \int_0^\infty d\lambda_1 \int_0^\infty E[X(t - \lambda_1)X(t - \lambda_2)]h(\lambda_1)h(\lambda_2)\,d\lambda_2 \qquad (8\text{–}9)$$

in which the subscripts on λ_1 and λ_2 have been introduced to keep the variables of integration distinct. The expected value inside the double integral is simply the autocorrelation function for the input random process; that is,

$$E[X(t - \lambda_1)X(t - \lambda_2)] = R_X(t - \lambda_1 - t + \lambda_2) = R_X(\lambda_2 - \lambda_1)$$

Hence, (8–9) becomes

$$\overline{Y^2} = \int_0^\infty d\lambda_1 \int_0^\infty R_X(\lambda_2 - \lambda_1)h(\lambda_1)h(\lambda_2)\,d\lambda_2 \qquad (8\text{–}10)$$

Although (8–10) is usually not difficult to evaluate, since both $R_X(\tau)$ and $h(t)$ are likely to contain only exponentials, it is frequently very tedious to carry out the details. This is because such autocorrelation functions often have discontinuous derivatives at the origin (which in this case is $\lambda_1 = \lambda_2$) and thus the integral must be broken up into several ranges. This point is illustrated later. At the moment, however, it is instructive to consider a much simpler situation—one in which the input is a sample function from white noise. For this case, it was shown in Section 7–7 that

$$R_X(\tau) = S_0 \delta(\tau)$$

where S_0 is the two-sided spectral density of the white noise. Hence (8–10) becomes

$$\overline{Y^2} = \int_0^\infty d\lambda_1 \int_0^\infty S_0 \delta(\lambda_2 - \lambda_1) h(\lambda_1) h(\lambda_2) \, d\lambda_2 \qquad (8\text{–}11)$$

Integrating over λ_2 yields

$$\overline{Y^2} = S_0 \int_0^\infty h^2(\lambda) \, d\lambda \qquad (8\text{–}12)$$

Hence, for this case it is the area of the *square* of the impulse response that is significant.[2]

As a means of illustrating some of these ideas with a simple example, consider the single-section, low-pass *RC* circuit shown in Figure 8–2. The mean value of the output is, from (8–7),

$$\overline{Y} = \overline{X} \int_0^\infty b e^{-b\lambda} \, d\lambda = \overline{X} b \left. \frac{e^{-b\lambda}}{-b} \right|_0^\infty = \overline{X} \qquad (8\text{–}13)$$

This result is obviously correct, since it is apparent by inspection that the dc gain of this circuit is unity.

Next consider the mean-square value of the output when the input is white noise. From (8–12), this is

$$\overline{Y^2} = S_0 \int_0^\infty b^2 e^{-2b\lambda} \, d\lambda = b^2 S_0 \left. \frac{e^{-2b\lambda}}{-2b} \right|_0^\infty = \frac{b S_0}{2} \qquad (8\text{–}14)$$

Note that the parameter b, which is the reciprocal of the time constant, is also related to the half-power bandwidth of the system. In particular, this bandwidth B is

$$B = \frac{1}{2\pi RC} = \frac{b}{2\pi} \text{ Hz}$$

---

[2]It should be noted that, for some functions, this integral can diverge even when (8–4) is satisfied. This occurs, for instance, whenever $h(t)$ contains δ functions. The high-pass *RC* circuit is an example of this.

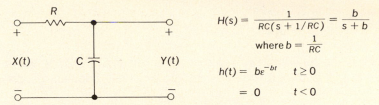

$$H(s) = \frac{1}{RC(s + 1/RC)} = \frac{b}{s+b}$$

where $b = \frac{1}{RC}$

$$h(t) = b\varepsilon^{-bt} \quad t \geq 0$$
$$= 0 \quad t < 0$$

Figure 8–2 Simple RC circuit and its impulse response.

so that (8–14) could be written as

$$\overline{Y^2} = \pi B S_0 \tag{8–15}$$

It is evident from the above that the mean-square value of the output of this system increases *linearly* with the bandwidth of the system. Such results are typical whenever the bandwidth of the input random process is large compared with the bandwidth of the system.

We might consider next a situation in which the input sample function is not from a white noise process. In this case the complete double integral of (8–10) must be evaluated. However, this is likely to be a tedious operation and is, in fact, just a special case of the more general problem of obtaining the autocorrelation function of the output. Since obtaining the complete autocorrelation function is only slightly more involved than finding just the mean-square value of the output, this task is postponed until the next section.

Exercise 8–3.1

A linear system has an impulse response of

$$h(t) = te^{-3t} u(t)$$

where $u(t)$ is the unit step function. The input to this system is a sample function from a white noise process having a two-sided spectral density of 4 V^2/Hz plus a dc component of 2 V.

a) Find the mean value of the output of the system.

b) Find the variance of the output.

c) Find the mean-square value of the output.

Answers: 0.0370, 0.0864, 0.2222

Exercise 8–3.2

White noise having a two-sided spectral density of 0.25 V^2/Hz is applied to the input of a finite-time integrator whose impulse response is

$$h(t) = 5[u(t) - u(t - 0.2)]$$

a) Find the mean value of the output of the system.

b) Find the mean-square value of the output.

Answers: 0, 1.25

8–4 Autocorrelation Function of System Output

A problem closely related to that of finding the mean-square value is the determination of the autocorrelation function at the output of the system. By definition, this autocorrelation function is

$$R_Y(\tau) = E[Y(t)Y(t + \tau)]$$

Following the same steps as in (8–9), except for replacing t by $t + \tau$ in one factor, the autocorrelation function may be written as

$$R_Y(\tau) = \int_0^\infty d\lambda_1 \int_0^\infty E[X(t - \lambda_1)X(t + \tau - \lambda_2)]h(\lambda_1)h(\lambda_2) \, d\lambda_2 \qquad \text{(8–16)}$$

In this case the expected value inside the integral is

$$E[X(t - \lambda_1)X(t + \tau - \lambda_2)] = R_X(t - \lambda_1 - t - \tau + \lambda_2) = R_X(\lambda_2 - \lambda_1 - \tau)$$

Hence, the output autocorrelation function becomes

$$R_Y(\tau) = \int_0^\infty d\lambda_1 \int_0^\infty R_X(\lambda_2 - \lambda_1 - \tau)h(\lambda_1)h(\lambda_2)d\lambda_2 \qquad \text{(8–17)}$$

Note the similarity between this result and that for the mean-square value. In particular, for $\tau = 0$, this reduces exactly to (8–10), as it must.

For the special case of white noise into the system, the expression for the output autocorrelation function becomes much simpler. Let

$$R_X(\tau) = S_0\delta(\tau)$$

as before, and substitute into (8–17). Thus,

$$R_Y(\tau) = \int_0^\infty d\lambda_1 \int_0^\infty S_0 \delta(\lambda_2 - \lambda_1 - \tau) h(\lambda_1) h(\lambda_2) \, d\lambda_2$$

$$= S_0 \int_0^\infty h(\lambda_1) h(\lambda_1 + \tau) \, d\lambda_1 \qquad (8\text{--}18)$$

Hence, for the white-noise case, the output autocorrelation function is proportional to the time correlation function of the impulse response.

This point can be illustrated by means of the linear system of Figure 8–2 and a white-noise input. Thus,

$$R_Y(\tau) = S_0 \int_0^\infty (be^{-b\lambda}) b e^{-b(\lambda + \tau)} \, d\lambda$$

$$= b^2 S_0 e^{-b\tau} \frac{e^{-2b\lambda}}{-2b} \Big|_0^\infty = \frac{bS_0}{2} e^{-b\tau} \qquad \tau \geq 0 \qquad (8\text{--}19)$$

This result is valid only for $\tau \geq 0$. When $\tau < 0$, the range of integration must be altered because the impulse response is always zero for negative values of the argument. The situation can be made clearer by means of the two sketches shown in Figure 8–3, which show the factors in the integrand of (8–18) for both ranges of τ. The integrand is zero, of course, when either factor is zero. When $\tau < 0$, the integral becomes

$$R_Y(\tau) = S_0 \int_{-\tau}^\infty (be^{-b\lambda}) b e^{-b(\lambda + \tau)} \, d\lambda$$

$$= b^2 S_0 e^{-b\tau} \frac{e^{-2b\lambda}}{-2b} \Big|_{-\tau}^\infty = \frac{bS_0}{2} e^{b\tau} \qquad \tau \leq 0 \qquad (8\text{--}20)$$

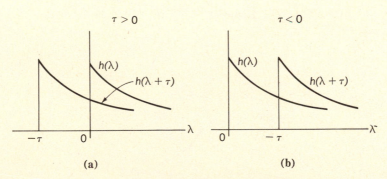

(a)   (b)

Figure 8–3 Factors in the integrand of (8–18) when the RC circuit of Figure 8–2 is used.

From (8–19) and (8–20), the complete autocorrelation function can be written as

$$R_Y(\tau) = \frac{bS_0}{2} e^{-b|\tau|} \qquad -\infty < \tau < \infty \qquad (8\text{–}21)$$

It is now apparent that the calculation for $\tau < 0$ was needless. Since the autocorrelation function is an even function of τ, the complete form could have been obtained immediately from the case for $\tau \geq 0$. This procedure is followed in the future.

It is desirable to consider at least one example in which the input random process is not white. In so doing, it is possible to illustrate some of the integration problems that develop and, at the same time, use the results to infer something about the validity and usefulness of the white-noise approximation. For this purpose, assume that the input random process to the RC circuit of Figure 8–2 has an autocorrelation function of the form

$$R_X(\tau) = \frac{\beta S_0}{2} e^{-\beta|\tau|} \qquad -\infty < \tau < \infty \qquad (8\text{–}22)$$

The coefficient $\beta S_0/2$ has been selected so that this random process has a spectral density at $\omega = 0$ of S_0; see (7–41) and Figure 7–7(b). Thus, at low frequencies, the spectral density is the same as the white-noise spectrum previously assumed.

In order to determine the appropriate ranges of integration, it is desirable to look at the autocorrelation function $R_X(\lambda_2 - \lambda_1 - \tau)$, as a function of λ_2 for $\tau > 0$. This is shown in Figure 8–4. Since λ_2 is always positive for the evaluation of (8–17), it is clear that the ranges of integration should be from 0 to $(\lambda_1 + \tau)$ and from $(\lambda_1 + \tau)$ to ∞. Hence, (8–17) may be written as

$$R_Y(\tau) = \int_0^\infty d\lambda_1 \int_0^{\lambda_1 + \tau} R_X(\lambda_2 - \lambda_1 - \tau)h(\lambda_1)h(\lambda_2) \, d\lambda_2$$

$$+ \int_0^\infty d\lambda_1 \int_{\lambda_1 + \tau}^\infty R_X(\lambda_2 - \lambda_1 - \tau)h(\lambda_1)h(\lambda_2) \, d\lambda_2$$

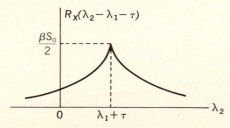

Figure 8–4 Autocorrelation function to be used in (8–17).

$$= \frac{b^2 \beta S_0}{2} \int_0^\infty e^{-(b+\beta)\lambda_1} \, d\lambda_1 \int_0^{\lambda_1+\tau} e^{-\beta\tau} e^{-(b-\beta)\lambda_2} \, d\lambda_2$$

$$+ \frac{b^2 \beta S_0}{2} \int_0^\infty e^{-(b-\beta)\lambda_1} \, d\lambda_1 \int_{\lambda_1+\tau}^\infty e^{\beta\tau} e^{-(b+\beta)\lambda_2} \, d\lambda_2 \qquad \text{(8–23)}$$

$$= \frac{b^2 \beta S_0}{-2(b-\beta)} e^{-\beta\tau} \int_0^\infty e^{-(b+\beta)\lambda_1} [e^{-(b-\beta)(\lambda_1+\tau)} - 1] \, d\lambda_1$$

$$- \frac{b^2 \beta S_0}{-2(b+\beta)} e^{\beta\tau} \int_0^\infty e^{-(b-\beta)\lambda_1} e^{-(b+\beta)(\lambda_1+\tau)} \, d\lambda_1$$

$$= \frac{b^2 \beta S_0}{2(b-\beta)} \left(-\frac{e^{-b\tau}}{2b} + \frac{e^{-\beta\tau}}{b+\beta} \right) + \frac{b^2 \beta S_0}{2(b+\beta)} \left(\frac{e^{-b\tau}}{2b} \right)$$

$$= \frac{b^2 \beta S_0}{2(b^2-\beta^2)} \left(e^{-\beta\tau} - \frac{\beta}{b} e^{-b\tau} \right) \qquad \tau > 0$$

From symmetry, the expression for $\tau < 0$ can be written directly. The final result is

$$R_Y(\tau) = \frac{b^2 \beta S_0}{2(b^2-\beta^2)} \left(e^{-\beta|\tau|} - \frac{\beta}{b} e^{-b|\tau|} \right) \qquad \text{(8–24)}$$

In order to compare this result with the previously obtained result for white noise at the input, it is only necessary to let β approach infinity. In this case,

$$\lim_{\beta \to \infty} R_Y(\tau) = \frac{bS_0}{2} e^{-b|\tau|} \qquad \text{(8–25)}$$

which is exactly the same as (8–21). Of greater interest, however, is the case when β is large compared to b but still finite. This corresponds to the physical situation in which the bandwidth of the input random process is large compared to the bandwidth of the system. In order to make this comparison, write (8–24) as

$$R_Y(\tau) = \frac{bS_0}{2} e^{-b|\tau|} \left[\frac{1}{1-(b^2)/\beta^2} \right] \left[1 - \frac{b}{\beta} e^{-(\beta-b)|\tau|} \right] \qquad \text{(8–26)}$$

The first factor in (8–26) is the autocorrelation function of the output when the input is white noise. The second factor is the one by which the true autocorrelation of the system output differs from the white-noise approximation. It is clear that as β becomes large compared to b, this factor approaches unity.

The point to this discussion is that there are many practical situations in which the input noise has a bandwidth that is much greater than the system bandwidth, and in these cases it is quite reasonable to use the white-noise approximation. In doing so, there is a great saving in labor without much loss in accuracy; for example, in a high-gain amplifier with a bandwidth of 10 MHz, the most important source of noise is shot noise in the first stage, which may have a bandwidth of 1000 MHz. Hence, the factor b/β in (8–26), assuming that this form applies, will be only 0.01, and the error in using the white-noise approximation will not exceed 1 percent.

Exercise 8–4.1

For the white noise and finite-time integrator of Exercise 8–3.2 find the value of the autocorrelation function of the system output at

   a) $\tau = 0$

   b) $\tau = 0.1$

   c) $\tau = 0.21$

   Answers: 0, 0.625, 1.25

Exercise 8–4.2

A linear system has an impulse response of

$$h(t) = 3e^{-3t}u(t)$$

The input to this system is a sample function from a random process having an autocorrelation function of

$$R_X(\tau) = 2e^{-4|\tau|}$$

Find the value of the autocorrelation function of the output of the system for

   a) $\tau = 0$

   b) $\tau = 0.5$

   c) $\tau = 1.$

   Answers: 0.1236, 0.4170, 0.8571

8–5 Crosscorrelation Between Input and Output

When a sample function from a random process is applied to the input of a linear system, the output must be related in some way to the input. Hence, they will be correlated, and the nature of the crosscorrelation function is important. In fact, it will be shown very shortly that this relationship can be used to provide a practical technique for measuring the impulse response of any linear system.

One of the crosscorrelation functions for input and output is defined by

$$R_{XY}(\tau) = E[X(t)Y(t + \tau)] \qquad (8\text{--}27)$$

which, in integral form, becomes

$$R_{XY}(\tau) = E\left[X(t)\int_0^\infty X(t + \tau - \lambda)h(\lambda)\,d\lambda\right] \qquad (8\text{--}28)$$

Since $X(t)$ is not a function of λ, it may be moved inside the integral and then the expectation may be moved inside. Thus,

$$R_{XY}(\tau) = \int_0^\infty E[X(t)X(t + \tau - \lambda)]h(\lambda)\,d\lambda$$

$$= \int_0^\infty R_X(\tau - \lambda)h(\lambda)\,d\lambda \qquad (8\text{--}29)$$

Hence, this crosscorrelation function is just the convolution of the input autocorrelation function and the impulse response of the system.

The other crosscorrelation function is

$$R_{YX}(\tau) = E[X(t + \tau)Y(t)] = E\left[X(t + \tau)\int_0^\infty X(t - \lambda)h(\lambda)\,d\lambda\right]$$

$$= \int_0^\infty E[X(t + \tau)X(t - \lambda)]h(\lambda)\,d\lambda \qquad (8\text{--}30)$$

$$= \int_0^\infty R_X(\tau + \lambda)h(\lambda)\,d\lambda$$

Since the autocorrelation function in (8–30) is symmetrical about $\lambda = -\tau$ and the impulse response is zero for negative values of λ, this crosscorrelation function will *always* be different from $R_{XY}(\tau)$. They will, however, have the same value at $\tau = 0$.

A simple example will serve to illustrate the calculation of this type of crosscorrelation function. If we consider the system of Figure 8–2 and an input from a random process having the autocorrelation function given by (8–22), the crosscorrelation function can be expressed as

$$R_{XY}(\tau) = \int_0^\tau \left[\frac{\beta S_0}{2} e^{-\beta(\tau - \lambda)} \right] (be^{-b\lambda})\, d\lambda$$

$$+ \int_\tau^\infty \left[\frac{\beta S_0}{2} e^{-\beta(\lambda - \tau)} \right] [be^{-b\lambda}]\, d\lambda \quad \text{(8–31)}$$

when $\tau > 0$. The integration is now straightforward and yields

$$R_{XY}(\tau) = \beta b S_0 \left[\frac{\beta}{\beta^2 - b^2} e^{-b\tau} - \frac{1}{2(\beta - b)} e^{-\beta\tau} \right] \quad \tau > 0 \quad \text{(8–32)}$$

For $\tau < 0$ the integration is even simpler.

$$R_{XY}(\tau) = \int_0^\infty \left[\frac{\beta S_0}{2} e^{-\beta(\lambda - \tau)} \right] [be^{-b\lambda}]\, d\lambda \quad \text{(8–33)}$$

Carrying out this integration leads to

$$R_{XY}(\tau) = \frac{\beta b S_0}{2(\beta + b)} e^{\beta\tau} \quad \tau < 0 \quad \text{(8–34)}$$

The other crosscorrelation function can be obtained from

$$R_{YX}(\tau) = R_{XY}(-\tau) \quad \text{(8–35)}$$

The above results become even simpler when the input to the system is considered to be a sample function of white noise. For this case,

$$R_X(\tau) = S_0 \delta(\tau)$$

and

$$R_{XY}(\tau) = \int_0^\infty S_0 \delta(\tau - \lambda) h(\lambda)\, d\lambda = S_0 h(\tau) \quad \tau \geq 0$$
$$= 0 \quad \tau < 0 \quad \text{(8–36)}$$

Likewise,

$$R_{YX}(\tau) = \int_0^\infty S_0 \delta(\tau + \lambda) h(\lambda)\, d\lambda = 0 \quad \tau > 0$$
$$= S_0 h(-\tau) \quad \tau \leq 0 \quad \text{(8–37)}$$

It is the result shown in (8–36) that leads to the procedure for measuring the impulse response, which will be discussed next.

Consider the block diagram shown in Figure 8–5. The input signal $X(t)$ is a sample function from a random process whose bandwidth is large compared to the bandwidth of the system to be measured. In practice, a bandwidth ratio of 10 to 1 gives very good results. For purposes of analysis this input will be assumed to be white.

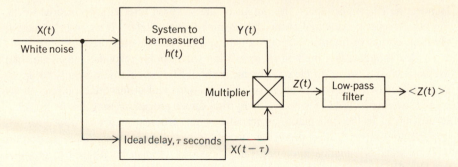

Figure 8–5 Method for measuring the impulse response of a linear system.

In addition to being applied to the system under test, this input signal is also delayed by τ seconds. If the complete impulse response is desired, then τ must be variable over a range from zero to a value at which the impulse response has become negligibly small. Several different techniques exist for obtaining such delay. An analog technique employs a magnetic recording drum on which the playback head can be displaced by a variable amount around the drum from the recording head. More modern techniques, however, would sample the signals at a rate that is at least twice the signal bandwidth and then delay the samples in a charge-coupled delay line or a switched capacitor delay line. Alternatively, the samples might be quantized into a finite number of amplitude levels (see Sec. 2–7) and then delayed by means of shift registers. For purposes of the present discussion, we simply assume the output of the delay device to be $X(t - \tau)$.

The system output $Y(t)$ and the delay unit output are then multiplied to form $Z(t) = X(t - \tau)Y(t)$, which is then passed through a low-pass filter. If the bandwidth of the lowpass filter is sufficiently small, its output will be mostly just the dc component of $Z(t)$, with a small random component added to it. For an ergodic input process, $Z(t)$ will be ergodic[3] and the dc component of $Z(t)$ (that is, its time average) will be the same as its expected value. Thus,

$$\langle Z(t) \rangle \simeq E[Z(t)] = E[Y(t)X(t - \tau)] = R_{XY}(\tau) \qquad (8\text{–}38)$$

since in the stationary case

$$E[Y(t)X(t - \tau)] = E[X(t)Y(t + \tau)] = R_{XY}(\tau) \qquad (8\text{–}39)$$

But from (8–36), it is seen that

$$\langle Z(t) \rangle \simeq S_0 h(\tau) \qquad \tau \geq 0$$

$$\simeq 0 \qquad \tau < 0$$

---

[3]This is true for a time-invariant system and a fixed delay τ.

Hence, the dc component at the output of the lowpass filter is proportional to the impulse response evaluated at the τ determined by the delay. If τ can be changed, then the complete impulse response of the system can be measured.

At first thought, this method of measuring the impulse response may seem like the hard way to solve an easy problem; it should be much easier simply to apply an impulse (or a reasonable approximation thereto) and observe the output. However, there are at least two reasons why this direct procedure may not be possible or desirable. In the first place, an impulse with sufficient area to produce an observable output may also drive the system into a region of *nonlinear* operation well outside its intended operating range. Second, it may be desired to monitor the impulse response of the system continuously while it is in normal operation. Repeated applications of impulses may seriously affect this normal operation. In the crosscorrelation method, however, the random input signal can usually be made small enough to have a negligible effect on the operation.

Some practical engineering situations in which this method has been successfully used include automatic control systems, chemical process control, and measurement of aircraft characteristics in flight. One of the more exotic applications is the continuous monitoring of the impulse response of a nuclear reactor in order to observe how close it is to critical–that is, unstable. It is also being used to measure the dynamic response of large buildings to earth tremors or wind gusts.

Exercise 8–5.1

For the white noise input and system impulse response of Exercise 8–4.1, evaluate both crosscorrelation functions at the same values of τ.

Answers: 0, 0, 0, 0, 1.25, 1.25

Exercise 8–5.2

For the input noise and system inpulse response of Exercise 8–4.2, evaluate both crosscorrelation functions for the same values of τ.

Answers: −0.4167, −0.1731, 0.0157, 0.1160, 0.8571, 0.8571

8–6 Examples of Time-Domain System Analysis

A simple RC circuit responding to a random input having an exponential autocorrelation function was analyzed in Section 8–4 and was found to involve an appreciable amount of labor. Actually, systems and inputs such as these are usually handled more conveniently by the frequency-domain methods that are discussed later in this chapter. Hence, it seems desirable to look at some situation in which time-domain methods are easier. These situations occur when the impulse response and autocorrelation function have a simple form over a finite time interval.

The system chosen for this example is the *finite-time integrator,* whose impulse response is shown in Figure 8–6(a). The input is assumed to have an autocorrelation function of the form shown in Figure 8–6(b). This autocorrelation function might come from the random binary process discussed in Section 6–2, for example.

For the particular input specified, the output of the finite-time integrator will have zero mean, since $\overline{X}$ is zero. In the more general case, however, the mean value of the output would be, from (8–7),

$$\overline{Y} = \overline{X} \int_0^T \frac{1}{T}dt = \overline{X} \tag{8–40}$$

Since the input process is not white, (8–10) must be used to determine the mean-square value of the output. Thus,

$$\overline{Y^2} = \int_0^T d\lambda_1 \int_0^T R_X(\lambda_2 - \lambda_1)\left(\frac{1}{T}\right)^2 d\lambda_2 \tag{8–41}$$

As an aid in evaluating this integral, it is helpful to sketch the integrand as shown in Figure 8–7 and note that the mean-square value is just the *volume* of the region indicated. Since this volume is composed of the volumes of two right pyramids,

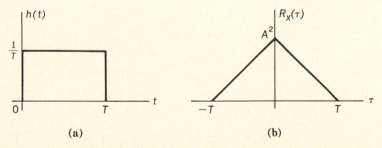

(a)   (b)

Figure 8–6 (a) Impulse response of finite-time integrator and (b) input autocorrelation function.

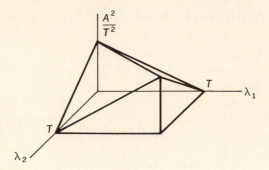

Figure 8–7   Integrand of (8–41)

each having a base of A^2/T^2 by $\sqrt{2}\,T$ and an altitude of $T/\sqrt{2}$, the total volume is seen to be

$$\overline{Y^2} = 2\left\{\frac{1}{3}\right\}\left(\frac{A^2}{T^2}\right)(\sqrt{2}\,T)\left(\frac{T}{\sqrt{2}}\right) = \frac{2}{3}A^2 \tag{8–42}$$

It is also possible to obtain the autocorrelation function of the output by using (8–17). Thus,

$$R_Y(\tau) = \int_0^T d\lambda_1 \int_0^T R_X(\lambda_2 - \lambda_1 - \tau)\left(\frac{1}{T}\right)^2 d\lambda_2 \tag{8–43}$$

It is left as an exercise for the reader to show that this has the shape shown in Figure 8–8 and is composed of segments of cubics.

It may be noted that the results become even simpler when the input random process can be treated as if it were white noise. Thus, using the special case derived in (8–12), the mean-square value of the output would be

$$\overline{Y^2} = S_0 \int_0^T \left(\frac{1}{T}\right)^2 d\lambda = \frac{S_0}{T} \tag{8–44}$$

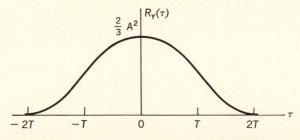

Figure 8–8   Autocorrelation function of the output of the finite-time integrator.

where S_0 is the spectral density of the input white noise. Furthermore, from the special case derived in (8–18), the output autocorrelation function can be sketched by inspection, as shown in Figure 8–9, since it is just the time correlation of the system impulse response with itself. Note that this result indicates another way in which a random process having a triangular autocorrelation function might be generated.

The second example utilizes the result of (8–14) to determine the amount of filtering required to make a good measurement of a small dc voltage in the presence of a large noise signal. Such a situation might arise in any system that attempts to measure the crosscorrelation between two signals that are only slightly correlated. Specifically, it is assumed that the signal appearing at the input to the RC circuit of Figure 8–2 has the form

$$X(t) = A + N(t)$$

where the noise $N(t)$ has an autocorrelation function of

$$R_N(\tau) = 10e^{-1000|\tau|}$$

It is desired to measure A with an rms error of 1 percent when A itself is on the order of 1 and it is necessary to determine the time-constant of the RC filter required to achieve this degree of accuracy.

Although an exact solution could be obtained by using the results of the exact analysis that culminated in (8–24), this approach is needlessly complicated. If it is recognized that the variance of the noise at the output of the filter must be very much smaller than that at the input, then it is clear that the bandwidth of the filter must also be very much smaller than the bandwidth of the input noise. Under these conditions the white-noise assumption for the input must be a very good approximation.

The first step in using this approximation is to find the spectral density of the noise in the vicinity of $\omega = 0$, since only frequency components in this region will be passed by the RC filter. Although this spectral density can be obtained directly by analogy to (8–22), the more general approach is employed here. It was shown in (7–40) that the spectral density is related to the autocorrelation function by

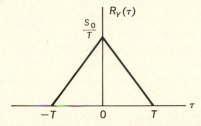

Figure 8–9 Output autocorrelation function with white-noise input.

$$S_N(\omega) = \int_{-\infty}^{\infty} R_N(\tau)e^{-j\omega\tau}\, d\tau$$

At $\omega = 0$, this becomes

$$S_N(0) = \int_{-\infty}^{\infty} R_N(\tau)\, d\tau = 2\int_0^{\infty} R_N(\tau)\, d\tau \qquad (8\text{--}45)$$

Hence, the spectral density of the assumed white-noise input would be this same value, that is, $S_N = S_N(0)$. Note that (8–45) is a general result that does not depend upon the form of the autocorrelation function. In the particular case being considered here, it follows that

$$S_N = 2(10)\int_0^{\infty} e^{-1000\tau}\, d\tau = \frac{20}{1000} = 0.02$$

From (8–14) it is seen that the mean-square of the filter output, $N_o(t)$, will be

$$\overline{N_o^2} = \frac{bS_N}{2} = \frac{b(0.02)}{2} = 0.01b$$

In order to achieve the desired accuracy of 1 percent it is necessary that

$$\sqrt{\overline{N_o^2}} \le (0.01)(1.0)$$

when A is 1.0, since the dc gain of this filter is unity. Thus,

$$\overline{N_o^2} = 0.01b \le 10^{-4}$$

so that

$$b \le 10^{-2}$$

Since $b = 1/RC$, it follows that

$$RC \ge 10^2$$

in order to obtain the desired accuracy.

It was noted that crosscorrelation of the input and output of a linear system yields an estimate of the system impulse response when the input has bandwidth that is large compared to the bandwidth of the system. Usually the crosscorrelation operation is carried out by sampling the input time function and the output time function, delaying the samples of the input, and then averaging the product of the delayed samples of the input and the samples of the output. A block diagram of a system that accomplishes this is shown in Figure 8–10. In order to analyze this method in more detail, let samples of the input time function, $X(t)$, be designated as

$$X_k = X(k\Delta t) \qquad k = 1, 2, \ldots, N$$

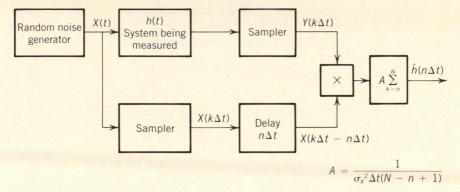

$$A = \frac{1}{\sigma_x^2 \Delta t(N - n + 1)}$$

Figure 8–10 Block diagram of a system that will estimate the impulse response of a linear system.

where Δt is the time interval between samples. In a similar way, let samples of the output time function be designated as

$$Y_k = Y(k\Delta t) \qquad k = 1, 2, \ldots, N$$

An estimate of the nth sample of the crosscorrelation function between the input and output is obtained from

$$\hat{R}_{XY}(n\Delta\tau) = \frac{1}{N - n + 1} \sum_{k=n}^{N} X_{k-n}Y_k \qquad n = 0, 1, 2, \ldots, M \ll N$$

In order to relate this estimate to an estimate of the system impulse response, it is necessary to relate the variance of the samples of $X(t)$ to the spectral density of the random process from which they came. If the bandwidth of the input process is sufficiently large that samples spaced Δt seconds apart may be assumed to be statistically independent, these samples can be imagined to have come from a bandlimited white process whose bandwidth is $1/2\Delta t$. Since the variance of such a white bandlimited process is just $2S_0 W$, it follows that $S_0 = \sigma_x^2 \Delta t$. It doesn't matter what the actual spectral density is. *Independent* samples from any process are indistinguishable from independent samples from a white bandlimited process having the same variance. Thus, an estimate of the system impulse response is, from (8–36), given by

$$\hat{h}(n\Delta t) = \frac{1}{\sigma_x^2 \Delta t} \hat{R}_{XY}(n\Delta t)$$

$$= \frac{1}{\sigma_x^2 \Delta t(N - n + 1)} \sum_{k=n}^{N} X_{k-n}Y_k \qquad (8\text{–}46)$$

$$n = 0, 1, 2, \ldots, M \ll N$$

By taking the expected value of (8–46) it is straightforward to show that this is an unbiased estimate of the system impulse response. Furthermore, it can be shown that the variance of this estimate is bounded by

$$\text{Var}\,[\hat{h}(n\Delta t)] \leq \frac{2}{N} \sum_{k=0}^{M} h^2(k\Delta t) \tag{8–47}$$

Often it is more convenient to replace the summation in (8–47) by

$$\sum_{k=0}^{M} h^2(k\Delta t) \leq \frac{1}{\Delta t} \int_0^{\infty} h^2(t)\, dt \tag{8–48}$$

Note that the bound on the variance of the estimate does not depend upon which sample of the impulse response is being estimated.

The above results are useful in determining how many samples of the input and output are necessary to achieve a given accuracy in the estimation of an impulse response. In order to illustrate this, assume it is desired to estimate an impulse response of the form

$$h(t) = 5e^{-5t} \sin 20t\, u(t)$$

with an error of less than 1 percent of the maximum value of the impulse response. Since the maximum value of this impulse response is about 3.376 at $t = 0.0785$, the variance of the estimate should be less than $(0.01 \times 3.376)^2 = 0.0011$. Furthermore,

$$\int_0^{\infty} h^2(t)\, dt = 1.25$$

Thus, from (8–47) and (8–48), the number of samples required to achieve the desired accuracy is bounded by

$$N \geq \frac{2 \times 1.25}{0.0011} \geq 2193$$

The selection of Δt is governed by the desired number of points, M, at which $h(t)$ is to be estimated and by the length of the time interval over which $h(t)$ has a significant magnitude. To illustrate this, suppose that in the above example it is desired to estimate the impulse response at 50 points over the range in which it is greater than 1 percent of its maximum value. Since the sine function can never be greater than 1.0, it follows that $5e^{-5t} \geq 0.01 \times 3.376$ implies that the greatest delay interval that must be considered is about 1 second. Thus, a value $\Delta t = 1/50 = 0.02$ seconds should be adequate. The bandwidth of an ideal white bandlimited source that would provide independent samples at intervals 0.02 seconds apart is 25 Hz, but a more practical source should probably have half-power bandwidth of about 250 Hz to guarantee the desired independence.

Exercise 8–6.1

White noise having a two-sided spectral density of 0.01 is applied to the input of a finite-time integrator having an impulse response of

$$h(t) = \frac{1}{3}[u(t) - u(t - 3)]$$

Find the value of the autocorrelation function of the output at

a) $\tau = 0$

b) $\tau = 1$

c) $\tau = 2.$

Answers: 0.1111, 0.2222, 0.3333

Exercise 8–6.2

A dc signal plus noise has sample functions of the form

$$X(t) = A + N(t)$$

where $N(t)$ has an autocorrelation function of the form

$$R_N(\tau) = 1 - \frac{|\tau|}{0.01} \qquad |\tau| \le 0.01$$

A finite-time integrator is used to estimate the value of A with an rms error or less than 0.02. If the impulse response of this integrator is

$$h(t) = \frac{1}{T}[u(t) - u(t - T)]$$

find the value of T required to accomplish this.

Answer: 25

8–7   Analysis in the Frequency Domain

The most common method of representing linear systems in the frequency domain is in terms of the system function $H(\omega)$ or the transfer function $H(s)$, which are the Fourier and Laplace transforms, respectively, of the system impulse response.

If the input to a system is $x(t)$ and the output $y(t)$, then the Fourier transforms of these quantities are related by

$$Y(\omega) = X(\omega)H(\omega) \tag{8–49}$$

and the Laplace transforms are related by

$$Y(s) = X(s)H(s) \tag{8–50}$$

provided the transforms exist. Neither of these forms is suitable when $X(t)$ is a sample function from a stationary random process. As discussed in Section 7–1, the Fourier transform of a sample function from a stationary random process generally never exists. In the case of the one-sided Laplace transform the input-output relationship is defined only for time functions existing for $t > 0$, and such time functions can never be sample functions from a stationary random process.

One approach to this problem is to make use of the spectral density of the process and to carry out the analysis using a truncated sample function in which the limit $T \to \infty$ is not taken until after the averaging operations are carried out. This procedure is valid and leads to correct results. There is, however, a much simpler procedure that can be used. In Section 7–6 it was shown that the spectral density of a stationary random process is the Fourier transform of the autocorrelation function of the process. Therefore, using the results we have already obtained for the correlation function of the output of a linear time-invariant system, we can obtain the corresponding results for the spectral density by carrying out the required transformations. When the basic relationship has been obtained, it will be seen that there is a close analogy between computations involving nonrandom signals and those involving random signals.

8–8   Spectral Density at the System Output

The spectral density of a process is a measure of how the average power of the process is distributed with respect to frequency. No information regarding the phases of the various frequency components is contained in the spectral density. The relationship between the spectral density $S_X(\omega)$ and the autocorrelation function $R_X(\tau)$ for a stationary process was shown to be

$$S_X(\omega) = \mathcal{F}\{R_X(\tau)\} \tag{8–51}$$

Using this relationship and (8–17), which relates the output autocorrelation function $R_Y(\tau)$ to the input correlation function $R_X(\tau)$ by means of the system impulse response, we have

$$R_Y(\tau) = \int_0^\infty d\lambda_1 \int_0^\infty R_X(\lambda_2 - \lambda_1 - \tau)h(\lambda_1)h(\lambda_2)\,d\lambda_2$$

$$S_Y(\omega) = \mathscr{F}\{R_Y(\tau)\}$$

$$= \int_{-\infty}^{\infty} \left[\int_0^{\infty} d\lambda_1 \int_0^{\infty} R_X(\lambda_2 - \lambda_1 - \tau)h(\lambda_1)h(\lambda_2) \, d\lambda_2 \right] e^{-j\omega\tau} \, d\tau$$

Interchanging the order of integration and carrying out the indicated operations gives

$$S_Y(\omega) = \int_0^{\infty} d\lambda_1 \int_0^{\infty} h(\lambda_1)h(\lambda_2) \, d\lambda_2 \int_{-\infty}^{\infty} R_X(\lambda_2 - \lambda_1 - \tau)e^{-j\omega\tau} \, d\tau$$

$$= \int_0^{\infty} d\lambda_1 \int_0^{\infty} h(\lambda_1)h(\lambda_2)S_X(\omega)e^{-j\omega(\lambda_2-\lambda_1)} \, d\lambda_2$$

$$= S_X(\omega) \int_0^{\infty} h(\lambda_1)e^{j\omega\lambda_1} \, d\lambda_1 \int_0^{\infty} h(\lambda_2)e^{-j\omega\lambda_2} \, d\lambda_2 \qquad \text{(8–52)}$$

$$= S_X(\omega) \, H(-\omega)H(\omega)$$

$$= S_X(\omega)|H(\omega)|^2$$

In arriving at (8–52) use was made of the property that $R_X(-\tau) = R_X(\tau)$.

From (8–52) it is seen that the output spectral density is related to the input spectral density by the power transfer function, $|H(\omega)|^2$. This result can also be expressed in terms of the complex frequency s as

$$S_Y(s) = S_X(s)H(s)H(-s) \qquad \text{(8–53)}$$

where $S_Y(s)$ and $S_X(s)$ are obtained from $S_Y(\omega)$ and $S_X(\omega)$ by substituting $-s^2 = \omega^2$, and where $H(s)$ is obtained from $H(\omega)$ by substituting $s = j\omega$. It is this form that will be used in further discussions of frequency analysis methods.

It is clear from (8–53) that the quantity $H(s)H(-s)$ plays the same role in relating input and output spectral densities as $H(s)$ does in relating input and output transforms. This similarity makes the use of frequency domain techniques for systems with rational transfer funtions very convenient when the input is a sample function from a *stationary random process*. However, this same technique is *not* always applicable when the input process is *nonstationary*, even though the definition for the spectral density of such processes is the same as we have employed. A detailed study of this matter is beyond the scope of the present discussion but the reader would do well to question any application of (8–53) for nonstationary processes.

Since the spectral density of the system has now been obtained, it is a simple matter to determine the mean-square value of the output. This is simply

$$\overline{Y^2} = \frac{1}{2\pi j} \int_{-j\infty}^{j\infty} H(s)H(-s)S_X(s) \, ds \qquad \text{(8–54)}$$

and may be evaluated by either of the methods discussed in Section 7–5.

In order to illustrate some of the methods, consider the RC circuit shown in Figure 8–11 and assume that its input is a sample function from a white-noise process having a spectral density of S_0. The spectral density at the output is simply

$$S_Y(s) = \frac{b}{s + b} \cdot \frac{b}{-s + b} \cdot S_0 = \frac{-b^2 S_0}{s^2 - b^2} \qquad (8\text{--}55)$$

The mean-square value of the output can be obtained by using the integral I_1, tabulated in Table 7–1, Section 7–5. In order to do this, it is convenient to write (8–55) as:

$$S_Y(s) = \frac{(b\sqrt{S_0})(b\sqrt{S_0})}{(s + b)(-s + b)}$$

from which it is clear that $n = 1$, and

$$c(s) = b\sqrt{S_0} = c_0$$

$$d(s) = s + b$$

Thus

$$d_0 = b$$

$$d_1 = 1$$

and

$$\overline{Y^2} = I_1 = \frac{c_0{}^2}{2 d_0 d_1} = \frac{b^2 S_0}{2b} = \frac{b S_0}{2} \qquad (8\text{--}56)$$

As a slightly more complicated example, let the input spectral density be

$$S_X(s) = \frac{-\beta^2 S_0}{s^2 - \beta^2} \qquad (8\text{--}57)$$

This spectral density, which corresponds to the autocorrelation function used in Section 8–4, has been selected so that its value at zero frequency is S_0. The spectral density at the output of the RC circuit is now

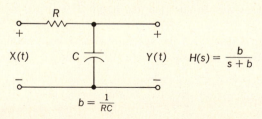

Figure 8–11   A simple RC circuit.

$$S_Y(s) = \frac{b}{s + b} \cdot \frac{b}{-s + b} \cdot \frac{-\beta^2 S_0}{s^2 - \beta^2}$$

$$= \frac{b^2 \beta^2 S_0}{(s^2 - b^2)(s^2 - \beta^2)} \tag{8-58}$$

The mean-square value for this output will be evaluated by using the integral I_2 tabulated in Table 7–1. Thus,

$$S_Y(s) = \frac{c(s)c(-s)}{d(s)d(-s)} = \frac{(b\beta\sqrt{S_0})(b\beta\sqrt{S_0})}{[s^2 + (b + \beta)s + b\beta][s^2 - (b + \beta)s + b\beta]} \tag{8-59}$$

it is clear that $n = 2$, and

$$c_0 = b\beta\sqrt{S_0}$$

$$c_1 = 0$$

$$d_0 = b\beta$$

$$d_1 = b + \beta$$

$$d_2 = 1$$

Hence,

$$\overline{Y^2} = I_2 = \frac{c_0^2 d_2 + c_1^2 d_0}{2d_0 d_1 d_2} = \frac{b^2 \beta^2 S_0}{2b\beta(b + \beta)} = \frac{b\beta S_0}{2(b + \beta)} \tag{8-60}$$

since $c_1 = 0$.

It is also of interest to look once again at the results when the input random process has a bandwidth much greater than the system bandwidth; that is, when $\beta \gg b$. From (8–58) it is clear that

$$S_Y(s) = \frac{-b^2 S_0}{(s^2 - b^2)(1 - s^2/\beta^2)} \tag{8-61}$$

and as β becomes large this spectral density approaches that for the white-input-noise case given by (8–55). In the case of the mean-square value, (8–60) may be written as

$$\overline{Y^2} = \frac{bS_0}{2(1 + b/\beta)} \tag{8-62}$$

which approaches the white-noise result of (8–56) when β is large.

Comparison of the foregoing examples with similar ones employing time-domain methods should make it evident that when the input spectral density and the system transfer function are rational, frequency domain methods are usually simpler. In fact, the more complicated the system, the greater the advantage of such methods. When either the input spectral density or the system transfer function is not rational, this conclusion may not hold.

Exercise 8–8.1

White noise having a two-sided spectral density of 2 V²/Hz is applied to the input of a linear system having an impulse response of

$$h(t) = te^{-3t} u(t)$$

a)   Find the value of the output spectral density at $\omega = 0$.

b)   Find the value of the output spectral density at $\omega = 3$.

c)   Find the mean-square value of the output.

Answers: 0.00617, 0.0185, 0.0247

Exercise 8–8.2

Find the mean-square value of the output of the system in Exercise 8–8.1 if the input has a spectral density of

$$S_X(\omega) = \frac{1800}{\omega^2 + 900}$$

Answer: 0.0185

8–9   Cross-Spectral Densities Between Input and Output

The cross-spectral densities between a system input and output are not widely used, but it is well to be aware of their existence. The derivation of these quantities would follow the same general pattern as shown above, but only the end results are quoted here. Specifically, they are

$$S_{XY}(s) = H(s)S_X(s) \qquad (8\text{–}63)$$

and

$$S_{YX}(s) = H(-s)S_X(s) \qquad (8\text{–}64)$$

The cross-spectral densities are related to the crosscorrelation functions between input and output in exactly the same way as ordinary spectral densities and autocorrelation functions are related. Thus,

$$S_{XY}(s) = \int_{-\infty}^{\infty} R_{XY}(\tau)e^{-s\tau}\, d\tau$$

and

$$S_{YX}(s) = \int_{-\infty}^{\infty} R_{YX}(\tau)e^{-s\tau}\, d\tau$$

Likewise, the inverse two-sided Laplace transform can be used to find the crosscorrelation functions from the cross-spectral densities, but these relations will not be repeated here. As noted in Section 7–8, it is not necessary that cross-spectral densities be even functions of ω, or that they be real and positive.

To illustrate the above results, we consider again the circuit of Figure 8–11 with an input of white noise having a two-sided spectral density of S_0. From (8–63) and (8–64) the two cross-spectral densities are

$$S_{XY}(s) = \frac{bS_0}{s + b}$$

and

$$S_{YX}(s) = \frac{bS_0}{-s + b}$$

If these are expressed as functions of ω by letting $s = j\omega$, it is obvious that the cross-spectral densities are not real, even, positive functions of ω. Clearly, similar results can be obtained for any other input spectral density.

Exercise 8–9.1

White noise having a two-sided spectral density of 0.5 V²/Hz is applied to the input of a finite-time integrator whose impulse response is

$$h(t) = [u(t) - u(t - 1)]$$

Find the values of both cross-spectral densities at

a) $\omega = 0$

b) $\omega = 0.5$

c) $\omega = 1.$

Answers: 0.5, 0.4794 $\pm j$0.1224, 0.4207 $\pm j$0.2298

Exercise 8–9.2

$X(t)$ and $Y(t)$ are from independent random processes having identical spectral densities of

$$S_X(s) = S_Y(s) = \frac{-1}{s^2 - 1}$$

a) Find both cross-spectral densities, $S_{XY}(s)$ and $S_{YX}(s)$.

b) Find both cross-spectral densities, $S_{UV}(s)$ and $S_{VU}(s)$ where $U(t) = X(t) + Y(t)$ and $V(t) = X(t) - Y(t)$.

Answers: 0, 0, 0, 0

8–10 Examples of Frequency-Domain Analysis

Frequency-domain methods tend to be most useful when dealing with conventional filters and random processes that have rational spectral densities. However, it is often possible to make the calculations even simpler, without introducing much error, by idealizing the filter characteristics and assuming the input processes to be white. An important concept in doing this is that of *equivalent-noise bandwidth*.

The equivalent-noise bandwidth, B, of a system is defined to be the bandwidth of an ideal filter that has the same maximum gain and the same mean-square value at its output as the actual system when the input is white noise. This concept is illustrated in Figure 8–12 for both lowpass and bandpass systems. It is clear that the rectangular power transfer function of the ideal filter must have the same *area*

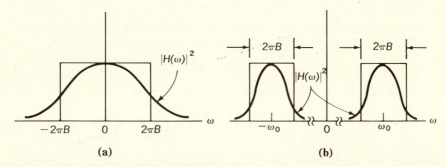

(a)   (b)

Figure 8–12 Equivalent-noise bandwidth of systems: (a) lowpass system and (b) bandpass system.

as the power transfer function of the actual system if they are to produce the same mean-square outputs with the same white-noise input. Thus, in the low pass case, the equivalent-noise bandwidth is given by

$$B = \frac{1}{4\pi|H(0)|^2} \int_{-\infty}^{\infty} |H(\omega)|^2 \, d\omega = \frac{1}{4\pi j|H(0)|^2} \int_{-j\infty}^{j\infty} H(s)H(-s) \, ds \quad \text{Hz} \quad \textbf{(8–65)}$$

If the input to the system is white noise with a spectral density of S_0, the mean-square value of the output is given by

$$\overline{Y^2} = 2S_0 B |H(0)|^2 \tag{8–66}$$

In the band pass case, $|H(0)|^2$ is replaced by $|H(\omega_0)|^2$ in both (8–65) and (8–66).

As a simple illustration of the calculation of equivalent noise bandwidth, consider the RC circuit of Figure 8–11. Since the integral of (8–65) has already been evaluated in obtaining the mean-square value of (8–56), it is easiest to use this result and (8–66). Thus,

$$\overline{Y^2} = \frac{bS_0}{2} = 2S_0 B |H(0)|^2$$

Since $|H(0)|^2 = 1$, it follows that

$$B = \frac{b}{4} = \frac{1}{4RC} \tag{8–67}$$

It is of interest to compare the equivalent-noise bandwidth of (8–67) with the more familiar half-power bandwidth. For a lowpass system, such as this RC circuit, the half-power bandwidth is defined to be that frequency at which the magnitude of the transfer function drops to $1/\sqrt{2}$ of its value at zero frequency. For this RC filter the half-power bandwidth is simply

$$B_{1/2} = \frac{1}{2\pi RC}$$

Hence, the equivalent-noise bandwidth is just $\pi/2$ times the half-power bandwidth for this particular circuit. If the transfer function of a system has steeper sides, then the equivalent-noise bandwidth and the half-power bandwidth are more nearly equal.

It is also possible to express the equivalent-noise bandwidth in terms of the system impulse response rather than the transfer function. Note first that

$$H(0) = \int_0^{\infty} h(t) \, dt$$

Then apply Parseval's theorem to the integral in (8–65).

$$\int_{-\infty}^{\infty} h(t)^2 dt = \frac{1}{2\pi} \int_{-\infty}^{\infty} |H(\omega)|^2 d\omega$$

Using these relations, the equivalent-noise bandwidth becomes

$$B = \frac{\displaystyle\int_0^\infty h^2(t)\ dt}{2\left[\displaystyle\int_0^\infty h(t)\ dt\right]^2} \tag{8-68}$$

The time-domain representation of equivalent-noise bandwidth may be simpler to use than the frequency-domain representation for systems having nonrational transfer functions. To illustrate this, consider the finite-time integrator defined as usual by the impulse response

$$h(t) = \frac{1}{T}[u(t) - u(t - T)]$$

Thus,

$$\int_0^\infty h(t)dt = \frac{1}{T}\,T = 1$$

and

$$\int_0^\infty h^2(t)dt = \frac{1}{T^2}T = \frac{1}{T}$$

Hence, the equivalent-noise bandwidth is

$$B = \frac{1/T}{2(1)^2} = \frac{1}{2T}$$

It is also of interest to relate this equivalent-noise bandwidth to the half-power bandwidth of the finite-time integrator. From the Fourier transform of $h(t)$ the transfer function becomes

$$|H(\omega)| = \frac{\sin \omega T}{\omega T}$$

and this has a half-power point at $B_{1/2} = 0.221/T$. Thus,

$$B = 2.26 B_{1/2}$$

One advantage of using equivalent-noise bandwidth is that it becomes possible to describe the noise response of even very complicated systems with only two numbers, B and $|H(\omega_0)|$.

Furthermore, these numbers can be measured quite readily in an experimental system. For example, suppose that a receiver in a communication system is measured and found to have a voltage gain of 10^6 at the frequency to which it is tuned and an equivalent-noise bandwidth of 10 kHz. The noise at the input to this receiver, from shot noise and thermal agitation, has a bandwidth of several hundred megahertz and, hence, can be assumed to be white over the bandwidth of the receiver. Suppose this noise has a spectral density of 2×10^{-20} V^2/Hz. (This is a realistic value for the input circuit of a high-quality receiver.) What should the effective value of the input signal be in order to achieve an output signal-to-noise power ratio of 100? The answer to this question would be very difficult to find if every stage of the receiver had to be analyzed exactly. It is very easy, however, using the equivalent-noise bandwidth since

$$(S/N)_0 = \frac{|H(\omega_0)|^2 \overline{X^2}}{2N_0 B |H(\omega_0)|^2} = \frac{\overline{X^2}}{2N_0 B} \qquad (8\text{–}69)$$

if $\overline{X^2}$ is the mean-square value of the input signal and N_0 is the spectral density of the input noise. Thus,

$$\frac{\overline{X^2}}{2N_0 B} = 100$$

and

$$\overline{X^2} = 2N_0 B(100) = 2(2 \times 10^{-20})(10^4)(100)$$
$$= 4 \times 10^{-14}$$

from which

$$\sqrt{\overline{X^2}} = 2 \times 10^{-7} \text{ V}$$

is the effective signal voltage being sought. Note that the actual value of receiver gain, although specified, was not needed to find the output signal-to-noise ratio.

It should be emphasized that the equivalent-noise bandwidth is useful only when one is justified in assuming that the spectral density of the input random process is white. If the input spectral density changes appreciably over the range of frequencies passed by the system, then significant errors can result from employing the concept.

The final example of frequency-domain analysis will consider the feedback system shown in Figure 8–13. This system might be a control system for positioning a radar antenna, in which $x(t)$ is the input control angle (assumed to be random since target position is unknown in advance) and $y(t)$ is the angular position of the antenna in response to this voltage. The disturbance $n(t)$ might represent the effects of wind blowing on the antenna, thus producing random perturbations on the angular position. The transfer function of the amplifier and motor within the feedback loop is

$$H(s) = \frac{A}{s(s + 1)}$$

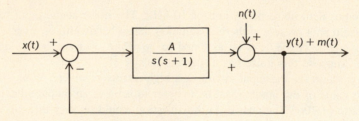

Figure 8–13 An automatic control system.

The transfer function relating $X(s) = \mathscr{L}[x(t)]$ and $Y(s) = \mathscr{L}[y(t)]$ can be obtained by letting $n(t) = 0$ and noting that

$$Y(s) = H(s)[X(s) - Y(s)]$$

since the input to the amplifier is the difference between the input control signal and the system output. Hence,

$$H_c(s) = \frac{Y(s)}{X(s)} = \frac{H(s)}{1 + H(s)} \tag{8–70}$$

$$= \frac{A}{s^2 + s + A}$$

If the spectral density of the input control signal (now considered to be a sample function from a random process) is

$$S_X(s) = \frac{-2}{s^2 - 1}$$

then the spectral density of the output is

$$S_Y(s) = S_X(s)H_c(s)H_c(-s) \tag{8–71}$$

$$= \frac{-2A^2}{(s^2 - 1)(s^2 + s + A)(s^2 - s + A)}$$

The mean-square value of the output is given by

$$\overline{Y^2} = \frac{2A^2}{2\pi j} \int_{-j\infty}^{j\infty} \frac{ds}{[s^3 + 2s^2 + (A + 1)s + A][-s^3 + 2s^2 - (A + 1)s + A]}$$

$$= 2A^2 I_3$$

in which

$$c_0 = 1, \qquad c_1 = 0, \qquad c_2 = 0$$

$$d_0 = A, \qquad d_1 = A + 1, \qquad d_2 = 2, \qquad d_3 = 1$$

From Table 7–1, this becomes

$$\overline{Y^2} = \frac{2A}{A + 2} \tag{8-72}$$

The transfer function relating $N(s) = \mathcal{L}[n(t)]$ and $M(s) = \mathcal{L}[m(t)]$ is not the same as (8–70) because the disturbance enters the system at a different point. It is apparent, however, that

$$M(s) = N(s) - H(s)M(s)$$

from which

$$H_n(s) = \frac{M(s)}{N(s)} = \frac{1}{1 + H(s)} \tag{8-73}$$

$$= \frac{s(s + 1)}{s^2 + s + A}$$

Let the interfering noise have a spectral density of

$$S_N(s) = \delta(s) - \frac{1}{s^2 - 0.25}$$

This corresponds to an input disturbance that has an average value as well as a random variation. The spectral density of the output disturbance becomes

$$S_M(s) = S_N(s)H_n(s)H_n(-s) \tag{8-74}$$

$$= \left[\delta(s) - \frac{1}{s^2 - 0.25} \right] \frac{s^2(s^2 - 1)}{(s^2 + s + A)(s^2 - s + A)}$$

The mean-square value of the output disturbance comes from

$$\overline{M^2} = \frac{1}{2\pi j} \int_{-j\infty}^{j\infty} \left[\delta(s) - \frac{1}{s^2 - 0.25} \right] \left[\frac{s^2(s^2 - 1)}{(s^2 + s + A)(s^2 - s + A)} \right] ds$$

Since the integrand vanishes at $s = 0$, the integral over $\delta(s)$ does not contribute anything to the mean-square value. The remaining terms are

$$\overline{M^2} = \frac{1}{2\pi j} \int_{-j\infty}^{j\infty} \frac{s(s + 1)(-s)(-s + 1)}{[s^3 + 1.5s^2 + (A + 0.5)s + 0.5A] \times} ds$$
$$[-s^3 + 1.5s^2 - (A + 0.5)s + 0.5A]$$

$$= I_3$$

The constants required for Table 7–1 are

$$c_0 = 0 \qquad c_1 = 1 \qquad c_2 = 1$$

$$d_0 = 0.5A \qquad d_1 = (A + 0.5) \qquad d_2 = 1.5 \qquad d_3 = 1$$

and the mean-square value becomes

$$\overline{M^2} = \frac{A + 1.5}{2A + 1.5} \tag{8-75}$$

The amplifier gain A has not been specified in the example in order that the effects of changing this gain can be made evident. It is clear from (8–72) and (8–75) that the desired signal mean-square value increases with larger values of A while the undesired noise mean-square value decreases. Thus, one would expect that large values of A would be desirable if output signal-to-noise ratio is the important criterion. In actuality, the dynamic response of the system to rapid input changes may be more important and this would limit the value of A that can be used.

Exercise 8–10.1

Find the equivalent-noise bandwidth of the transfer function

$$|H(\omega)| = \frac{1}{\left[1 + \dfrac{(\omega - \omega_0)^2}{B_{1/2}^2}\right]^{1/2}}$$

Answer: $\dfrac{\pi}{2} B_{1/2}$

Exercise 8–10.2

Find the equivalent-noise bandwidth of the system whose impulse response is

$$h(t) = (1 - t)[u(t) - u(t - 1)]$$

Answer: 2/3

PROBLEMS

8–2.1 A deterministic random signal has sample functions of

$$X(t) = M + B \cos(20t + \theta)$$

in which M is a random variable having a Gaussian probability density function with a mean of 5 and a variance of 64, B is a random variable having a Rayleigh probability density function with a mean-square value of 32, and θ is a random variable that is uniformly distributed from 0 to 2π. All three random variables are mutually independent. This sample function is the input to a system having an impulse response of

$$h(t) = 10e^{-10t} u(t)$$

a) Write an expression for the output of the system.

b) Find the mean value of the output.

c) Find the mean-square value of the output.

8–2.2 Repeat Problem 8–2.1 if the system impulse response is

$$h(t) = \delta(t) - 10e^{-10t} u(t)$$

8–3.1 A finite-time integrator has an impulse response of

$$h(t) = 1 \qquad 0 \le t \le 0.5$$
$$= 0 \qquad \text{elsewhere}$$

The input to this system is white noise with a two-sided spectral density of $10V^2/Hz$.

a) Find the mean value of the output.

b) Find the mean-square value of the output.

c) Find the variance of the output.

8–3.2 Repeat Problem 8–3.1 if the input to the finite-time integrator is a sample function from a stationary random process having an autocorrelation function of

$$R_X(\tau) = 16e^{-2|\tau|}$$

8–3.3 A sample function from a random process having an autocorrelation function of

$$R_X(\tau) = 16e^{-2|\tau|} + 16$$

is the input to a linear system having an impulse response of

$$h(t) = \delta(t) - 2e^{-2t} u(t)$$

a) Find the mean value of the output.

b) Find the mean-square value of the output.

c) Find the variance of the output.

8–3.4

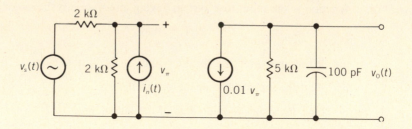

The above circuit models a single-stage transistor amplifier including an internal noise source for the transistor. The input signal is

$$v_s(t) = 0.1 \cos 2000\pi t$$

and $i_n(t)$ is a white-noise current source having a spectral density of 2×10^{-16} A²/Hz that models the internal transistor noise. Find the ratio of the mean-square output signal to the mean-square output noise.

8–4.1

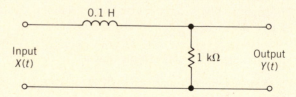

White noise having a spectral density of 10^{-4} V²/Hz is applied to the input of the above circuit.

a) Find the autocorrelation function of the output.

b) Find the mean-square value of the output.

8–4.2 Repeat Problem 8–4.1 if the input to the system is a sample function from a stationary random process having an autocorrelation function of

$$R_X(\tau) = 2e^{-5000|\tau|}$$

8–4.3 The objective of this problem is to demonstrate a general result that frequently arises in connection with many systems analysis problems. Consider the arbitrary triangular waveform shown on page 323.

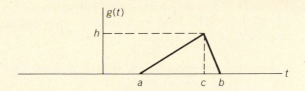

Show that

$$\int_{-\infty}^{\infty} g^2(t)\, dt = \left(\frac{1}{3}\right)h^2(b - a)$$

for any triangle in which $a \le c \le b$.

8–4.4 Consider a linear system having an impulse response of

$$h(t) = [1 - t][u(t) - u(t - 1)]$$

The input to this system is a sample function from a random process having an autocorrelation function of

$$R_X(\tau) = 2\delta(\tau) + 9$$

a) Find the mean value of the output.

b) Find the mean-square value of the output.

c) Find the autocorrelation function of the output.

8–5.1 For the system and input of Problem 8–3.1 find both crosscorrelation functions between input and output.

8–5.2 For the system and input of Problem 8–3.2 find both crosscorrelation functions between input and output.

8–5.3 For the system and input of Problem 8–4.4 find both crosscorrelation functions between input and output.

8–5.4

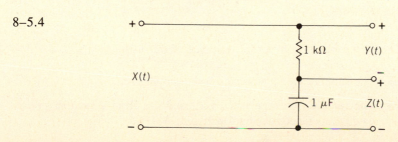

The input $X(t)$ to the above circuit is white noise having a spectral density of 0.1 V²/Hz. Find the crosscorrelation function between the two outputs, $R_{YZ}(\tau)$, for all τ.

8–6.1   A finite-time integrator has an impulse response of

$$h(t) = \frac{1}{T}[u(t) - u(t - T)]$$

The input to this integrator is a sample function from a stationary random process having an autocorrelation function of

$$R_X(\tau) = A^2\left[1 - \frac{|\tau|}{T}\right] \qquad |\tau| \le T$$

$$= 0 \qquad\qquad\quad \text{elsewhere}$$

a)   Find the mean-square value of the output.

b)   Find the autocorrelation function of the output.

8–6.2

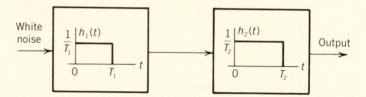

White noise having a spectral density of 0.001 V²/Hz is applied to the input of two finite-time integrators connected in cascade as shown in the figure above. Find the variance of the output if

a) $T_1 = T_2 = 0.1$

b) $T_1 = 0.1$ and $T_2 = 0.01$

c) $T_1 = 0.1$ and $T_2 = 1.0$

8–6.3   It is desired to estimate the mean value of a stationary random process by averaging N samples from the process. That is, let

$$\hat{\bar{X}} = \frac{1}{N}\sum_{n=1}^{N} X_n$$

Derive a general result for the variance of this estimate if:

a)   The samples are uncorrelated from one another.

b)   The samples are separated by Δt and are from a random process having an autocorrelation function of $R_X(\tau)$.

8–6.4   It is desired to estimate the impulse response of a system by sampling the input and output of the system and crosscorrelating the samples. The input samples are independent and have a variance of 2.0. The system impulse response is of the form

$$h(t) = 10te^{-20t}\, u(t)$$

and 60 samples of $h(t)$ are to be estimated in the range in which the impulse response is greater than 2 percent of its maximum value.

a)   Find the time separation between samples.

b)   Find the number of samples required to estimate the impulse response with an rms error less than 1 percent of the maximum value of the impulse response.

c)   Find the total amount of time required to make the measurements.

8–7.1

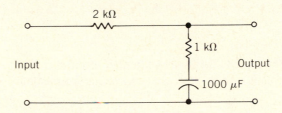

a)   Determine the transfer function, $H(s)$, for the system shown above.

b)   If the input to the system has a Laplace transform of

$$X(s) = \frac{s}{s + 4}$$

find $|Y(s)|^2$ where $Y(s)$ is the Laplace transform of the output of the system.

8–7.2   A three-pole Butterworth filter has poles as shown in the sketch below.

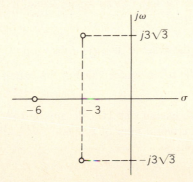

The filter gain at $\omega = 0$ is unity.

a) Write the transfer function $H(\omega)$ for this filter.

b) Write the power transfer function $|H(\omega)|^2$ for this filter.

c) Find $|H(s)|^2$ for this filter.

8–8.1 Find the spectral density of the output of the system of Problem 8–7.1 if the input is:

a) A sample function of white noise having a spectral density of 0.5 V^2/Hz.

b) A sample function from a random process having a spectral density of

$$S_X(\omega) = \frac{\omega^2}{\omega^4 + 5\omega^2 + 4}$$

8–8.2 The input to the Butterworth filter of Problem 8–7.2 is a sample function from a random process having an autocorrelation function of

$$R_X(\tau) = 10e^{-|\tau|}$$

a) Find the spectral density of the output as a function of ω.

b) Find the value of the spectral density at $\omega = 0$.

8–8.3 A linear system has a transfer function of

$$H(s) = \frac{s}{s^2 + 15s + 50}$$

White noise having a mean-square value of 1.2 V^2/Hz is applied to the input.

a) Write the spectral density of the output.

b) Find the mean-square value of the output.

8–8.4 White noise having a spectral density of $0.8 V^2/Hz$ is applied to the input of the Butterworth filter of Problem 8–7.2. Find the mean-square value of the output.

8–9.1 For the system and input of Problem 8–8.2 find both cross-spectral densities for the input and output.

8–9.2

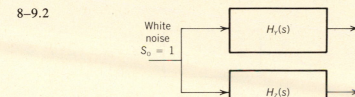

Derive general expressions for the cross-spectral densities $S_{YZ}(s)$ and $S_{ZY}(s)$ for the system shown above.

8–9.3 In the system of Problem 8–9.2 let

$$H_Y(s) = \frac{1}{s + 1}$$

and

$$H_Z(s) = \frac{s}{s + 1}$$

Evaluate both cross-spectral densities between $Y(t)$ and $Z(t)$.

8–10.1 a) Find the equivalent-noise bandwidth of the three-pole Butterworth filter of Problem 8–7.2.
   b) Find the half-power power bandwidth of the Butterworth filter and compare it to the equivalent-noise bandwidth.

8–10.2

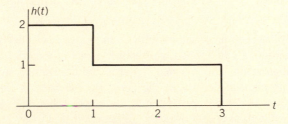

a) Find the equivalent-noise bandwidth of the system whose impulse response is shown above.

    b) If the input to this system is white noise having a spectral density of $2V^2/Hz$, find the mean-square value of the output.

    c) Repeat part (b) using the integral of $h^2(t)$.

8–10.3

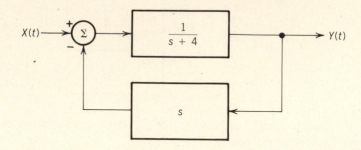

    a) Find the closed-loop transfer function of the control system shown above.

    b) If the input to this control system system is a sample function from a stationary process having a spectral density of

$$S_X(s) = \frac{10}{s + 2}$$

    find the mean-square value of the output.

    c) For the input of part (b), find the mean-square value of the error $X(t) - Y(t)$.

8–10.4   A tuned amplifier has a gain of 40 dB at a frequency of 10.7 MHz and a half-power bandwidth of 1 MHz. The response curve has a shape equivalent to that of a single-stage parallel RLC circuit. It is found that thermal noise at the input to the amplifier produces an rms value at the output of 0.1 V. Find the spectral density of the input thermal noise.

8–10.5   It has been proposed to measure the range to a reflecting object by transmitting a bandlimited white-noise signal at a carrier frequency of f_0 and then adding the received signal to the transmitted signal and measuring the spectral density of the sum. Periodicities in the amplitude of the spectral density are related to the range. Using the system model shown and assuming that α^2 is negligible compared to α, investigate the possibility of this approach. What effect that would adversely affect the measurement has been omitted in the system model?

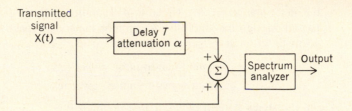

8–10.6   It is frequently useful to approximate the shape of a filter by a Gaussian function of frequency. Determine the standard deviation of a Gaussian shaped lowpass filter that has a maximum gain of unity and a halfpower bandwidth of W Hz. Find the equivalent-noise bandwidth of a Gaussian shaped filter in terms of its half-power bandwidth and in terms of its standard deviation.

8–10.7   The thermal agitation noise generated by a resistance can be closely approximated as white noise having a spectral density of $2kTR$ V^2/Hz, where $k = 1.37 \times 10^{-23}$ W-s/°K is the Boltzmann constant, T is the absolute temperature in degrees Kelvin, and R is the resistance in ohms. Any physical resistance in an amplifier is paralleled by a capacitance, so that the equivalent circuit is as shown.

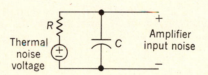

a.   Calculate the mean-square value of the amplifier input noise and show that it is independent of R.

b.   Explain this result on a physical basis.

c.   Show that the maximum noise power available (that is, with a matched load) from a resistance is kTB watts, where B is the equivalent-noise bandwidth over which the power is measured.

8–10.8   Any signal at the input of an amplifier is always accompanied by noise. The minimum noise theoretically possible is the thermal noise present in the resistive component of the input impedance as described in Problem 8–10.7. In general the amplifier will add additional noise in the process of amplifying the signal. The amount of noise is measured in terms of the deterioration of the signal-to-noise ratio of the signal when it is passed through the amplifier. A common method of specifying this characteristic of an amplifier is in terms of a noise figure F, defined as

$$F = \frac{\text{input signal-to-noise power ratio}}{\text{output signal-to-noise power ratio}}$$

a. Using the above definition, show that the overall noise figure for two cascaded amplifiers is $F = F_1 + (F_2 - 1)/G_1$ where the individual amplifiers have power gains of G_1 and G_2 and noise figures of F_1 and F_2 respectively.
b. A particular wide-band video amplifier has a single time constant roll-off with a half-power bandwidth of 100 MHz, a gain of 100 dB, a noise figure of 13 dB, and input and output impedances of 300 Ω. Find the rms output noise voltage when the input signal is zero.
c. Find the amplitude of the input sine wave required to give an output signal-to-noise power ratio of 10 dB.

References

See the references for Chapter 1. Of particular interest for material of this chapter are the books by Davenport and Root and by Lanning and Battin. For a review of system analysis see the following text.

McGillem, C. D. and G. R. Cooper, *Continuous and Discrete Signal and System Analysis.* Second Edition, New York: Holt, Rinehart and Winston, 1984.

CHAPTER 9

Optimum Linear Systems

9–1 Introduction

It was pointed out previously that almost any practical system has some sort of random disturbance introduced into it in addition to the desired signal. The presence of this random disturbance means that the system output is never quite what it should be, and may deviate very considerably from its desired value. When this occurs, it is natural to ask if the system can be modified in any way to reduce the effects of the disturbances. Usually it turns out that it is possible to select a system impulse response or transfer function that minimizes some attribute of the output disturbance. Such a system is said to be *optimum*.

The study of optimum systems for various types of desired signals and various types of noise disturbance is very involved because of the many different situations that can be specified. The literature on the subject is quite extensive and the methods used to determine the optimum system are quite general, quite powerful, and quite beyond the scope of the present discussion. Nevertheless, it is desirable to introduce some of the terminology and a few of the basic concepts in order that the student be aware of some of the possibilities and be in a better position to read the literature.

One of the first steps in the study of optimum systems is a precise definition of what constitutes optimality. Since many different criteria of optimality exist, it is necessary to use some care in selecting an appropriate one. This problem is discussed in the following section.

After a criterion has been selected, the next step is to specify the nature of the system to be considered. Again there are many possibilities, and the ease of carrying out the optimization may be critically dependent upon the choice. Section 9–3 considers this problem briefly.

Once the optimum system has been determined, there remains the problem of evaluating its performance. In some cases this is relatively easy, while in other cases it may be more difficult than actually determining the optimum system. No general treatment of the evaluation problem will be given here; each case considered is handled separately.

In an actual engineering problem, the final step is to decide if the optimum system can be built economically, or whether it will be necessary to approximate it. If it turns out, as it often does, that it is not possible to build the true optimum system, then it is reasonable to question the value of the optimization techniques. Strangely enough, however, it is frequently useful and desirable to carry out the optimizing exercise even though there is no intention of attempting to construct the optimum system. The reason is that the optimum performance provides a yardstick against which the performance of any actual system can be compared. Since the optimum performance cannot be exceeded, this comparison clearly indicates whether any given system needs to be improved or whether its performance is already so close to the optimum that further effort on improving it would be uneconomical. In fact, it is probably this type of comparison that provides the greatest motivation for studying optimum systems since it is only rarely that the true optimum system can actually be constructed.

9–2   Criteria of Optimality

Since there are many different criteria of optimality that might be selected, it is necessary to establish some guidelines as to what constitutes a reasonable criterion. In the first place, it is necessary that the criterion satisfy certain requirements, such as:

1. The criterion must have physical significance and not lead to a trivial result. For example, if the criterion were that of minimizing the output noise power, the obvious result would be a system having *zero* output for both signal and noise. This is clearly a trivial result. On the other hand, a criterion of minimizing the output noise power subject to the constraint of maintaining a given output signal power might be quite reasonable.
2. The criterion must lead to a unique solution. For example, the criterion that the *average error* of the output signal be zero can be satisfied by many systems, not all equally good in regard to the *variance* of the error.
3. The criterion should result in a mathematical form that is capable of being solved. This requirement turns out to be a very stringent one and is the primary reason why so few criteria have found practical application. As a

consequence, the criterion is often selected primarily on this basis even though some other criterion might be more desirable in a given situation.

The choice of a criterion is often influenced by the nature of the input signal—that is, whether it is deterministic or random. The reason for this is that the purpose of the system is usually different for these two types of signals. For example, if the input signal is deterministic, then its form is known and the purpose in observing it is to determine such things as whether it is present or not, the time at which it occurs, how big it is, and so on. On the other hand, when the signal is random its form is unknown and the purpose of the system is usually to determine its form as nearly as possible. In either of these cases there are a number of criteria that might make sense. However, only one criterion for each case is discussed here, and the one selected is the one that is most common and most easily handled mathematically.

In the case of deterministic signals, the criterion of optimality used here is to *maximize* the output signal-to-noise power ratio at some specified time. This criterion is particularly useful when the purpose of the system is to detect the presence of a signal of known shape or to measure the time at which such a signal occurs. There is some flexibility in this criterion with respect to choosing the time at which the signal-to-noise ratio is to be maximized, but reasonable choices are usually apparent from the nature of the signal.

In the case of random signals, the criterion of optimality used here is to *minimize* the mean-square value of the difference between the actual system output and the actual value of the signal being observed. This criterion is particularly useful when the purpose of the system is to observe an unknown signal for purposes of measurement or control. The difference between the output of the system and the true value of the signal consists of two components. One component is the *signal error* and represents the difference between the input and output when there is no input noise. The second component is the output noise, which also represents an error in the output. The total error is the sum of these components, and the quantity to be minimized is the mean-square value of this total error.

Several examples serve to clarify the criteria discussed above and to illustrate situations in which they might arise. Maximum output signal-to-noise ratio is a very commonly used criterion for radar systems. Radar systems operate by periodically transmitting very short bursts of radio-frequency energy. The received signal is simply one or more replicas of the transmitted signal that are created by being reflected from any objects that the transmitted signal illuminates. Thus, the form of the received signals is known exactly. The things that are not known about received signals are the number of reflections, the time delay between the transmitted and received signals, the amplitude, and even whether there is a received signal or not. It can be shown, by methods that are beyond the scope of this text, that the probability of detecting a weak radar signal in the presence of noise or other interference is greatest when the signal-to-noise ratio is greatest.

Thus, the criterion of maximizing the signal-to-noise ratio is an appropriate one with respect to the task the radar system is to perform.

A similar situation arises in digital communication systems. In such a system the message to be transmitted is converted to a sequence of binary symbols, say 0 and 1. Each of the two binary symbols is then represented by time function having a specified form. For example, a negative rectangular pulse might represent a 0 and a positive rectangular pulse represent a 1. At the receiver, it is important that a correct decision be made when each pulse is received as to whether it is positive or negative, and this decision may not be easy to make if there is a large amount of noise present. Again, the probability of making the correct decision is maximized by maximizing the signal-to-noise ratio.

On the other hand, there are many signals of interest in which the form of the signal is not known before it is observed, and the signals can only be observed in the presence of noise. For example, in an analog communication system the messages, such as speech or music, are not converted to binary signals but are transmitted in their original form after an appropriate modulation process. At the receiver, it is desired to recover these messages in a form that is as close to the original message as possible. In this case, minimizing the mean-square error between the received message and the transmitted message is the appropriate criterion. Another situation in which this is the appropriate criterion is the measurement of biological signals such as is done for electrotroencephalograms and electrocardiograms. Here it is important that an accurate representation of the signal be obtained and that the effects of noise be minimized as much as possible.

The above discussion may be summarized more succinctly by the following statements that are true in general:

a) To determine the presence or absence of signal of known form, use the maximum output signal-to-noise ratio criterion.

b) To determine the form of a signal that is known to be present, use the minimum mean-square error criterion.

There are, of course, situations that are not encompassed by either of the above general rules, but treatment of such situations is beyond the scope of this book.

Exercise 9–2.1

For each of the following situations, state whether the criterion of optimality should be (1) maximum signal-to-noise ratio or (2) minimum mean-square error.

a) Picking up noise signals from distant radio stars

b) Listening to music from a conventional record

c) Listening to music from a digital recording

d) Communication links between computers

e) Using a cordless telephone

f) Detecting flaws in large castings with an ultrasonic flaw detector

Answers: 1, 1, 1, 1, 2, 2

9–3   Restrictions on the Optimum System

It is usually necessary to impose some sort of restriction on the type of system that will be permitted. The most common restriction is that the system must be *causal,*[1] since this is a fundamental requirement of physical realizability. It frequently is true that a noncausal system, which can respond to *future* values of the input, could do a better job of satisfying the chosen criterion than any physically realizable system. A noncausal system usually cannot be built, however, and does not provide a fair comparison with real systems, so is usually inappropriate. A possible exception to this rule arises when the data available to the system is in recorded form so that future values can, in fact, be utilized.

Another common assumption is that the system is *linear*. The major reason for this assumption is that it is usually not possible to carry out the analytical solution for the optimum nonlinear system. In many cases, particularly those involving Gaussian noise, it is possible to show that there is no nonlinear system that will do a better job than the optimum linear system. However, in the more general case, the linear system may not be the best. Nevertheless, the difficulty of determining the optimum nonlinear system is such that it is usually not feasible to hunt for it.

With present day technology it is becoming more and more common to approximate an analog system with a digital system. Such an implementation may eliminate the need for large capacitances and inductances and, thus, reduce the physical size of the optimum system. Furthermore, it may be possible to implement, in an economical fashion, systems that would be too complex and too costly to build as analog systems. It is not the intent of the discussion here to consider the implementation of such digital systems since this is a subject that is too vast to be dealt with in a single chapter. The reader should be aware, however, that while digital approximations to very complex system functions are indeed possible, there

---

[1] By *causal* we mean that the system impulse response satisfies the condition $h(t) = 0$, $t < 0$ [see equation (8–3)]. In addition, the *stability* condition of equation (8–4) is also assumed to apply.

are always errors that arise due to both the discrete-time nature of the operation and the necessary quantization of amplitudes. Thus, the discussion of errors in analog systems in the following sections is not adequate for all of the sources of error that arise in a digital system.

Once a reasonable criterion has been selected, and the system restricted to being causal and linear, then it is usually possible to find the impulse response or transfer function that optimizes the criterion. However, in some cases it may be desirable to further restrict the system to a particular form. The reason for such a restriction is usually that it guarantees a system having a given complexity (and, hence, cost) while the more general optimization may yield a system that is costly or difficult to approximate. An example of this specialized type of otpimization will be considered in the next section.

9—4   Optimization by Parameter Adjustment

As suggested by the title, this method of optimization is carried out by specifying the form of the system to be used and then by finding the values of the components of that system that optimize the selected criterion. This procedure has the obvious advantage of yielding a system whose complexity can be predetermined and, hence, has its greatest application in cases in which the complexity of the system is of critical importance because of size, weight, or cost considerations. The disadvantage is that the performance of this type of optimum system will never be quite as good as that of a more general system whose form is not specified in advance. Any attempt to improve performance by picking a slightly more complicated system to start out with leads to analytical problems in determining the optimum values of more than one parameter (because the simultaneous equations that must be solved are seldom linear), although computer solutions are quite possible. As a practical matter, analytical solutions are usually limited to a single parameter. Two different examples are discussed in order to illustrate the basic ideas.

As a first example, assume that the signal consists of a rectangular pulse as shown in Figure 9–1(a) and that this signal is combined with white noise having a spectral density of N_o. Since the form of the signal is known, the objective of the system is to detect the presence of the signal. As noted earlier, a reasonable criterion for this purpose is to find that system maximizing the output signal-to-noise power ratio at some instant of time. That is, if the output signal is $s_o(t)$, and the mean-square value of the output noise is $\overline{M^2}$, then it is desired to find the system that will maximize the ratio $s_o^2(t_o)/\overline{M^2}$, where t_o is the time chosen for this to be a maximum.

In the method of parameter adjustment, the form of the system is specified and in this case is assumed to be a simple RC circuit, as shown in Figure 9–1(b). The parameter to be adjusted is the time constant of the filter—or, rather, the recipro-

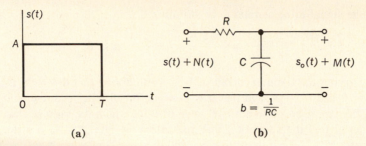

(a)   (b)

Figure 9–1   Signal and system for maximizing signal-to-noise ratio: (a) signal to be detected and (b) specified form of optimum system.

cal of this time constant. One of the first steps is to select a time t_o at which the signal-to-noise ratio is a maximum. An appropriate choice for t_o becomes apparent when the output signal component is considered. This output signal is given by

$$s_o(t) = A[1 - e^{-bt}] \qquad\qquad 0 \le t < T \qquad\qquad (9\text{–}1)$$
$$\quad\;\; = A[1 - e^{-bT}]e^{-b(t-T)} \qquad T \le t < \infty$$

and is sketched in Figure 9–2. This result is arrived at by any of the conventional methods of system analysis. It is clear from this sketch that the output signal component has its largest value at time T. Hence, it is reasonable to choose $t_o = T$, and thus

$$s_o(t_o) = A(1 - e^{-bT}) \qquad\qquad\qquad (9\text{–}2)$$

The mean-square value of the output noise from this type of circuit has been considered several times before and has been shown to be

$$\overline{M^2} = \frac{bN_o}{2} \qquad\qquad\qquad (9\text{–}3)$$

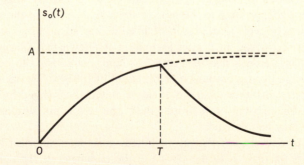

Figure 9–2   Signal component at the RC filter output.

Hence, the signal-to-noise ratio to be maximized is

$$\frac{s_o^2(t_o)}{\overline{M^2}} = \frac{A^2(1 - e^{-bT})^2}{bN_o/2} \qquad b \geq 0 \qquad (9\text{-}4)$$

Before carrying out the maximization, it is worth noting that this ratio is zero for both $b = 0$ and $b = \infty$, and that it is positive for all other positive values of b. Hence, there must be some positive value of b for which the ratio is a maximum.

In order to find the value of b that maximizes the ratio, (9–4) is differentiated with respect to b and the derivative equated to zero. Thus,

$$\frac{d[s_o^2(t_o)/\overline{M^2}]}{db} = \frac{2A^2}{N_o} \frac{[2b(1 - e^{-bT})Te^{-bT} - (1 - e^{-bT})^2]}{b^2} = 0 \qquad (9\text{-}5)$$

This can be simplified to yield the nontrivial equation

$$2bT + 1 = e^{bT} \qquad (9\text{-}6)$$

This equation is easily solved for bT by trial-and-error methods and leads to

$$bT \doteq 1.256 \qquad (9\text{-}7)$$

from which the optimum time constant is

$$RC = \frac{T}{1.256} \qquad (9\text{-}8)$$

This, then, is the value of time constant that should be used in the RC filter in order to maximize the signal-to-noise ratio at time T.

The next step in the procedure is to determine how good the filter actually is. This is easily done by substituting the optimum value of bT, as given by (9–7), into the signal-to-noise ratio of (9–4). When this is done, it is found that

$$\left[\frac{s_o^2(t_o)}{\overline{M^2}}\right]_{max} = \frac{0.8145A^2T}{N_o} \qquad (9\text{-}9)$$

It may be noted that the *energy* of the pulse is A^2T, so that the maximum signal-to-noise ratio is proportional to the ratio of the signal energy to the noise spectral density. This is typical of *all* cases of maximizing signal-to-noise ratio in the presence of white noise. In a later section, it is shown that if the form of the optimum system were not specified as it was here, but allowed to be general, the constant of proportionality would be 1.0 instead of 0.8145. The reduction in signal-to-noise ratio encountered in this example may be considered as the price that must be paid for insisting on a simple filter. The loss is not serious here, but in other cases it may be appreciable.

The final step, which is frequently omitted, is to determine just how sensitive the signal-to-noise ratio is to the choice of the parameter b. This is most easily done by sketching the proportionality constant in (9–4) as a function of b. Since this constant is simply

$$K = \frac{2(1 - e^{-bT})^2}{bT} \tag{9–10}$$

the result is as shown in Figure 9–3. It is clear from this sketch that the output signal-to-noise ratio does not change rapidly with b in the vicinity of the maximum so that it is not very important to have precisely the right value of time constant in the optimum filter.

The fact that this particular system is not very sensitive to the value of the parameter should not be construed to imply that this is always the case. If, for example, the signal were a sinusoidal pulse and the system a resonant circuit, the performance is critically dependent upon the resonant frequency and, hence, upon the values of inductance and capacitance.

The second example of optimization by parameter adjustment will consider a random signal and employ the minimum mean-square error criterion. In this example the system will be an ideal lowpass filter, rather than a specified circuit configuration, and the parameter to be adjusted will be the bandwidth of that filter.

Assume that the signal $X(t)$ is a sample function from a random process having a spectral density of

$$S_X(\omega) = \frac{A^2}{\omega^2 + (2\pi f_a)^2} \tag{9–11}$$

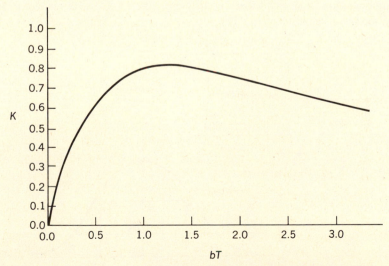

Figure 9–3 Output signal-to-noise ratio as function of the parameter bT.

Added to this signal is white noise $N(t)$ having a spectral density of N_o. These are illustrated in Figure 9–4 along with the power transfer characteristic of the ideal lowpass filter.

Since the filter is an ideal lowpass filter, the error in the output *signal component*, $E(t) = X(t) - Y(t)$, will be due entirely to that portion of the signal spectral density falling outside the filter pass band. Its mean-square value can be obtained by integrating the signal spectral density over the region *outside* of $\pm 2\pi B$. Because of symmetry, only one side need be evaluated and then doubled. Hence

$$
\begin{aligned}
\overline{E^2} &= \frac{2}{2\pi} \int_{2\pi B}^{\infty} \frac{A^2}{\omega^2 + (2\pi f_a)^2} \, d\omega \\
&= \frac{2A^2}{4\pi^2 f_a} \left(\frac{\pi}{2} - \tan^{-1} \frac{B}{f_a} \right)
\end{aligned}
$$
(9–12)

The noise out of the filter, $M(t)$; has a mean-square value of

$$
\overline{M^2} = \frac{1}{2\pi} \int_{-2\pi B}^{2\pi B} N_o \, d\omega = 2BN_o
$$
(9–13)

The total mean-square error is the sum of these two (since signal and noise are statistically independent) and is the quantity that is to be minimized by selecting B. Thus,

$$
\overline{E^2} + \overline{M^2} = \frac{2A^2}{4\pi^2 f_a} \left(\frac{\pi}{2} - \tan^{-1} \frac{B}{f_a} \right) + 2BN_o
$$
(9–14)

The minimization is accomplished by differentiating (9–14) with respect to B and setting the result equal to zero. Thus,

$$
\frac{2A^2}{4\pi^2 f_a} \left[\frac{-1/f_a}{1 + (B/f_a)^2} \right] + 2N_o = 0
$$

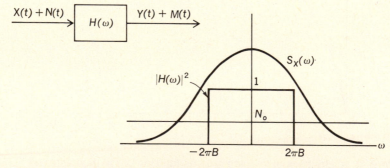

Figure 9–4 Signal and noise spectral densities, filter characteristic.

from which it follows that

$$B = \left(\frac{A^2}{4\pi^2 N_o} - f_a^2\right)^{1/2} \tag{9-15}$$

is the optimum value. The actual value of the minimum mean-square error can be obtained by substituting this value into (9–14).

The form of (9–15) is not easy to interpret. A somewhat simpler form can be obtained by noting that the mean-square value of the signal is

$$\overline{X^2} = \frac{A^2}{4\pi f_a}$$

and that the mean-square value of that portion of the *noise* contained within the equivalent-noise bandwidth of the *signal* is just

$$\overline{N_X^2} = \pi f_a N_o$$

since the equivalent-noise bandwidth of the signal is $(\pi/2)f_a$. Hence, (9–15) can be written as

$$B = f_a\left(\frac{\overline{X^2}}{\overline{N_X^2}} - 1\right)^{1/2} \qquad \overline{X^2} > \overline{N_X^2} \tag{9-16}$$

and sketched as in Figure 9–5.

It is of interest to note from Figure 9–5 that the optimum bandwidth of the filter is zero when the mean-square value of the signal into the filter is equal to the mean-square value of the noise within the equivalent-noise bandwidth of the signal. Under these circumstances there is no signal and no noise at the output of filter. Thus, the minimum mean-square error is just the mean-square value of the signal. For smaller values of signal mean-square value the optimum bandwidth

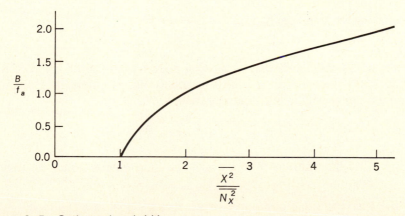

Figure 9–5 Optimum bandwidth.

remains at zero and the minimum mean-square error is still the mean-square value of the signal.

The example just discussed is not a practical example as it stands because it uses an ideal lowpass filter, which cannot be realized with a finite number of elements. This filter was chosen for reasons of analytical convenience rather than practicality. However, a practical filter with a transfer function that drops off much more rapidly than the rate at which the signal spectral density drops off would produce essentially the same result as the ideal lowpass filter. Thus, this simple analysis can be used to find the optimum bandwidth of a practical lowpass filter with a sharp cutoff. One should not conclude, however, that this simplified approach would work for *any* practical filter. For example, if a simple RC circuit, such as that shown in Figure 9–1(b), were used, the optimum filter bandwidth is quite different from that given by (9–15). This is illustrated in Exercise 9–4.2 below.

In each of the examples just discussed only *one* parameter of the system was adjusted in order to optimize the desired criterion. The procedure for adjusting two or more parameters is quite similar. That is, the quantity to be maximized or minimized is differentiated with respect to *each* of the parameters to be adjusted and the derivatives set equal to zero. This yields a set of simultaneous equations that can, in principle, be solved to yield the desired parameter values. As a practical matter, however, this procedure is only rarely possible because the equations are usually nonlinear and analytical solutions are not known. Computer solutions may be obtained frequently, but there may be unresolved questions concerning uniqueness.

Exercise 9–4.1

A rectangular pulse defined by

$$s(t) = 2[u(t) - u(t - 1)]$$

is combined with white noise having a spectral density of 2 V^2/Hz. It is desired to maximize the signal-to-noise ratio at output of a finite-time integrator whose impulse response is

$$h(t) = \frac{1}{T}[u(t) - u(t - T)]$$

a)   Find the value of T that maximizes the output signal-to-noise ratio.

b)   Find the value of the maximum output signal-to-noise ratio.

c) If the integration time, T, is changed by 10% on either side of the optimum value, find the percent drop in output signal-to-noise ratio.

Answers: 1, 2, 9.09, 10

Exercise 9–4.2

A signal having a spectral density of

$$S_X(\omega) = \frac{40}{\omega^2 + 2.25}$$

and white noise having a spectral density of 1 V^2/Hz are applied to a lowpass RC filter having a transfer function of

$$H(\omega) = \frac{b}{j\omega + b}$$

a) Find the value of b that minimizes the mean-square error between the input signal and the total filter output.

b) Find the value of the minimum mean-square error.

c) Find the value of mean-square error that is approached as b approaches 0.

Answers: 4.82, 5.57, 13.33

9–5   Systems That Maximize Signal-to-Noise Ratio

This section will consider systems that maximize signal-to-noise ratio at a specified time, when the form of the signal is known. The form of the system is *not* specified; the only restrictions on the system being that it must be *causal* and *linear*.

The notation is illustrated in Figure 9–6. The signal $s(t)$ is deterministic and assumed to be known (except, possibly, for amplitude and time of occurrence). The noise $N(t)$ is assumed to be white with a spectral density of N_o. Although the case of nonwhite noise is not considered here (except for a brief mention at the end of this section), the same general procedure can be used for it also. The output signal-to-noise ratio is defined to be $s_o^2(t_o)/\overline{M^2}$, and the time t_o is to be selected.

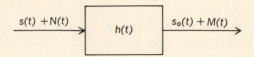

Figure 9–6 Notation for optimum filter.

The objective is to find the form of $h(t)$ that maximizes this output signal-to-noise ratio.

In the first place, the output signal is given by

$$s_o(t) = \int_0^\infty h(\lambda)s(t - \lambda) \, d\lambda \qquad (9\text{–}17)$$

and the mean-square value of the output noise is, for a white noise input, given by

$$\overline{M^2} = N_o \int_0^\infty h^2(\lambda) \, d\lambda \qquad (9\text{–}18)$$

Hence, the signal-to-noise ratio at time t_o is

$$\frac{s_o{}^2(t_o)}{\overline{M^2}} = \frac{\left[\int_0^\infty h(\lambda)s(t_o - \lambda) \, d\lambda \right]^2}{N_o \int_0^\infty h^2(\lambda) \, d\lambda} \qquad (9\text{–}19)$$

In order to maximize this ratio it is convenient to use the *Schwarz inequality*. This inequality states that for any two functions, say $f(t)$ and $g(t)$, that

$$\left[\int_a^b f(t)g(t) \, dt \right]^2 \leq \int_a^b f^2(t) \, dt \int_a^b g^2(t) \, dt \qquad (9\text{–}20)$$

Furthermore, the *equality* holds if and only if $f_;(t) = kg(t)$, where k is independent of t.

Using the Schwarz inequality on (9–19) leads to

$$\frac{s_o{}^2(t_o)}{\overline{M^2}} \leq \frac{\int_0^\infty h^2(\lambda) \, d\lambda \int_0^\infty s^2(t_o - \lambda) \, d\lambda}{N_o \int_0^\infty h^2(\lambda) \, d\lambda} \qquad (9\text{–}21)$$

From this it is clear that the maximum value of the signal-to-noise ratio occurs when the equality holds, and that this maximum value is just

$$\left[\frac{s_o{}^2(t_o)}{\overline{M^2}} \right]_{\text{max}} = \frac{1}{N_o} \int_0^\infty s^2(t_o - \lambda) \, d\lambda \qquad (9\text{–}22)$$

since the integrals of $h^2(\lambda)$ cancel out. Furthermore, the condition that is required for the equality to hold is

$$h(\lambda) = ks(t_o - \lambda)u(\lambda) \qquad (9\text{--}23)$$

Since the k is simply a gain constant that does not affect the signal-to-noise ratio, it can be set equal to any value; a convenient value is $k = 1$. The $u(\lambda)$ has been added to guarantee that the system is causal. Note that the desired impulse response is simply the signal waveform run backwards in time and delayed by t_o seconds.

The right side of (9–22) can be written in slightly different form by letting $t = t_o - \lambda$. Upon making this change of variable, the integral becomes

$$\int_0^\infty s^2(t_o - \lambda) \, d\lambda = \int_{-\infty}^{t_o} s^2(t) \, dt = \varepsilon(t_o) \qquad (9\text{--}24)$$

and it is clear that this is simply the energy in the signal up to the time the signal-to-noise ratio is to be maximized. This signal energy is designated as $\varepsilon(t_o)$. To summarize, then:

1. The output signal-to-noise ratio at time t_o is maximized by a filter whose impulse response is

$$h(t) = s(t_o - t)u(t) \qquad (9\text{--}25)$$

2. The value of the maximum signal-to-noise ratio is

$$\left[\frac{s_o{}^2(t_o)}{\overline{M^2}} \right]_{\text{max}} = \frac{\varepsilon(t_o)}{N_o} \qquad (9\text{--}26)$$

where $\varepsilon(t_o)$ is the energy in $s(t)$ up to the time t_o.

The filter defined by (9–25) is usually referred to as a *matched filter*.

As a first example of this procedure, consider again the case of the rectangular signal pulse, as shown in Figure 9–7(a), and find the $h(t)$ that will maximize the signal-to-noise ratio at $t_o = T$. The reversed and translated signal is shown (for an arbitrary t_o) in Figure 9–7(b). The resulting impulse response for $t_o = T$ is shown in Figure 9–7(c) and is represented mathematically by

$$\begin{aligned} h(t) &= A \qquad 0 \le t \le T \\ &= 0 \qquad \text{elsewhere} \end{aligned} \qquad (9\text{--}27)$$

The maximum signal-to-noise ratio is given by

$$\left[\frac{s_o{}^2(t_o)}{\overline{M^2}} \right]_{\text{max}} = \frac{\varepsilon(t_o)}{N_o} = \frac{A^2 T}{N_o} \qquad (9\text{--}28)$$

This result may be compared with (9–9).

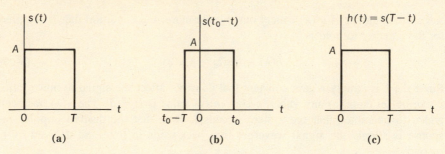

Figure 9–7 Matched filter for a rectangular pulse: (a) signal, (b) reversed, translated signal, and (c) optimum filter for $t_o = T$.

In order to see the effect of changing the value of t_o, the sketches of Figure 9–8 are presented. The sketches show $s(t_o - t)$, $h(t)$, and the output signal $s_o(t)$, all for the same input $s(t)$ shown in Figure 9–7(a). It is clear from these sketches that making $t_o < T$ decreases the maximum signal-to-noise ratio because not all of the energy of the pulse is available at time t_o. On the other hand, making $t_o > T$ does not further increase the output signal-to-noise ratio, since all of the pulse energy is available by time T. It is also clear that the signal out of the matched filter does not have the same shape as the input signal. Thus, the matched filter is *not* suitable if the objective of the filter is to recover a nearly undistorted rectangular pulse.

As a second example of matched filters, it is of interest to consider a signal having finite energy but infinite time duration. Such a signal might be

$$s(t) = Ae^{-bt}u(t) \tag{9–29}$$

as shown in Figure 9–9. For some arbitrarily selected t_o, the optimum matched filter is

$$h(t) = Ae^{-b(t_o - t)}u(t) \tag{9–30}$$

and is also shown. The maximum signal-to-noise ratio depends upon t_o, since the available energy increases with t_o. In this case it is

$$\left[\frac{s_o^2(t_o)}{\overline{M^2}}\right]_{max} = \frac{\varepsilon(t_o)}{N_o} = \frac{\displaystyle\int_0^{t_o} A^2 e^{-2bt}\,dt}{N_o} = \frac{A^2}{2bN_o}[1 - e^{-2bt_o}] \tag{9–31}$$

It is clear that this approaches a limiting value of $A^2/2bN_o$ as t_o is made large. Hence, the choice of t_o is governed by how close to this limit one wants to come—remembering that larger values of t_o generally represent a more costly system.

The third and final illustration of the matched filter considers signals having

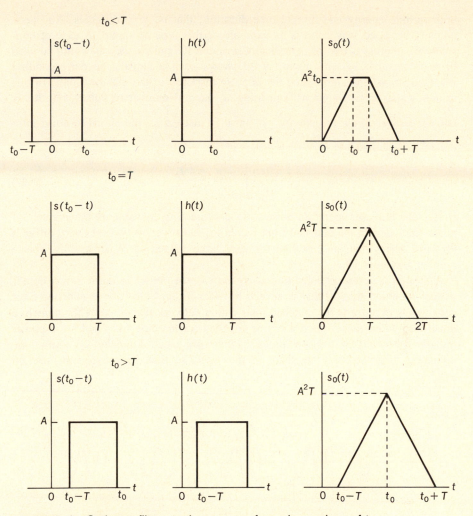

Figure 9–8 Optimum filters and responses for various values of t_0.

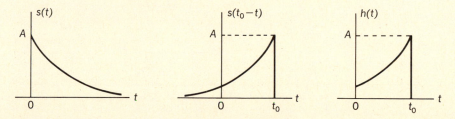

Figure 9–9 Matched filter for an exponential signal.

both infinite energy and infinite time duration, that is, power signals. Any peri-odically repeated waveform is an example of this type of signal. A case of con-siderable interest is that of periodically repeated RF pulses such as would be used in a pulse radar system. Figure 9–10 shows such a signal, the corresponding reversed, translated signal, and the impulse response of the matched filter. In this sketch, t_o has been shown as containing an integral number of pulses, but this is not necessary. Since the energy per pulse is just $\frac{1}{2}A^2 t_p$, the signal-to-noise ratio out of a filter matched to N such pulses is

$$\left[\frac{s_o^2(t_o)}{M^2}\right]_{max} = \frac{NA^2 t_p}{2N_o} \tag{9–32}$$

It is clear that this signal-to-noise ratio continues to increase as the number of pulses included in the matched filter increases. However, it becomes very difficult to build filters that are matched for very large values of N, so usually N is a number less than 10.

Although it is not intended to discuss the case of nonwhite noise in any detail, it may be noted that all that is needed in order to apply the above matched filter concepts is to precede the matched filter with a network that converts the nonwhite noise into white noise. Such a device is called a *whitening* filter and has a power transfer function that is the reciprocal of the noise spectral density. Of course, the whitening filter changes the shape of the signal so that the subsequent matched

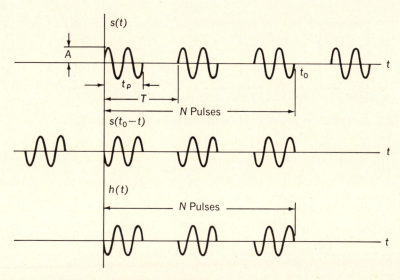

Figure 9–10   Matched filter for N pulses.

filter has to be matched to this new signal shape rather than to the original signal shape.

An interesting phenomenon know as *singular detection* may arise sometimes for certain combinations of input signal and nonwhite noise spectral density. Suppose, for example, that the nonwhite noise has a spectral density of

$$S_N(\omega) = \frac{1}{\omega^2 + 1}$$

The power transfer function of the whitening filter that converts this spectral density to white noise is

$$|H(\omega)|^2 = \frac{1}{S_N(\omega)} = \omega^2 + 1 = (j\omega + 1)(-j\omega + 1)$$

Thus, the voltage transfer function of the whitening filter is

$$H(\omega) = j\omega + 1$$

and this corresponds to an impulse response of

$$h(t) = \delta(t) + \dot{\delta}(t)$$

Hence, for any input signal $s(t)$, the output of the whitening filter is $s(t) + \dot{s}(t)$. If the input signal is a rectangular pulse as shown in Figure 9–7(a), the output of the whitening filter will contain two δ functions because of the differentiation action of the filter. Since any δ function contains infinite energy, it can always be detected regardless of how small the input signal might be. The same result would occur for any input signal having a discontinuity in amplitude. This is clearly not a realistic situation and arises because the input signal is modeled in an idealistic way. Actual signals can never have discontinuities and, hence, singular detection never actually occurs. Nevertheless, the fact that the analysis suggests this possibility emphasizes the importance of using realistic mathematical models for signals when matched filters for nonwhite noise are considered.

Exercise 9–5.1

A signal of the form

$$s(t) = 1.5t[u(t) - u(t - 2)]$$

is to be detected in the presence of white noise having a spectral density of 0.15 V²/Hz using a matched filter.

a) Find the smallest value of t_0 that will yield the maximum output signal-to-noise ratio.

b) For this value of t_0 find the value of the matched-filter impulse response at $t = 0, 1$, and 2.

c) Find the maximum output signal-to-noise ratio.

Answers: 0, 1.5, 2, 3, 5

Exercise 9–5.2

A signal of the form

$$s(t) = 5e^{-(t+2)} u(t + 2)$$

is to be detected in the presence of white noise having a spectral density of 0.25 V^2/Hz using a matched filter.

a) For $t_0 = 2$ find the value of the impulse response of the matched filter at $t = 0, 2, 4$.

b) Find the maximum output signal-to-noise ratio that can be achieved if $t_0 = \infty$.

c) Find the value of t_0 that should be used to achieve an output signal-to-noise ratio that is 0.95 of that achieved in part (b).

Answers: −0.502, 0.0916, 0.677, 5, 50

9–6   Systems That Minimize Mean-Square Error

This section considers systems that minimize the mean-square error between the total system output and the input signal component when that signal is from a stationary random process. The form of the system is not specified in advance, but it is restricted to be linear and causal.

It is convenient to use s-plane notation in carrying out this analysis, although it can be done in the time-domain as well. The notation is illustrated in Figure 9–11, in which the random input signal $X(t)$ is assumed to have a spectral density of $S_X(s)$, while the input noise $N(t)$ has a spectral density of $S_N(s)$. The output spectral densities for these components are $S_Y(s)$ and $S_M(s)$, respectively. There is

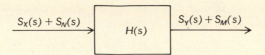

Figure 9–11 Notation for the optimum system.

no particular simplification from assuming the input noise to be white (as there was in the case of the matched filter), so it will not be done here.

The error in the signal component, produced by the system, is defined as before by

$$E(t) = X(t) - Y(t)$$

and its Laplace transform is

$$F_E(s) = F_X(s) - F_Y(s) = F_X(s) - H(s)F_X(s) = F_X(s)[1 - H(s)] \quad \text{(9–33)}$$

Hence, $1 - H(s)$ is the transfer function relating the *signal error* to the *input signal*, and the mean-square value of the signal error is given by

$$\overline{E^2} = \frac{1}{2\pi j} \int_{-j\infty}^{j\infty} S_X(s)[1 - H(s)][1 - H(-s)] \, ds \quad \text{(9–34)}$$

The noise appearing at the system output is $M(t)$, and its mean-square value is

$$\overline{M^2} = \frac{1}{2\pi j} \int_{-j\infty}^{j\infty} S_N(s)H(s)H(-s) \, ds \quad \text{(9–35)}$$

The total mean-square error is $\overline{E^2} + \overline{M^2}$ (since signal and noise are statistically independent) and may be expressed as

$$\overline{E^2} + \overline{M^2} = \frac{1}{2\pi j} \int_{-j\infty}^{j\infty} \{S_X(s)[1 - H(s)][1 - H(-s)]$$
$$+ S_N(s)H(s)H(-s)\} ds \quad \text{(9–36)}$$

The objective now is to find the form of $H(s)$ that minimizes (9–36).

If there were no requirement that the system be causal, finding the optimum value of $H(s)$ would be very simple. In order to do this, rearrange the terms in (9–36) as

$$\overline{E^2} + \overline{M^2} = \frac{1}{2\pi j} \int_{-j\infty}^{j\infty} \{[S_X(s) + S_N(s)]H(s)H(-s)$$
$$- S_X(s)H(s) - S_X(s)H(-s) + S_X(s)\} ds \quad \text{(9–37)}$$

Since $[S_X(s) + S_N(s)]$ is also a spectral density, it must have the same symmetry properties and, hence, can be factored into one factor having poles and zeros in the left half plane and another factor having the same poles and zeros in the right half plane. Thus, it can always be written as

$$S_X(s) + S_N(s) = F_i(s)F_i(-s) \qquad (9\text{--}38)$$

Substituting this into (9–37), and again rearranging terms, leads to

$$\overline{E^2} + \overline{M^2} = \frac{1}{2\pi j} \int_{-j\infty}^{j\infty} \left\{ \left[F_i(s)H(s) - \frac{S_X(s)}{F_i(-s)} \right] \left[F_i(-s)H(-s) - \frac{S_X(s)}{F_i(s)} \right] \right.$$
$$\left. + \frac{S_X(s)S_n(s)}{F_i(s)F_i(-s)} \right\} ds \qquad (9\text{--}39)$$

It may now be noted that the last term of (9–39) does not involve $H(s)$. Hence, the minimum value of $\overline{E^2} + \overline{M^2}$ will occur when the two factors in the first term of (9–39) are zero (since the product of these factors cannot be negative.) This implies, therefore, that

$$H(s) = \frac{S_X(s)}{F_i(s)F_i(-s)} = \frac{S_X(s)}{S_X(s) + S_N(s)} \qquad (9\text{--}40)$$

should be the optimum transfer function. This would be true except that (9–40) is also symmetrical in the s-plane and, hence, cannot represent a causal system.

Since the $H(s)$ defined by (9–40) is not causal, the first inclination is to simply use the left half plane poles and zeros of (9–40) to define a causal system. This would appear to be analogous to eliminating the negative time portion of $s(t_o - t)$ in the matched filter of the previous section. Unfortunately, the problem is not quite that simple, because in this case the total random process at the system input, $X(t) + N(t)$, is not white. If it were white, its autocorrelation function would be a δ function and, hence, all future values of the input would be uncorrelated with the present and past values. Thus, a system that could not respond to future inputs (that is, a causal system) would not be ignoring any information that might lead to a better estimate of the signal. It appears, therefore, that the first step in obtaining a causal system should be to transform the spectral density of signal plus noise into white noise. Hence, a whitening filter is needed.

From (9–38), it is apparent that if one had a filter with a transfer function of

$$H_1(s) = \frac{1}{F_i(s)} \qquad (9\text{--}41)$$

then the output of this filter would be white, because

$$[S_X(s) + S_N(s)]H_1(s)H_1(-s) = \frac{S_X(s) + S_N(s)}{F_i(s)F_i(-s)} = 1$$

Furthermore, $H_1(s)$ would be causal because $F_i(s)$ by definition has only left half plane poles and zeros. Thus, $H_1(s)$ is the whitening filter for the input signal plus noise.

Next, look once more at the factor in (9–39) that was set equal to zero; that is,

$$F_i(s)H(s) - \frac{S_X(s)}{F_i(-s)}$$

The source of the right half plane poles is the second term of this factor, but that term can be broken up (by means of a partial fraction expansion) into the *sum* of one term having only left half plane poles and one having only right half plane poles. Thus, write this factor as

$$F_i(s)H(s) - \frac{S_X(s)}{F_i(-s)} = F_i(s)H(s) - \left[\frac{S_X(s)}{F_i(-s)}\right]_L - \left[\frac{S_X(s)}{F_i(-s)}\right]_R \quad \textbf{(9–42)}$$

where the sub L implies left half plane poles only and the sub R implies right half plane poles only. It is now clear that it is not possible to make this entire factor zero with a causal $H(s)$, and that the smallest value that it can have is obtained by making the difference between the first two terms of the right side of (9–42) equal to zero. That is, let

$$F_i(s)H(s) - \left[\frac{S_X(s)}{F_i(-s)}\right]_L = 0$$

or

$$H(s) = \frac{1}{F_i(s)}\left[\frac{S_X(s)}{F_i(-s)}\right]_L \quad \textbf{(9–43)}$$

Note that the first factor of (9–43) is $H_1(s)$, the whitening filter. Thus, the elimination of the *noncausal* parts of the second factor represents the best that can be done in minimizing the total mean-square error.

The optimum filter, which minimizes total mean-square error, is often referred to as the *Wiener filter*. It can be considered as a cascade of two parts, as shown in Figure 9–12. The first part is the whitening filter $H_1(s)$, while the second part, $H_2(s)$, does the actual filtering. Often $H_1(s)$ and $H_2(s)$ have common factors that cancel to yield an $H(s)$ that is simpler than might be expected (and easier to build than either factor).

As an example of the Wiener filter, consider a signal having a spectral density of

$$S_X(s) = \frac{-1}{s^2 - 1}$$

and noise with a spectral density of

$$S_N(s) = \frac{-1}{s^2 - 4}$$

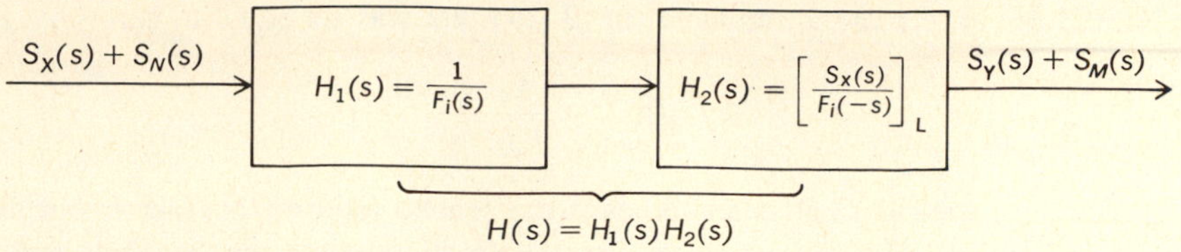

$$S_X(s) + S_N(s) \rightarrow \boxed{H_1(s) = \frac{1}{F_i(s)}} \rightarrow \boxed{H_2(s) = \left[\frac{S_X(s)}{F_i(-s)}\right]_L} \rightarrow S_Y(s) + S_M(s)$$

$$H(s) = H_1(s)H_2(s)$$

$$F_i(s)F_i(-s) = S_X(s) + S_N(s)$$

Figure 9–12   The optimum Wiener filter.

Thus,

$$F_i(s)F_i(-s) = S_X(s) + S_N(s) = \frac{-1}{s^2 - 1} + \frac{-1}{s^2 - 4} = \frac{-(2s^2 - 5)}{(s^2 - 1)(s^2 - 4)}$$

from which it follows that

$$F_i(s) = \frac{\sqrt{2}\,(s + \sqrt{2.5})}{(s + 1)(s + 2)} \tag{9–44}$$

Therefore, the whitening filter is

$$H_1(s) = \frac{1}{F_i(s)} = \frac{(s + 1)(s + 2)}{\sqrt{2}\,(s + \sqrt{2.5})} \tag{9–45}$$

The second filter section is obtained readily from

$$\frac{S_X(s)}{F_i(-s)} = \frac{-1}{s^2 - 1} \cdot \frac{(-s + 1)(-s + 2)}{\sqrt{2}\,(-s + \sqrt{2.5})} = \frac{s - 2}{\sqrt{2}\,(s + 1)(s - \sqrt{2.5})}$$

which may be broken up by means of a partial fraction expansion into

$$\frac{S_X(s)}{F_i(-s)} = \frac{0.822}{s + 1} - \frac{0.115}{s - \sqrt{2.5}}$$

Hence,

$$H_2(s) = \left[\frac{S_X(s)}{F_i(-s)}\right]_L = \frac{0.822}{s + 1} \tag{9–46}$$

The final optimum filter is

$$H(s) = H_1(s)H_2(s) = \left[\frac{(s + 1)(s + 2)}{\sqrt{2}\,(s + \sqrt{2.5})}\right]\left[\frac{0.822}{s + 1}\right] = \frac{0.582(s + 2)}{s + \sqrt{2.5}} \tag{9–47}$$

Note that the final optimum filter is simpler than the whitening filter and can be built as an RC circuit.

The remaining problem is that of evaluating the performance of the optimum filter; that is, to determine the actual value of the minimum mean-square error. This problem is greatly simplified by recognizing that in an optimum system of this sort, the error that remains must be uncorrelated with the actual output of the system. If this were not true it would be possible to perform some further linear operation on the output and obtain a still smaller error. Thus, the minimum mean-square error is simply the difference between the mean-square value of the input *signal* component and the mean-square value of the *total* filter output. That is,

$$(\overline{E^2 + M^2})_{\min} \tag{9-48}$$

$$= \frac{1}{2\pi j} \int_{-j\infty}^{j\infty} S_X(s)\,ds - \frac{1}{2\pi j} \int_{-j\infty}^{j\infty} [S_X(s) + S_N(s)]H(s)H(-s)\,ds$$

when $H(s)$ is as given by (9-43).

The above result can be used to evaluate the minimum mean-square error that is achieved by the Wiener filter described by (9-47). The first integral in (9-48) is evaluated easily by using either Table 7-1 or by summing the residues. Thus,

$$\frac{1}{2\pi j} \int_{-j\infty}^{j\infty} S_X(s)\,ds = \frac{1}{2\pi j} \int_{-j\infty}^{j\infty} \frac{-1}{s^2 - 1}\,ds = 0.5$$

The second integral is similarly evaluated as

$$\frac{1}{2\pi j} \int_{-j\infty}^{j\infty} [S_X(s) + S_N(s)]H(s)H(-s)\,ds$$

$$= \frac{1}{2\pi j} \int_{-j\infty}^{j\infty} \frac{-(2s^2 - 5)}{(s^2 - 1)(s^2 - 4)} \cdot \frac{(0.582)^2(s^2 - 4)}{(s^2 - 2.5)}\,ds$$

$$= \frac{1}{2\pi j} \int_{-j\infty}^{j\infty} \frac{-2(0.582)^2}{(s^2 - 1)}\,ds = 0.339$$

The minimum mean-square error now becomes

$$(\overline{E^2 + M^2})_{\min} = 0.5 - 0.339 = 0.161$$

It is of interest to compare this value with the mean-square error that would result if no filter were used. With no filtering there would be no signal error and the total mean-square error would be the mean-square value of the noise. Thus

$$(\overline{E^2 + M^2}) = \overline{N^2} = \frac{1}{2\pi j} \int_{-j\infty}^{j\infty} \frac{-1}{s^2 - 4}\,ds = 0.25$$

and it is seen that the use of the filter has substantially reduced the total error. This reduction would have been even more pronounced had the input noise had a wider bandwidth.

Exercise 9–6.1

A random signal has a spectral density of

$$S_X(s) = \frac{-1}{s^2 - 1}$$

and is combined with noise having a spectral density of

$$S_N(s) = \frac{s^2}{s^2 - 1}$$

Find the minimum mean-square error between the input signal and the total filter output that can be achieved with any linear, causal filter.

     Answer: 0.375

Exercise 9–6.2

A random signal has a spectral density of

$$S_X(s) = \frac{-2s^2}{s^4 - 13s^2 + 36}$$

and is combined with white noise having a spectral density of 1.0. Find the poles and zeros of the optimum causal Wiener filter that minimizes the mean-square error between the input signal and the total filter output.

    Answers: $0,\ -\sqrt{3},\ -2\sqrt{3}$

_____ PROBLEMS _____

9–2.1   For each of the situations listed below, indicate whether the appropriate criterion of optimality is maximum signal-to-noise ratio or minimum mean-square error.

    a)   An automatic control system subject to random disturbances

b)   An aircraft flight-control system

c)   A pulse radar system

d)   A police speed radar system

e)   A particle detector for measuring nuclear radiation

f)   A passive sonar system for detecting underwater sounds

9–2.2   A signal consisting of a steady-state sinusoid having a peak value of 2 V
and a frequency of 80 Hz is combined with white noise having a spectral
density of 0.01 V²/Hz. A single-section RC filter having a transfer function of

$$H(\omega) = \frac{b}{b + j\omega}$$

is used to extract the signal from the noise.

a)   Determine the output signal-to-noise ratio if the half-power bandwidth of the filter is 10 Hz.

b)   Repeat if the filter half-power bandwidth is 100 Hz.

c)   Repeat if the filter half-power bandwidth is 1000 Hz.

9–3.1   The impulse response, $h(t)$, of the system shown below is causal and the
input noise $N(t)$ is zero-mean, Gaussian and white.

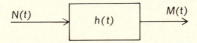

a)   Prove that the output $M(t)$ is *independent* of $N(t + \tau)$ for all $\tau > 0$
(that is, future values of the input) but is *not* independent of
$N(t + \tau)$ for $\tau \leq 0$ (that is, past and present values of the input).

b)   Prove that the statement in (a) is not true if the system is noncausal.

9–4.1   a)   For the signal and noise of Problem 9–2.2, find the filter half-power
bandwidth that maximizes the output signal-to-noise ratio.

b)   Find the value of the maximum output signal-to-noise ratio.

9–4.2 The signal $s(t)$ below is combined with white noise having a spectral density of 2 V^2/Hz. It is desired to maximize the signal-to-noise at the output of the RC filter, also shown below, at $t = 0.01$ seconds. Find the value of RC in the filter that achieves this result.

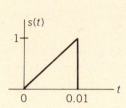
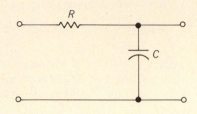

9–4.3 A random signal having a spectral density of

$$S_X(\omega) = 2 \qquad |\omega| \leq 10$$
$$= 0 \qquad \text{elsewhere}$$

is observed in the presence of white noise having a spectral density of 2V^2/Hz. Both are applied to the input of a lowpass RC filter having a transfer function of

$$H(\omega) = \frac{b}{j\omega + b}$$

a) Find the value of b that minimizes the mean-square error between the input signal and the total filter output.

b) Find the value of the minimum mean-square error.

9–4.4 A random signal having a spectral density of

$$S_X(\omega) = 1 - \frac{|\omega|}{10} \qquad |\omega| \leq 10$$
$$= 0 \qquad \text{elsewhere}$$

is observed in the presence of white noise having a spectral density of 0.1V^2/Hz. Both are applied to the input of an ideal lowpass filter whose transfer function is

$$H(\omega) = 1 \qquad |\omega| \leq 2\pi W$$
$$= 0 \qquad \text{elsewhere}$$

a)   Find the value of W that maximizes the ratio of signal power to noise power at the output of the filter.

b)   Find the value of W that minimizes the mean-square error between the input signal and the total filter output.

9–5.1

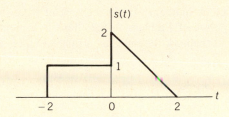

a)   The signal shown above is combined with white noise having a spectral density of 0.1 V^2/Hz. Find the impulse response of the causal filter that will maximize the output signal-to-noise ratio at $t_0 = 2$.

b)   Find the value of the maximum output signal-to-noise ratio.

c)   Repeat (a) and (b) for $t_0 = 0$.

9–5.2   A signal has the form

$$s(t) = te^{-t} u(t)$$

and is combined with white noise having a spectral density of 0.005 V^2/Hz.

a)   What is the largest output signal-to-noise ratio that can be achieved with any linear filter?

b)   For what observation time t_0 should a matched filter be constructed to achieve an output signal-to-noise ratio that is 0.9 of that determined in (a)?

9–5.3   A power signal consists of rectangular pulses having an amplitude of 1 V and a duration of 1 millisecond repeated periodically at a rate of 100 pulses per second. This signal is observed in the presence of white noise having a spectral density of 0.001 V^2/Hz.

a)   If a causal filter is to be matched to N successive pulses, find the output signal-to-noise ratio that can be achieved as a function of N.

$(4)^2$

    b) How many pulses must the filter be matched to in order to achieve an output signal-to-noise ratio of 100?

    c) Sketch a block diagram showing how such a matched filter might be constructed using a finite-time integrator and a transversal filter.

9–5.4 Below is a block diagram of another type of filter that might be used to extract the pulses of Problem 9–5.3. This is a recursive filter that does not distort the shape of the pulse as a matched filter does.

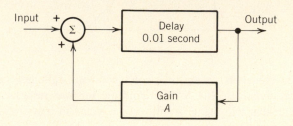

    a) Find the largest value the gain parameter A can have in order for the filter to be stable.

    b) Find a relationship between the output signal-to-noise ratio and the gain parameter A.

    c) Find the value of A that is required to achieve an output signal-to-noise ratio of 100.

9–5.5 The diagram below represents a particle detector connected to an amplifier with a matched filter in its output. The particle detector may be modeled as having a source impedance of 1 MΩ and producing an open circuit voltage for each particle of

$$s(t) = 10^{-4}e^{-10^5 t}u(t)$$

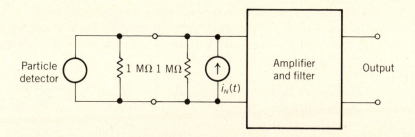

The input circuit of the amplifier may be modeled as a 1 MΩ resistor in parallel with a current source for which the current is from a white noise source having a spectral density of 10^{-26} A^2/Hz. The amplifier may be assumed to have an output impedance that is negligibly small compared to the input impedance of the filter.

a) Find the impulse response of the filter that will maximize the output signal-to-noise ratio at the time at which the output of the particle detector drops to one-hundredth of its maximum value.

b) Find the value of the maximum output signal-to-noise ratio.

9–6.1 A signal is from a stationary random process having a spectral density of

$$S_X(\omega) = \frac{16}{\omega^2 + 16}$$

and is combined with white noise having a spectral density of 0.1 V^2/Hz.

a) Find the transfer function of the noncausal linear filter that will minimize the mean-square error between the input signal and the total filter output.

b) Find the value of the minimum mean-square error that is achieved with this filter.

9–6.2 Repeat Problem 9–6.1 using a causal linear filter.

9–6.3 A signal is from a stationary random process having a spectral density of

$$S_X(\omega) = \frac{4}{\omega^2 + 4}$$

and is combined with noise having a spectral density of

$$S_M(\omega) = \frac{\omega^2}{\omega^2 + 4}$$

a) Find the transfer function of the causal linear filter that minimizes the mean-square error between the input signal and the total filter output.

b) Find the value of the minimum mean-square error.

9–6.4

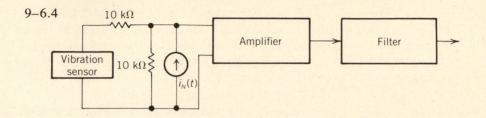

The block diagram above illustrates a system for measuring vibrations with a sensitive vibration sensor having an internal impedance of 10,000 Ω. The open-circuit signal produced by this sensor comes from a stationary random process having a spectral density of

$$S_X(\omega) = \frac{10^{-4}\omega^2}{\omega^4 + 13\omega^2 + 36} \text{ V}^2/\text{Hz}$$

The output of the vibration sensor is connected to the input of a broadband amplifier whose input circuit may be modeled by a resistance of 10,000 Ω in parallel with a noise current source in which the current is a sample function from a white noise source having a spectral density of 10^{-8} A^2/Hz. The amplifier has a voltage gain of 10 and has an output impedance that is negligibly small compared to the input impedance of the causal linear filter connected to the output.

a) Find the transfer function of the output filter that will minimize the mean-square error between the signal into the filter and the total output from the filter. Normalize the filter gain so that its maximum gain is unity.

b) Find the ratio of the minimum mean-square error to the mean-square value of the signal into the filter.

References

See the References for Chapter 1. Of particular interest for the material of this chapter are the books by Davenport and Root and by Lanning and Battin.

Appendices

APPENDIX A

Mathematical Tables

Table A–1   Trigonometric Identities

$$\sin (A \pm B) = \sin A \cos B \pm \cos A \sin B$$

$$\cos (A \pm B) = \cos A \cos B \mp \sin A \sin B$$

$$\cos A \cos B = \frac{1}{2} [\cos (A + B) + \cos (A - B)]$$

$$\sin A \sin B = \frac{1}{2} [\cos (A - B) - \cos (A + B)]$$

$$\sin A \cos B = \frac{1}{2} [\sin (A + B) + \sin (A - B)]$$

$$\sin A + \sin B = 2 \sin \frac{1}{2} (A + B) \cos \frac{1}{2} (A - B)$$

$$\sin A - \sin B = 2 \sin \frac{1}{2} (A - B) \cos \frac{1}{2} (A + B)$$

$$\cos A + \cos B = 2 \cos \frac{1}{2} (A + B) \cos \frac{1}{2} (A - B)$$

$$\cos A - \cos B = -2 \sin \frac{1}{2} (A + B) \sin \frac{1}{2} (A - B)$$

$$\sin 2A = 2 \sin A \cos A$$

$$\cos 2A = 2 \cos^2 A - 1 = 1 - 2 \sin^2 A = \cos^2 A - \sin^2 A$$

$$\sin \frac{1}{2} A = \sqrt{\frac{1}{2} (1 - \cos A)}$$

$$\cos \frac{1}{2} A = \sqrt{\frac{1}{2} (1 + \cos A)}$$

Table A–1   (continued)

$$\sin^2 A = \frac{1}{2}(1 - \cos 2A)$$

$$\cos^2 A = \frac{1}{2}(1 + \cos 2A)$$

$$\sin x = \frac{e^{jx} - e^{-jx}}{2j} \quad \text{and} \quad \cos x = \frac{e^{jx} + e^{-jx}}{2}$$

$$e^{jx} = \cos x + j \sin x$$

$$A \cos(\omega t + \phi_1) + B \cos(\omega t + \phi_2) = C \cos(\omega t + \phi_3)$$

$$\text{where } C = \sqrt{A^2 + B^2 + 2AB \cos(\phi_2 - \phi_1)}$$

$$\phi_3 = \tan^{-1}\left[\frac{A \sin \phi_1 + B \sin \phi_2}{A \cos \phi_1 + B \cos \phi_2}\right]$$

$$\sin(\omega t + \phi) = \cos(\omega t + \phi - 90°)$$

Table A–2   Indefinite Integrals

$$\int \sin ax \, dx = -\frac{1}{a} \cos ax \qquad \int \cos ax \, dx = \frac{1}{a} \sin ax$$

$$\int \sin^2 ax \, dx = \frac{x}{2} - \frac{\sin 2ax}{4a}$$

$$\int x \sin ax \, dx = \frac{1}{a^2}(\sin ax - ax \cos ax)$$

$$\int x^2 \sin ax \, dx = \frac{1}{a^3}(2ax \sin ax + 2 \cos ax - a^2x^2 \cos ax)$$

$$\int \cos^2 ax \, dx = \frac{x}{2} + \frac{\sin 2ax}{4a}$$

$$\int x \cos ax \, dx = \frac{1}{a^2}(\cos ax + ax \sin ax)$$

$$\int x^2 \cos ax \, dx = \frac{1}{a^3}(2ax \cos ax - 2 \sin ax + a^2x^2 \sin ax)$$

$$\int \sin ax \sin bx \, dx = \frac{\sin(a-b)x}{2(a-b)} - \frac{\sin(a+b)}{2(a+b)} \qquad a^2 \neq b^2$$

Table A–2   (continued)

$$\int \sin ax \cos bx \, dx = -\left[\frac{\cos (a - b)x}{2(a - b)} + \frac{\cos (a + b)x}{2(a + b)} \right]$$

$$\int \cos ax \cos bx \, dx = \frac{\sin (a - b)x}{2(a - b)} + \frac{\sin (a + b)x}{2(a + b)}$$

$\left. \right\} \quad a^2 \neq b^2$

$$\int e^{ax} \, dx = \frac{1}{a} e^{ax}$$

$$\int x e^{ax} \, dx = \frac{e^{ax}}{a^2} (ax - 1)$$

$$\int x^2 e^{ax} \, dx = \frac{e^{ax}}{a^3} (a^2 x^2 - 2ax + 2)$$

$$\int e^{ax} \sin bx \, dx = \frac{e^{ax}}{a^2 + b^2} (a \sin bx - b \cos bx)$$

$$\int e^{ax} \cos bx \, dx = \frac{e^{ax}}{a^2 + b^2} (a \cos bx + b \sin bx)$$

Table A–3   Definite Integrals

$$\int_0^\infty x^n e^{-ax} \, dx = \frac{n!}{a^{n+1}} = \frac{\Gamma(n + 1)}{a^{n+1}}$$

where $\Gamma(u) = \int_0^\infty z^{u-1} e^{-z} \, dz$ (Gamma function)

$$\int_0^\infty e^{-r^2 x^2} \, dx = \frac{\sqrt{\pi}}{2r}$$

$$\int_0^\infty x e^{-r^2 x^2} \, dx = \frac{1}{2r^2}$$

$$\int_0^\infty x^2 e^{-r^2 x^2} \, dx = \frac{\sqrt{\pi}}{4r^3}$$

$$\int_0^\infty x^n e^{-r^2 x^2} \, dx = \frac{\Gamma[(n + 1)/2]}{2r^{n+1}}$$

$$\int_0^\infty \frac{\sin ax}{x} \, dx = \frac{\pi}{2}, 0, -\frac{\pi}{2} \quad \text{for} \quad a > 0, a = 0, a < 0$$

$$\int_0^\infty \frac{\sin^2 x}{x^2} \, dx = \frac{\pi}{2}$$

Table A–3   (continued)

$$\int_0^\infty \frac{\sin^2 ax}{x^2} \, dx = |a| \frac{\pi}{2}$$

$$\int_0^\pi \sin^2 mx \, dx = \int_0^\pi \sin^2 x \, dx = \int_0^\pi \cos^2 mx \, dx = \int_0^\pi \cos^2 x \, dx = \frac{\pi}{2},$$

$$m \text{ an integer}$$

$$\int_0^\pi \sin mx \sin nx \, dx = \int_0^\pi \cos mx \cos nx \, dx = 0 \quad m \neq n$$

$$m, n \text{ integers}$$

$$\int_0^\pi \sin mx \cos nx \, dx = \begin{cases} \dfrac{2m}{m^2 - n^2} & \text{if } m + n \text{ odd} \\ 0 & \text{if } m + n \text{ even} \end{cases}$$

Table A–4   Fourier Transforms

Description	$f(t)$	$F(\omega)$
Definition	$f(t) = \dfrac{1}{2\pi} \displaystyle\int_{-\infty}^\infty F(\omega)e^{j\omega t} \, d\omega$	$F(\omega) = \displaystyle\int_{-\infty}^\infty f(t)e^{-j\omega t} \, dt$
Reversal	$f(-t)$	$F(-\omega)$
Symmetry	$F(t)$	$2\pi f(-\omega)$
Scaling	$f(at)$	$\dfrac{1}{|a|} F\left(\dfrac{\omega}{a}\right)$
Delay	$f(t - t_0)$	$e^{-j\omega t_0}F(\omega)$
Complex conjugate	$f^*(t)$	$F^*(-\omega)$
Time differentiation	$\dfrac{d^n f(t)}{dt^n}$	$(j\omega)^n F(\omega)$
Frequency differentiation	$t^n f(t)$	$(j)^n \dfrac{d^n F(\omega)}{d\omega^n}$
Time integration	$\displaystyle\int_{-\infty}^t f(t) \, dt$	$\dfrac{1}{j\omega} F(\omega) + \pi F(0)\delta(\omega)$
Time convolution	$\displaystyle\int_{-\infty}^\infty f_1(\lambda)f_2(t - \lambda) \, d\lambda$	$F_1(\omega)F_2(\omega)$
Frequency convolution	$f_1(t)f_2(t)$	$\dfrac{1}{2\pi} \displaystyle\int_{-\infty}^\infty F_1(\xi)F_2(\omega - \xi) \, d\xi$
Parseval's theorem	$\displaystyle\int_{-\infty}^\infty f_1(t)f_2(t) \, dt = \dfrac{1}{2\pi} \displaystyle\int_{-\infty}^\infty F_1(\omega)F_2(-\omega) \, d\omega$	

Table A–4 (continued)

Description	$f(t)$	$F(\omega)$
Modulation	$e^{j\omega_0 t}f(t)$	$F(\omega - \omega_0)$
Unit impulse	$\delta(t)$	1
Unit step	$u(t)$	$\pi\delta(\omega) + \dfrac{1}{j\omega}$
Signum function	$\operatorname{sgn} t$	$\dfrac{2}{j\omega}$
Sine	$\sin \omega_0 t$	$-j\pi[\delta(\omega - \omega_0) - \delta(\omega + \omega_0)]$
Cosine	$\cos \omega_0 t$	$\pi[\delta(\omega - \omega_0) + \delta(\omega + \omega_0)]$

Rectangular pulse		$T\dfrac{\sin (\omega T/2)}{\omega T/2}$

Triangular pulse		$\dfrac{T}{2}\left(\dfrac{\sin (\omega T/4)}{\omega T/4}\right)^2$
Gaussian pulse	$e^{-\alpha^2 t^2}$	$\dfrac{\sqrt{\pi}}{\alpha} e^{-(\omega^2/4\alpha^2)}$
Fourier series	$\displaystyle\sum_{n=-\infty}^{\infty} \alpha_n e^{j(2\pi nt/T)}$	$2\pi \displaystyle\sum_{n=-\infty}^{\infty} \alpha_n \delta\left(\omega - \dfrac{2\pi n}{T}\right)$
Impulse train	$\displaystyle\sum_{n=-\infty}^{\infty} \delta(t - nT)$	$\dfrac{2\pi}{T} \displaystyle\sum_{n=-\infty}^{\infty} \delta\left(\omega - \dfrac{2\pi n}{T}\right)$
Constant	K	$2\pi K\delta(\omega)$

Table A–5 One-Sided Laplace Transforms

Description	$f(t)$	$F(s)$
Definition	$f(t) = \dfrac{1}{2\pi j} \displaystyle\int_{c-j\infty}^{c+j\infty} F(s)\, e^{st}\, ds$	$F(s) = \displaystyle\int_{0}^{\infty} f(t) e^{-st}\, dt$
Derivative	$f'(t) = \dfrac{df(t)}{dt}$	$sF(s) - f(0)$
2nd derivative	$f''(t) = \dfrac{d^2 f(t)}{dt^2}$	$s^2 F(s) - sf(0) - f'(0)$
Integral	$\displaystyle\int_{0}^{t} f(\xi)\, d\xi$	$\dfrac{1}{s} F(s)$
t multiplication	$tf(t)$	$-\dfrac{dF(s)}{ds}$
Division by t	$\dfrac{1}{t} f(t)$	$\displaystyle\int_{s}^{\infty} F(\xi)\, d\xi$
Delay	$f(t - t_0) u(t - t_0)$	$e^{-st_0} F(s)$
Exponential decay	$e^{-at} f(t)$	$F(s + a)$
Scale change	$f(at) \quad a > 0$	$\dfrac{1}{a} F\left(\dfrac{s}{a}\right)$
Convolution	$\displaystyle\int_{0}^{t} f_1(\lambda) f_2(t - \lambda)\, d\lambda$	$F_1(s) F_2(s)$
Initial value	$f(0^+)$	$\displaystyle\lim_{s \to \infty} sF(s)$
Final value	$f(\infty)$	$\displaystyle\lim_{s \to 0} sF(s)$ [$F(s)$-Left-half plane poles only]
Impulse	$\delta(t)$	1
Step	$u(t)$	$\dfrac{1}{s}$
Ramp	$tu(t)$	$\dfrac{1}{s^2}$
nth order ramp	$t^n u(t)$	$\dfrac{n!}{s^{n+1}}$
Exponential	$e^{-\alpha t} u(t)$	$\dfrac{1}{s + \alpha}$
Damped ramp	$te^{-\alpha t} u(t)$	$\dfrac{1}{(s + \alpha)^2}$
Sine wave	$\sin(\beta t) u(t)$	$\dfrac{\beta}{s^2 + \beta^2}$
Cosine wave	$\cos(\beta t) u(t)$	$\dfrac{s}{s^2 + \beta^2}$
Damped sine	$e^{-\alpha t} \sin(\beta t) u(t)$	$\dfrac{\beta}{(s + \alpha)^2 + \beta^2}$
Damped cosine	$e^{-\alpha t} \cos(\beta t) u(t)$	$\dfrac{s + \alpha}{(s + \alpha)^2 + \beta^2}$

APPENDIX B

Frequently Encountered Probability Distributions

There are a number of probability distributions that occur quite frequently in the application of probability theory to practical problems. Mathematical expressions for the most common of these distributions are collected here along with their most important parameters.

The following notation is used throughout:

$\Pr(x)$—probability of the event x occurring.

$f_X(x)$—probability density function of the random variable X at the point x.

$\overline{X} = E\{X\}$—mean of the random variable X.

$\sigma_X^2 = E\{(X - \overline{X})^2\}$—variance of the random variable X.

$\phi(u) = \displaystyle\int_{-\infty}^{\infty} f_X(x)e^{jux}\,dx$—characteristic function of the random variable X.

Discrete Probability Functions

Bernoulli (Special case of Binomial)

$$\Pr(x) = \begin{cases} p & x = 1 \\ q = 1 - p & x = 0 \qquad 0 < p < 1 \\ 0 & \text{otherwise} \end{cases}$$

$$f_X(x) = p\delta(x - 1) + q\delta(x)$$

$$\overline{X} = p$$

$$\sigma_X^2 = pq$$

$$\phi(u) = 1 - p + pe^{ju}$$

371

Binomial

$$\Pr(x) = \begin{cases} \binom{n}{x} p^x q^{n-x} & x = 0, 1, 2, \ldots, n \\ 0 & \text{otherwise} \end{cases}$$

$$0 < p < 1 \qquad q = 1 - p \qquad n = 1, 2, \ldots$$

$$f_X(x) = \sum_{k=0}^{n} \binom{n}{k} p^k q^{n-k} \delta(x - k)$$

$$\overline{X} = np$$

$$\sigma_X^2 = npq$$

$$\phi(u) = [1 - p + pe^{ju}]^n$$

Pascal

$$\Pr(x) = \begin{cases} \binom{x-1}{n-1} p^n q^{x-n} & x = n, n+1, \ldots \\ 0 & \text{otherwise} \end{cases}$$

$$0 < p < 1 \qquad q = 1 - p \qquad n = 1, 2, 3, \ldots$$

$$\overline{X} = np^{-1}$$

$$\sigma_X^2 = nqp^{-2}$$

$$\phi(u) = p^n e^{jnu}[1 - qe^{ju}]^{-n}$$

Poisson

$$\Pr(x) = \frac{a^x e^{-a}}{x!} \qquad x = 0, 1, 2, \ldots$$

$$a > 0$$

$$\overline{X} = a$$

$$\sigma_X^2 = a$$

$$\phi(u) = e^{a(e^{ju} - 1)}$$

Continuous Distributions

Beta

$$f_X(x) = \begin{cases} \dfrac{(a+b-1)!}{(a-1)!(b-1)!} x^{a-1}(1-x)^{b-1} & 0 < x < 1 \\ 0 & \text{otherwise} \end{cases}$$

$$a > 0 \qquad b > 0$$

$$\bar{X} = \frac{a}{a+b}$$

$$\sigma_X^2 = \frac{ab}{(a+b)^2(a+b+1)}$$

Cauchy

$$f_X(x) = \frac{1}{\pi} \frac{a}{a^2 + (x-b)^2} \qquad -\infty < x < \infty$$

$$a > 0 \qquad -\infty < b < \infty$$

Mean and variance not defined

$$\phi(u) = e^{jbu - a|u|}$$

Chi-square

$$f_X(x) = \begin{cases} \left[\dfrac{1}{\Gamma\left(\dfrac{n}{2}\right)} 2^{-n/2} x^{(n/2)-1} e^{-x/2} \right] & x > 0 \\ & \text{otherwise} \end{cases}$$

$$n = 1, 2, \ldots$$

$$\bar{X} = n$$

$$\sigma_X^2 = 2n$$

$$\phi(u) = (1 - 2ju)^{-n/2}$$

Erlang

$$f_X(x) = \begin{cases} \dfrac{a^n x^{n-1} e^{-ax}}{(n-1)!} & x > 0 \\ 0 & \text{otherwise} \end{cases}$$

$$a > 0 \qquad n = 1, 2, \ldots$$

$$\overline{X} = na^{-1}$$
$$\sigma_X^2 = na^{-2}$$
$$\phi(u) = a^n(a - ju)^{-n}$$

Exponential

$$f_X(x) = \begin{cases} ae^{-ax} & x > 0 \\ 0 & \text{otherwise} \end{cases}$$
$$a > 0$$
$$\overline{X} = a^{-1}$$
$$\sigma_X^2 = a^{\pm 2}$$
$$\phi(u) = a(a - ju)^{-1}$$

Gamma

$$f_X(x) = \begin{cases} \dfrac{x^a e^{-x/b}}{a! b^{a+1}} & x > 0 \\ 0 & \text{otherwise} \end{cases}$$
$$a > -1 \quad b > 0$$
$$\overline{X} = (a + 1)b$$
$$\sigma_X^2 = (a + 1)b^2$$
$$\phi(u) = (1 - jbu)^{-a-1}$$

Laplace

$$f_X(x) = \frac{a}{2} e^{-a|x-b|} \qquad \begin{array}{l} -\infty < x < \infty \\ -\infty < b < \infty \end{array}$$
$$a > 0$$
$$\overline{X} = b$$
$$\sigma_X^2 = 2a^{-2}$$
$$\phi(u) = a^2 e^{jbu}(a^2 + u^2)^{-1}$$

Log-normal

$$f_X(x) = \begin{cases} \dfrac{\exp\{-[\ln(x - a) - b]^2/2\sigma^2\}}{\sqrt{2\pi}\,\sigma(x - a)} & x \geq a \\ 0 & \text{otherwise} \end{cases}$$
$$\sigma > 0 \qquad -\infty < a < \infty \qquad -\infty < b < \infty$$

$$\overline{X} = a + e^{b + .5\sigma^2}$$
$$\sigma_X^2 = e^{2b + \sigma^2}(e^{\sigma^2} - 1)$$

Maxwell

$$f_X(x) = \begin{cases} \sqrt{\dfrac{2}{\pi}}\, a^3 x^2 e^{-a^2 x^2/2} & x > 0 \\ 0 & \text{otherwise} \end{cases}$$
$$a > 0$$
$$\overline{X} = \sqrt{8/\pi}\, a^{-1}$$
$$\sigma_X^2 = \left(3 - \frac{8}{\pi}\right) a^{-2}$$

Normal

$$f_X(x) = \frac{1}{\sqrt{2\pi}\,\sigma_X}\, e^{-(x - \overline{X})^2/2\sigma_X^2} \qquad -\infty < x < \infty$$
$$\sigma_X > 0 \qquad -\infty < \overline{X} < \infty$$
$$\phi(u) = e^{ju\overline{X} - (u^2\sigma_X^2/2)}$$

Normal-bivariate

$$f_{X,Y}(x,y) = \frac{1}{2\pi\sigma_X\sigma_Y \sqrt{1 - \rho^2}} \exp\left\{ \frac{-1}{2(1 - \rho^2)} \left[\left(\frac{x - \overline{X}}{\sigma_X}\right)^2 + \left(\frac{y - \overline{Y}}{\sigma_Y}\right)^2 \right. \right.$$
$$\left. \left. - \frac{2\rho}{\sigma_X\sigma_Y} (x - \overline{X})(y - \overline{Y}) \right] \right\}$$
$$-\infty < x < \infty \qquad -\infty < y < \infty \qquad \sigma_X > 0 \qquad \sigma_Y > 0$$
$$-1 < \rho < 1$$

$$\phi(u,v) = \exp\left[ju\overline{X} + jv\overline{Y} - \frac{1}{2}(u^2\sigma_X^2 + 2\rho uv\sigma_X\sigma_Y + v^2\sigma_Y^2) \right]$$

Rayleigh

$$f_X(x) = \begin{cases} \dfrac{x}{a^2}\, e^{-x^2/2a^2} & x > 0 \\ 0 & \text{otherwise} \end{cases}$$
$$\overline{X} = a\sqrt{\frac{\pi}{2}}$$
$$\sigma_X^2 = \left(2 - \frac{\pi}{2}\right) a^2$$

Uniform

$$f_X(x) = \begin{cases} \dfrac{1}{b-a} & a < x < b \\ 0 & \text{otherwise} \end{cases}$$

$$-\infty < a < b < \infty$$

$$\bar{X} = \frac{a+b}{2}$$

$$\sigma_X^2 = \frac{(b-a)^2}{12}$$

$$\phi(u) = \frac{e^{jub} - e^{jua}}{ju[b-a)}$$

Weibull

$$f_X(x) = \begin{cases} abx^{b-1}e^{-ax^b} & x > 0 \\ 0 & \text{otherwise} \end{cases}$$

$$a > 0 \qquad b > 0$$

$$\bar{X} = \left(\frac{1}{a}\right)^{1/b} \Gamma(1 + b^{-1})$$

$$\sigma_X^2 = \left(\frac{1}{a}\right)^{2/b} \{\Gamma(1 + 2b^{-1}) - [\Gamma(1 + b^{-1})]^2\}$$

APPENDIX C

Binomial Coefficients

$$\binom{n}{m} = \frac{n!}{(n-m)!m!}$$

n	$\binom{n}{0}$	$\binom{n}{1}$	$\binom{n}{2}$	$\binom{n}{3}$	$\binom{n}{4}$	$\binom{n}{5}$	$\binom{n}{6}$	$\binom{n}{7}$	$\binom{n}{8}$	$\binom{n}{9}$
0	1									
1	1	1								
2	1	2	1							
3	1	3	3	1						
4	1	4	6	4	1					
5	1	5	10	10	5	1				
6	1	6	15	20	15	6	1			
7	1	7	21	35	35	21	7	1		
8	1	8	28	56	70	56	28	8	1	
9	1	9	36	84	126	126	84	36	9	1
10	1	10	45	120	210	252	210	120	45	10
11	1	11	55	165	330	462	462	330	165	55
12	1	12	66	220	495	792	924	792	495	220
13	1	13	78	286	715	1287	1716	1716	1287	715
14	1	14	91	364	1001	2002	3003	3432	3003	2002
15	1	15	105	455	1365	3003	5005	6435	6435	5005
16	1	16	120	560	1820	4368	8008	11440	12870	11440
17	1	17	136	680	2380	6188	12376	19448	24310	24310
18	1	18	153	816	3060	8568	18564	31824	43758	48620
19	1	19	171	969	3876	11628	27132	50388	75582	92378

$$\binom{n}{m} + \binom{n}{m+1} = \binom{n+1}{m+1}$$

Useful Relationships

$$\binom{n}{n-m} = \binom{n}{m}$$

APPENDIX *D*

Normal Probability Distribution Function

$$\Phi(x) = \frac{1}{\sqrt{2\pi}} \int_{-\infty}^{x} e^{-t^2/2} \, dt; \ \Phi(-x) = 1 - \Phi(x)$$

x	.00	.01	.02	.03	.04	.05	.06	.07	.08	.09
0.0	.5000	.5040	.5080	.5120	.5160	.5199	.5239	.5279	.5319	.5359
0.1	.5398	.5438	.5478	.5517	.5557	.5596	.5636	.5675	.5714	.5753
0.2	.5793	.5832	.5871	.5910	.5948	.5987	.6026	.6064	.6103	.6141
0.3	.6179	.6217	.6255	.6293	.6331	.6368	.6406	.6443	.6480	.6517
0.4	.6554	.6591	.6628	.6664	.6700	.6736	.6772	.6808	.6844	.6879
0.5	.6915	.6950	.6985	.7019	.7054	.7088	.7123	.7157	.7190	.7224
0.6	.7257	.7291	.7324	.7357	.7389	.7422	.7454	.7486	.7517	.7549
0.7	.7580	.7611	.7642	.7673	.7704	.7734	.7764	.7794	.7823	.7852

x	.00	.01	.02	.03	.04	.05	.06	.07	.08	.09
0.8	.7881	.7910	.7939	.7967	.7995	.8023	.8051	.8078	.8106	.8133
0.9	.8159	.8186	.8212	.8238	.8264	.8289	.8315	.8340	.8365	.8389
1.0	.8413	.8438	.8461	.8485	.8508	.8531	.8554	.8577	.8599	.8621
1.1	.8643	.8665	.8686	.8708	.8729	.8749	.8770	.8790	.8810	.8830
1.2	.8849	.8869	.8888	.8907	.8925	.8944	.8962	.8980	.8997	.9015
1.3	.9032	.9049	.9066	.9082	.9099	.9115	.9131	.9147	.9162	.9177
1.4	.9192	.9207	.9222	.9236	.9251	.9265	.9279	.9292	.9306	.9319
1.5	.9332	.9345	.9357	.9370	.9382	.9394	.9406	.9418	.9429	.9441
1.6	.9452	.9463	.9474	.9484	.9495	.9505	.9515	.9525	.9535	.9545
1.7	.9554	.9564	.9573	.9582	.9591	.9599	.9608	.9616	.9625	.9633
1.8	.9641	.9649	.9656	.9664	.9671	.9678	.9686	.9693	.9699	.9706
1.9	.9713	.9719	.9726	.9732	.9738	.9744	.9750	.9756	.9761	.9767
2.0	.9772	.9778	.9783	.9788	.9793	.9798	.9803	.9808	.9812	.9817
2.1	.9821	.9826	.9830	.9834	.9838	.9842	.9846	.9850	.9854	.9857
2.2	.9861	.9864	.9868	.9871	.9875	.9878	.9881	.9884	.9887	.9890
2.3	.9893	.9896	.9898	.9901	.9904	.9906	.9909	.9911	.9913	.9916
2.4	.9918	.9920	.9922	.9925	.9927	.9929	.9931	.9932	.9934	.9936
2.5	.9938	.9940	.9941	.9943	.9945	.9946	.9948	.9949	.9951	.9952
2.6	.9953	.9955	.9956	.9957	.9959	.9960	.9961	.9962	.9963	.9964
2.7	.9965	.9966	.9967	.9968	.9969	.9970	.9971	.9972	.9973	.9974
2.8	.9974	.9975	.9976	.9977	.9977	.9978	.9979	.9979	.9980	.9981
2.9	.9981	.9982	.9982	.9983	.9984	.9984	.9985	.9985	.9986	.9986
3.0	.9987	.9987	.9987	.9988	.9988	.9989	.9989	.9989	.9990	.9990
3.1	.9990	.9991	.9991	.9991	.9992	.9992	.9992	.9992	.9993	.9993
3.2	.9993	.9993	.9994	.9994	.9994	.9994	.9994	.9995	.9995	.9995
3.3	.9995	.9995	.9996	.9996	.9996	.9996	.9996	.9996	.9996	.9997
3.4	.9997	.9997	.9997	.9997	.9997	.9997	.9997	.9997	.9998	.9998
3.5	.9998	.9998	.9998	.9998	.9998	.9998	.9998	.9998	.9998	.9998
3.6	.9998	.9999	.9999	.9999	.9999	.9999	.9999	.9999	.9999	.9999
3.7	.9999	.9999	.9999	.9999	.9999	.9999	.9999	.9999	.9999	.9999
3.8	.9999	.9999	.9999	.9999	.9999	.9999	.9999	1.0000	1.0000	1.0000

APPENDIX *E*

The Q-Function

$$Q(x) = \frac{1}{\sqrt{2\pi}} \int_x^\infty e^{-u^2/2}\, du \qquad Q(-x) = 1 - Q(x)$$

x	0.00	0.01	0.02	0.03	0.04	0.05	0.06	0.07	0.08	0.09
0.0	0.5000	0.4960	0.4920	0.4880	0.4840	0.4801	0.4761	0.4721	0.4681	0.4641
0.1	0.4602	0.4562	0.4522	0.4483	0.4443	0.4404	0.4364	0.4325	0.4286	0.4247
0.2	0.4207	0.4168	0.4129	0.4090	0.4052	0.4013	0.3974	0.3936	0.3897	0.3859
0.3	0.3821	0.3783	0.3745	0.3707	0.3669	0.3632	0.3594	0.3557	0.3520	0.3483
0.4	0.3446	0.3409	0.3372	0.3336	0.3300	0.3264	0.3228	0.3192	0.3156	0.3121
0.5	0.3085	0.3050	0.3015	0.2981	0.2946	0.2912	0.2877	0.2843	0.2810	0.2776
0.6	0.2743	0.2709	0.2676	0.2643	0.2611	0.2578	0.2546	0.2514	0.2483	0.2451
0.7	0.2420	0.2389	0.2358	0.2327	0.2297	0.2266	0.2236	0.2206	0.2177	0.2148
0.8	0.2119	0.2090	0.2061	0.2033	0.2005	0.1977	0.1949	0.1922	0.1894	0.1867
0.9	0.1841	0.1814	0.1788	0.1762	0.1736	0.1711	0.1685	0.1660	0.1635	0.1611
1.0	0.1587	0.1562	0.1539	0.1515	0.1492	0.1469	0.1446	0.1423	0.1401	0.1379
1.1	0.1357	0.1335	0.1314	0.1292	0.1271	0.1251	0.1230	0.1210	0.1190	0.1170
1.2	0.1151	0.1131	0.1112	0.1093	0.1075	0.1056	0.1038	0.1020	0.1003	0.0985
1.3	0.0968	0.0951	0.0934	0.0918	0.0901	0.0885	0.0869	0.0853	0.0838	0.0823
1.4	0.0808	0.0793	0.0778	0.0764	0.0749	0.0735	0.0721	0.0708	0.0694	0.0681
1.5	0.0668	0.0655	0.0643	0.0630	0.0618	0.0606	0.0594	0.0582	0.0571	0.0559
1.6	0.0548	0.0537	0.0526	0.0516	0.0505	0.0495	0.0485	0.0475	0.0465	0.0455
1.7	0.0446	0.0436	0.0427	0.0418	0.0409	0.0401	0.0392	0.0384	0.0375	0.0367
1.8	0.0359	0.0351	0.0344	0.0336	0.0329	0.0322	0.0314	0.0307	0.0301	0.0294
1.9	0.0287	0.0281	0.0274	0.0268	0.0262	0.0256	0.0250	0.0244	0.0239	0.0233
2.0	0.2275E-01	0.2222E-01	0.2169E-01	0.2118E-01	0.2068E-01	0.2018E-01	0.1970E-01	0.1923E-01	0.1876E-01	0.1831E-01
2.1	0.1786E-01	0.1743E-01	0.1700E-01	0.1659E-01	0.1618E-01	0.1578E-01	0.1539E-01	0.1500E-01	0.1463E-01	0.1426E-01
2.2	0.1390E-01	0.1355E-01	0.1321E-01	0.1287E-01	0.1255E-01	0.1222E-01	0.1191E-01	0.1160E-01	0.1130E-01	0.1101E-01
2.3	0.1072E-01	0.1044E-01	0.1017E-01	0.9903E-02	0.9642E-02	0.9387E-02	0.9137E-02	0.8894E-02	0.8656E-02	0.8424E-02
2.4	0.8198E-02	0.7976E-02	0.7760E-02	0.7549E-02	0.7344E-02	0.7143E-02	0.6947E-02	0.6756E-02	0.6569E-02	0.6387E-02
2.5	0.6210E-02	0.6037E-02	0.5868E-02	0.5703E-02	0.5543E-02	0.5386E-02	0.5234E-02	0.5085E-02	0.4940E-02	0.4799E-02
2.6	0.4661E-02	0.4527E-02	0.4396E-02	0.4269E-02	0.4145E-02	0.4025E-02	0.3907E-02	0.3793E-02	0.3681E-02	0.3573E-02
2.7	0.3467E-02	0.3364E-02	0.3264E-02	0.3167E-02	0.3072E-02	0.2980E-02	0.2890E-02	0.2803E-02	0.2718E-02	0.2635E-02
2.8	0.2555E-02	0.2477E-02	0.2401E-02	0.2327E-02	0.2256E-02	0.2186E-02	0.2118E-02	0.2052E-02	0.1988E-02	0.1926E-02
2.9	0.1866E-02	0.1807E-02	0.1750E-02	0.1695E-02	0.1641E-02	0.1589E-02	0.1538E-02	0.1489E-02	0.1441E-02	0.1395E-02

x	0.00	0.01	0.02	0.03	0.04	0.05	0.06	0.07	0.08	0.09
3.0	0.1350E-02	0.1306E-02	0.1264E-02	0.1223E-02	0.1183E-02	0.1144E-02	0.1107E-02	0.1070E-02	0.1035E-02	0.1001E-02
3.1	0.9676E-03	0.9354E-03	0.9043E-03	0.8740E-03	0.8447E-03	0.8164E-03	0.7888E-03	0.7622E-03	0.7364E-03	0.7114E-03
3.2	0.6871E-03	0.6637E-03	0.6410E-03	0.6190E-03	0.5977E-03	0.5770E-03	0.5571E-03	0.5377E-03	0.5190E-03	0.5009E-03
3.3	0.4834E-03	0.4665E-03	0.4501E-03	0.4342E-03	0.4189E-03	0.4041E-03	0.3897E-03	0.3758E-03	0.3624E-03	0.3495E-03
3.4	0.3369E-03	0.3248E-03	0.3131E-03	0.3018E-03	0.2909E-03	0.2803E-03	0.2701E-03	0.2602E-03	0.2507E-03	0.2415E-03
3.5	0.2326E-03	0.2241E-03	0.2158E-03	0.2078E-03	0.2001E-03	0.1926E-03	0.1854E-03	0.1785E-03	0.1718E-03	0.1653E-03
3.6	0.1591E-03	0.1531E-03	0.1473E-03	0.1417E-03	0.1363E-03	0.1311E-03	0.1261E-03	0.1213E-03	0.1166E-03	0.1121E-03
3.7	0.1078E-03	0.1036E-03	0.9961E-04	0.9574E-04	0.9201E-04	0.8842E-04	0.8496E-04	0.8162E-04	0.7841E-04	0.7532E-04
3.8	0.7235E-04	0.6948E-04	0.6673E-04	0.6407E-04	0.6152E-04	0.5906E-04	0.5669E-04	0.5442E-04	0.5223E-04	0.5012E-04
3.9	0.4810E-04	0.4615E-04	0.4427E-04	0.4247E-04	0.4074E-04	0.3908E-04	0.3748E-04	0.3594E-04	0.3446E-04	0.3304E-04
4.0	0.3167E-04	0.3036E-04	0.2910E-04	0.2789E-04	0.2673E-04	0.2561E-04	0.2454E-04	0.2351E-04	0.2252E-04	0.2157E-04
4.1	0.2066E-04	0.1978E-04	0.1894E-04	0.1814E-04	0.1737E-04	0.1662E-04	0.1591E-04	0.1523E-04	0.1458E-04	0.1395E-04
4.2	0.1335E-04	0.1277E-04	0.1222E-04	0.1168E-04	0.1118E-04	0.1069E-04	0.1022E-04	0.9774E-05	0.9345E-05	0.8934E-05
4.3	0.8540E-05	0.8163E-05	0.7802E-05	0.7456E-05	0.7124E-05	0.6807E-05	0.6503E-05	0.6212E-05	0.5934E-05	0.5668E-05
4.4	0.5413E-05	0.5169E-05	0.4935E-05	0.4712E-05	0.4498E-05	0.4294E-05	0.4098E-05	0.3911E-05	0.3732E-05	0.3561E-05
4.5	0.3398E-05	0.3241E-05	0.3092E-05	0.2949E-05	0.2813E-05	0.2682E-05	0.2558E-05	0.2439E-05	0.2325E-05	0.2216E-05
4.6	0.2112E-05	0.2013E-05	0.1919E-05	0.1828E-05	0.1742E-05	0.1660E-05	0.1581E-05	0.1506E-05	0.1434E-05	0.1366E-05
4.7	0.1301E-05	0.1239E-05	0.1179E-05	0.1123E-05	0.1069E-05	0.1017E-05	0.9680E-06	0.9211E-06	0.8765E-06	0.8339E-06
4.8	0.7933E-06	0.7547E-06	0.7178E-06	0.6827E-06	0.6492E-06	0.6173E-06	0.5869E-06	0.5580E-06	0.5304E-06	0.5042E-06
4.9	0.4792E-06	0.4554E-06	0.4327E-06	0.4112E-06	0.3906E-06	0.3711E-06	0.3525E-06	0.3348E-06	0.3179E-06	0.3019E-06
5.0	0.2867E-06	0.2722E-06	0.2584E-06	0.2452E-06	0.2328E-06	0.2209E-06	0.2096E-06	0.1989E-06	0.1887E-06	0.1790E-06
5.1	0.1698E-06	0.1611E-06	0.1528E-06	0.1449E-06	0.1374E-06	0.1302E-06	0.1235E-06	0.1170E-06	0.1109E-06	0.1051E-06
5.2	0.9964E-07	0.9442E-07	0.8946E-07	0.8475E-07	0.8029E-07	0.7605E-07	0.7203E-07	0.6821E-07	0.6459E-07	0.6116E-07
5.3	0.5790E-07	0.5481E-07	0.5188E-07	0.4911E-07	0.4647E-07	0.4398E-07	0.4161E-07	0.3937E-07	0.3724E-07	0.3523E-07
5.4	0.3332E-07	0.3151E-07	0.2980E-07	0.2818E-07	0.2664E-07	0.2518E-07	0.2381E-07	0.2250E-07	0.2127E-07	0.2010E-07
5.5	0.1899E-07	0.1794E-07	0.1695E-07	0.1601E-07	0.1512E-07	0.1428E-07	0.1349E-07	0.1274E-07	0.1203E-07	0.1135E-07
5.6	0.1072E-07	0.1012E-07	0.9548E-08	0.9011E-08	0.8503E-08	0.8022E-08	0.7569E-08	0.7140E-08	0.6735E-08	0.6352E-08
5.7	0.5990E-08	0.5649E-08	0.5326E-08	0.5022E-08	0.4734E-08	0.4462E-08	0.4206E-08	0.3964E-08	0.3735E-08	0.3519E-08
5.8	0.3316E-08	0.3124E-08	0.2942E-08	0.2771E-08	0.2610E-08	0.2458E-08	0.2314E-08	0.2179E-08	0.2051E-08	0.1931E-08
5.9	0.1818E-08	0.1711E-08	0.1610E-08	0.1515E-08	0.1425E-08	0.1341E-08	0.1261E-08	0.1186E-08	0.1116E-08	0.1049E-08
6.0	0.9866E-09	0.9276E-09	0.8721E-09	0.8198E-09	0.7706E-09	0.7242E-09	0.6806E-09	0.6396E-09	0.6009E-09	0.5646E-09
6.1	0.5303E-09	0.4982E-09	0.4679E-09	0.4394E-09	0.4126E-09	0.3874E-09	0.3637E-09	0.3415E-09	0.3205E-09	0.3008E-09
6.2	0.2823E-09	0.2649E-09	0.2486E-09	0.2332E-09	0.2188E-09	0.2052E-09	0.1925E-09	0.1805E-09	0.1693E-09	0.1587E-09
6.3	0.1488E-09	0.1395E-09	0.1308E-09	0.1226E-09	0.1149E-09	0.1077E-09	0.1009E-09	0.9451E-10	0.8854E-10	0.8294E-10
6.4	0.7769E-10	0.7276E-10	0.6814E-10	0.6380E-10	0.5974E-10	0.5593E-10	0.5235E-10	0.4900E-10	0.4586E-10	0.4292E-10
6.5	0.4016E-10	0.3758E-10	0.3515E-10	0.3289E-10	0.3076E-10	0.2877E-10	0.2690E-10	0.2516E-10	0.2352E-10	0.2199E-10
6.6	0.2056E-10	0.1922E-10	0.1796E-10	0.1678E-10	0.1568E-10	0.1465E-10	0.1369E-10	0.1279E-10	0.1195E-10	0.1116E-10
6.7	0.1042E-10	0.9731E-11	0.9086E-11	0.8483E-11	0.7919E-11	0.7392E-11	0.6900E-11	0.6435E-11	0.6009E-11	0.5607E-11
6.8	0.5231E-11	0.4880E-11	0.4552E-11	0.4246E-11	0.3960E-11	0.3693E-11	0.3443E-11	0.3210E-11	0.2993E-11	0.2790E-11
6.9	0.2600E-11	0.2423E-11	0.2258E-11	0.2104E-11	0.1961E-11	0.1826E-11	0.1701E-11	0.1585E-11	0.1476E-11	0.1374E-11
7.0	0.1280E-11	0.1192E-11	0.1109E-11	0.1033E-11	0.9612E-12	0.8946E-12	0.8325E-12	0.7747E-12	0.7208E-12	0.6706E-12

Q(x)	x
1E-01	1.28115
1E-02	2.32635
1E-03	3.09023
1E-04	3.71902
1E-05	4.26489

Q(x)	x
1E-06	4.75342
1E-07	5.19934
1E-08	5.61200
1E-09	5.99781
1E-10	6.63134

APPENDIX F

Student's t
Distribution Function

$$F_T(t) = \int_{-\infty}^{t} \frac{\Gamma\left(\frac{v+1}{2}\right)}{\sqrt{v\pi}\ \Gamma\left(\frac{v}{2}\right)} \left(1 + \frac{x^2}{v}\right)^{-\frac{v+1}{2}} dx$$

v \ F	0.60	0.75	0.90	0.95	0.975	0.99	0.995	0.9995
1	0.325	1.000	3.078	6.314	12.71	31.82	63.66	636.6
2	0.289	0.816	1.886	2.920	4.303	6.965	9.925	31.60
3	0.277	0.765	1.638	2.353	3.182	4.541	5.841	12.94
4	0.271	0.741	1.533	2.132	2.776	3.747	4.604	8.611
5	0.267	0.727	1.476	2.015	2.571	3.365	4.032	6.859
6	0.265	0.718	1.440	1.943	2.447	3.143	3.707	5.959
7	0.263	0.711	1.415	1.895	2.365	2.998	3.499	5.405
8	0.262	0.706	1.397	1.860	2.306	2.896	3.355	5.041
9	0.261	0.703	1.383	1.833	2.262	2.821	3.250	4.781
10	0.260	0.700	1.372	1.812	2.228	2.764	3.169	4.587
11	0.260	0.697	1.363	1.796	2.201	2.718	3.106	4.437
12	0.259	0.695	1.356	1.782	2.179	2.681	3.055	4.318
13	0.259	0.694	1.350	1.771	2.160	2.650	3.012	4.221
14	0.258	0.692	1.345	1.761	2.145	2.624	2.977	4.140
15	0.258	0.691	1.341	1.753	2.131	2.602	2.947	4.073
16	0.258	0.690	1.337	1.746	2.120	2.583	2.921	4.015
17	0.257	0.689	1.333	1.740	2.110	2.567	2.898	3.965
18	0.257	0.688	1.330	1.734	2.101	2.552	2.878	3.922
19	0.257	0.688	1.328	1.729	2.093	2.539	2.861	3.883
20	0.257	0.687	1.325	1.725	2.086	2.528	2.845	3.850
21	0.257	0.686	1.323	1.721	2.080	2.518	2.831	3.819
22	0.256	0.686	1.321	1.717	2.074	2.508	2.819	3.792
23	0.256	0.685	1.319	1.714	2.069	2.500	2.807	3.767
24	0.256	0.685	1.318	1.711	2.064	2.492	2.797	3.745
25	0.256	0.684	1.316	1.708	2.060	2.485	2.787	3.725
26	0.256	0.684	1.315	1.706	2.055	2.479	2.779	3.707
27	0.256	0.684	1.314	1.703	2.052	2.473	2.771	3.690
28	0.256	0.683	1.313	1.701	2.048	2.467	2.763	3.674
29	0.256	0.683	1.311	1.699	2.045	2.462	2.756	3.659
30	0.256	0.683	1.310	1.697	2.042	2.457	2.750	3.646
40	0.255	0.681	1.303	1.684	2.021	2.423	2.704	3.551
60	0.254	0.679	1.296	1.671	2.000	2.390	2.660	3.460
120	0.254	0.677	1.289	1.658	1.980	2.358	2.617	3.373
∞	0.253	0.674	1.282	1.645	1.960	2.326	2.576	3.291

Computer Programs for Estimating Correlation Functions and Spectral Density

The following Fortran programs are useful for estimating autocorrelation functions and spectral densities from data that has been stored in a file called 'data.' It is assumed the data consist of sample values from a random time function that has been sampled at equally spaced intervals separated by $\Delta t = 1$ second. The total number of data points is mm and should be put in the file called 'data' with a format of (f10.7). For purposes of illustration, we have set nn = 31 and mm = 1000, but any other values can be used so long as nn $\ll$ mm.

The first program calculates the sample mean of the data and the autocorrelation function $R(k\Delta t)$ for k = 0, 1, . . . , nn. It also calculates an upper bound on the variance of the estimated autocorrelation function using the estimated data values. The second program also calculates the estimated autocorrelation function and then uses this estimate to make a Hamming-window estimate of the spectral density, $S(q\Delta\omega)$, q = 0, 1, . . . , nn.

Estimated Autocorrelation Function

```
          parameter (nn=31, mm=1000)
c
c         Variable declarations
c
          integer i1, i2, n
          real data(0:(mm−1)), r(0:nn), mean, var
```

```
c
        open(unit = 2,file = 'data',status = 'old')
        do 4 i1 = 0, (mm − 1)
4       read (2,5) data(i1)
5       format(f10.7)
c
c       Calculate the mean.
c
        do 10 i1 = 0, (mm − 1)
            mean = mean + data(i1)
10      continue
        mean = mean/mm
        print *,'The mean is',mean
c
c       Find the estimated autocorrelation function and
c           maximum variance.
c
        print *,' The autocorrelation values are:'
        do 20 n = 0,nn
            do 30 i2 = 0, (mm − n − 1)
                r(n) = r(n) + data(i2)*data(i2 + n)
30          continue
            r(n) = r(n)/(mm − n)
            print *,' R(',n,')',r(n)
            var = var + r(n)**2
20      continue
        var = (var*4.0−2.0*r(0))/mm
        print *, 'The maximum estimate variance is', var
        stop
        end
```

Estimated Spectral Density

```
        parameter (nn = 31, mm = 1000)
c
c       Variable declarations
```

```
c
          integer i1, i2, i5, n, q
          real data(0:(mm − 1)), r(0:nn), rs( − 1:(nn + 1)), hs(0:nn), pi
c
          pi = 4.0*atan(1.0)
          open(unit = 2,file = 'data',status = 'old')
          do 4 i1 = 0, (mm − 1)
4         read (2,5) data(i1)
5         format(f10.7)
c
c         Find the autocorrelation estimates and maximum
c              variance.
          print *,' The estimated correlation values are:'
          do 20 n = 0,nn
            do 30 i2 = 0,(mm − n − 1)
              r(n) = r(n) + data(i2)*data(i2 + n)
30          continue
            r(n) = r(n)/(mm − n)
          print *,' R(',n,')',r(n)
20        continue
c
c         Find the Hamming-window estimate of the
c              spectral density.
c
          print *,' The estimates of spectral density are:'
          do 40 q = 0,nn
            sum = 0.0
            do 50 i5 = 1, (nn − 1)
              sum = 2.0*r(i5)*cos(q*i5*pi/nn) + sum
50          continue
            rs(q) = r(0) + r(nn)*cos(q*i5*pi/nn) + sum
40        continue
          rs( − 1) = rs(1)
          rs(nn + 1) = 0.0
c
          do 60 q = 0,nn
```

```
            hs(q) = 0.54*rs(q) + 0.23*(rs(q + 1) + rs(q - 1))
            print *,'S(', q*pi/nn,') = ',hs(q)
60          continue
            stop
            end
```

Table of Correlation Function—Spectral Density Pairs

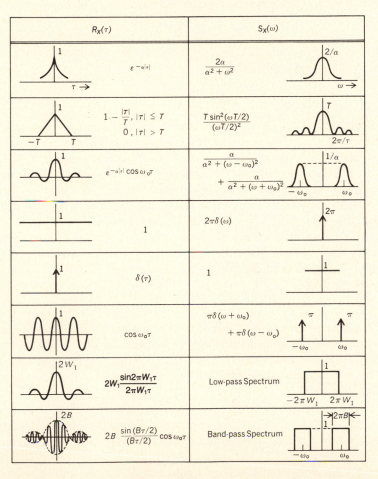

$R_X(\tau)$	$S_X(\omega)$
$\varepsilon^{-\alpha|\tau|}$	$\dfrac{2\alpha}{\alpha^2 + \omega^2}$
$1 - \dfrac{|\tau|}{T}, \ |\tau| \leq T$ $0, \ |\tau| > T$	$\dfrac{T \sin^2(\omega T/2)}{(\omega T/2)^2}$
$\varepsilon^{-\alpha|\tau|} \cos \omega_o \tau$	$\dfrac{\alpha}{\alpha^2 + (\omega - \omega_o)^2}$ $+ \dfrac{\alpha}{\alpha^2 + (\omega + \omega_o)^2}$
1	$2\pi\delta(\omega)$
$\delta(\tau)$	1
$\cos \omega_o \tau$	$\pi\delta(\omega + \omega_o)$ $+ \pi\delta(\omega - \omega_o)$
$2W_1 \dfrac{\sin 2\pi W_1 \tau}{2\pi W_1 \tau}$	Low-pass Spectrum
$2B \dfrac{\sin(B\tau/2)}{(B\tau/2)} \cos \omega_o \tau$	Band-pass Spectrum

APPENDIX I

Contour Integration

Integrals of the following types are frequently encountered in the analysis of linear systems

$$\frac{1}{2\pi j} \int_{c-j\infty}^{c+j\infty} F(s)e^{st} \, ds \tag{I-1}$$

$$\frac{1}{2\pi j} \int_{-\infty}^{\infty} S_X(s) \, ds \tag{I-2}$$

The integral of (I-1) is the inversion integral for the Laplace transform while (I-2) represents the mean-square value of a random process having a spectral density $S_X(s)$. Only in very special cases can these integrals be evaluated by elementary methods. However, because of the generally well-behaved nature of their integrands, these integrals can frequently be evaluated very simply by the method of residues. This method of evaluation is based on the following theorem from complex variable theory: if a function $F(s)$ is analytic on and interior to a closed contour C, except at a number of poles, then the integral of $F(s)$ around the contour is equal to $2\pi j$ times the sum of the residues at the poles within the contour. In equation form this becomes

$$\oint_C F(s) \, ds = 2\pi j \sum \text{residues at poles enclosed} \tag{I-3}$$

What is meant by the left-hand side of (I-3) is that the value of $F(s)$ at each point on the contour, C, is to be multiplied by the differential path length and summed over the complete contour. As indicated by the arrow, the contour is to be traversed counterclockwise. Reversing the direction introduces a minus sign on the right-hand side of (I-3).

In order to utilize (I-3) for the evaluation of integrals such as (I-1) and (I-2), two further steps are required: We must learn how to find the residue at a pole and then we must reconcile the closed contour in (I-3) with the apparently open paths of integration in (I-1) and (I-2).

Consider the problem of poles and residues first. A single valued function $F(s)$

is *analytic* at a point, $s = s_0$, if its derivative exists at every point in the neighborhood of (and including) s_0. A function is *analytic in a region* of the s-plane if it is analytic at every point in that region. If a function is analytic at every point in the neighborhood of s_0, but not at s_0 itself, then s_0 is called a *singular point*. For example, the function $F(s) = 1/(s - 2)$ has a derivative $F'(s) = -1/(s - 2)^2$. It is readily seen by inspection that this function is analytic everywhere except at $s = 2$, where it has a singularity. An *isolated singular point* is a point interior to a region throughout which the function is analytic except at that point. It is evident that the above function has an isolated singularity at $s = 2$. The most frequently encountered singularity is the *pole*. If a function $F(s)$ becomes infinite at $s = s_0$ in such a manner that by multiplying $F(s)$ by a factor of the form $(s - s_0)^n$, where n is a positive integer, the singularity is removed, then $F(s)$ is said to have a pole of order n at $s = s_0$. For example, the function $1/\sin s$ has a pole at $s = 0$ and can be written as

$$F(s) = \frac{1}{\sin s} = \frac{1}{s - s^2/3! + s^5/5! - \cdots}$$

Multiplying by s [that is, the factor $(s - s_0)$] we obtain

$$\phi(s) = \frac{s}{s - s^3/3! + s^5/5! + \cdots} = \frac{1}{1 - s^2/3! + s^4/5! + \cdots}$$

which is seen to be well-behaved near $s = 0$. It may therefore be concluded that $1/\sin s$ has a simple (that is, first order) pole at $s = 0$.

It is an important property of analytic functions that they can be represented by convergent power series throughout their region of analyticity. By a simple extension of this property it is possible to represent functions in the vicinity of a singularity. Consider a function $F(s)$ having an nth order pole at $s = s_0$. Define a new function $\phi(s)$ such that

$$\phi(s) = (s - s_0)^n F(s) \tag{I-4}$$

Now $\phi(s)$ will be analytic in the region of s_0 since the singularity of $F(s)$ has been removed. Therefore, $\phi(s)$ can be expanded in a Taylor series as follows:

$$\phi(s) = A_{-n} + A_{-n+1}(s - s_0) + A_{-n+2}(s - s_0)^2 \tag{I-5}$$

$$+ \cdots + A_{-1}(s - s_0)^{n-1} + \sum_{k=0}^{\infty} B_k (s - s_0)^{n+k}$$

Substituting (I-5) into (I-4) and solving for $F(s)$ gives

$$F(s) = \frac{A_{-n}}{(s - s_0)^n} + \frac{A_{-n+1}}{(s - s_0)^{n-1}} + \cdots + \frac{A_{-1}}{s - s_0} + \sum_{k=0}^{\infty} B_k (s - s_0)^k \tag{I-6}$$

This expansion is valid in the vicinity of the pole at $s = s_0$. The series converges in a region around s_0 that extends out to the nearest singularity. Equation (I-6) is called the *Laurent expansion* or *Laurent series* for $F(s)$ about the singularity at

$s = s_0$. There are two distinct parts to the series: The first, called the *principal part*, consists of the terms containing $(s - s_0)$ raised to negative powers; the second, sometimes called the Taylor part, consists of terms containing $(s - s_0)$ raised to zero or positive powers. It should be noted that the second part is analytic throughout the s-plane (except at infinity) and assumes the value B at $s = s_0$. If there were no singularity in $F(s)$, only the second part of the expansion would be present and would just be the Taylor series expansion. The coefficient of $(s - s_0)^{-1}$, which is A_{-1} in (I-6), is called the *residue* of $F(s)$ of the pole at $s = s_0$.

Formally the coefficients of the Laurent series can be determined from the usual expression for the Taylor series expansion for the function $\phi(s)$ and the subsequent division by $(s - s_0)^n$. For most cases of engineering interest, simpler methods can be employed. Due to the uniqueness of the properties of analytic functions it follows that any series of the proper form [that is, the form given in (I-6)] must, in fact, be the Laurent series. When $F(s)$ is a ratio of two polynomials in s, a simple procedure for finding the Laurent series is as follows: Form $\phi(s) = (s - s_0)^n F(s)$; let $s - s_0 = v$ or $s = v + s_0$; expand $\phi(v + s_0)$ around $v = 0$ by dividing the denominator into the numerator; and replace v by $s - s_0$. As an example consider the following

$$F(s) = \frac{2}{s^2(s^2 - 1)}$$

Let it be required to find the Laurent series for $F(s)$ in the vicinity of $s = -1$:

$$\phi(s) = \frac{2}{s^2(s - 1)}$$

Let $s = v - 1$

$$\phi(v - 1) = \frac{2}{(v^2 - 2v + 1)(v - 2)} = \frac{2}{v^3 - 4v^2 - 3v - 2}$$

$$
\begin{array}{r}
-1 + \dfrac{3v}{2} - \dfrac{v^2}{4} \\[4pt]
-2 - 3v - 4v^2 + v^3 \overline{\smash{\big)}\, 2 } \\[2pt]
\underline{2 + 3v + 4v^2 - v^3} \\[2pt]
- 3v - 4v^2 + v^3 \\[2pt]
\underline{- 3v - \dfrac{9v^2}{2} - 6v^3 + \dfrac{3v^4}{2}} \\[2pt]
\dfrac{v^2}{2} + 7v^3 - \dfrac{3v^4}{2} \\[2pt]
\underline{\dfrac{v^2}{2} + \dfrac{3v^3}{4} + v^4 - \dfrac{v^5}{4}}
\end{array}
$$

$$\phi(v - 1) = -1 + \frac{3}{2}v - \frac{1}{4}v^2 - \cdots$$

Replacing $v - 1$ by s gives

$$\phi(s) = -1 + \frac{3}{2}(s + 1) - \frac{1}{4}(s + 1)^2 - \cdots$$

$$F(s) = -\frac{1}{s + 1} + \frac{3}{2} - \frac{1}{4}(s + 1) - \cdots$$

The residue is seen to be -1.

A useful formula for finding the residue at an nth order pole, $s = s_0$, is as follows:[1]

$$K_{s_0} = \frac{\phi^{(n-1)}(s_0)}{(n - 1)!} \tag{I-7}$$

where $\phi(s) = (s - s_0)^n F(s)$. This formula is valid for $n = 1$ and is not restricted to rational functions.

When $F(s)$ is not a ratio of polynomials it is permissible to replace transcendental terms by series valid in the vicinity of the pole. For example

$$F(s) = \frac{\sin s}{s^2} = \frac{1}{s^2}\left(s - \frac{s^3}{3!} + \frac{s^5}{5!} - \cdots\right)$$

$$= \frac{1}{s} - \frac{s}{3!} + \frac{s^3}{5!} - \cdots$$

In this instance the residue of the pole at $s = 0$ is 1.

There is a direct connection between the Laurent series and the partial fraction expansion of a function $F(s)$. In particular, if $H_i(s)$ is the principal part of the Laurent series at the pole $s = s_i$, then the partial fraction expansion of $F(s)$ can be written as

$$F(s) = H_1(s) + H_2(s) \cdots H_k(s) + q(s)$$

where the first k terms are the principal parts of the Laurent series about the k poles and $q(s)$ is a polynomial $a_0 + a_1 s + a_2 s^2 + \cdots a_m s^m$ representing the behavior of $F(s)$ for large s. The value of m is the difference of the degree of the numerator polynomial minus the degree of the denominator polynomial. In general $q(s)$ can be determined by dividing the denominator polynomial into the numerator polynomial until the remainder is of lower order than the denominator. The remainder can then be expanded in its principal parts.

With the question of how to determine residues out of the way, the only remaining question is how to relate the closed contour of (I-3) with the open (straight line) contours of (I-1) and (I-2). This is handled quite easily by restricting consideration to integrands that approach zero rapidly enough for large values of the variable so that there will be no contribution to the integral from distant por-

---

[1]Where $\phi^{(n-1)}(s_o)$ denotes the $(n - 1)$-derivative of $\phi(s)$, with respect to s, evaluated at $s = s_o$.

tions of the contour. Thus, although the specified path of integration in the s-plane may be from $s = c - j\infty$ to $c + j\infty$, the integral that will be evaluated will have a path of integration as shown in Figure I-1. The path of integration will be along the contour C_1 going from $c - jR_o$ to $c + jR_o$ and then along the contour C_2 which is a semicircle closing to the left. The integral can then be written as

$$\oint_{C_1+C_2} F(s)\, ds = \int_{C_1} F(s)\, ds + \int_{C_2} F(s)\, ds \tag{I-8}$$

If in the limit as $R_o \to \infty$ the contribution from the right-hand integral is zero, then we have

$$\int_{c-j\infty}^{c+j\infty} F(s)\, ds = \lim_{R_o\to\infty} \oint_{C_1+C_2} F(s)\, ds$$

$$= 2\pi j \,\Sigma \text{ residues at poles enclosed} \tag{I-9}$$

In any specific case the behavior of the integral over C_2 can be examined to verify that there is no contribution from this portion of the contour. However, the following two special cases will cover many of the situations in which this problem arises:

1. Whenever $F(s)$ is a rational function having a denominator whose order exceeds the numerator by at least two, then it may be shown readily that

$$\int_{c-j\infty}^{c+j\infty} F(s)\, ds = \oint_{C_1+C_2} F(s)\, ds$$

2. If $F_1(s)$ is analytic in the left half plane except for a finite number of poles and tends uniformly to zero as $|s| \to \infty$ with $\sigma < 0$ then for positive t the following is true (Jordan's lemma)

$$\lim_{R_o\to\infty} \int_{C_2} F_1(s)e^{st}\, ds = 0$$

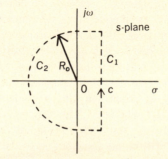

Figure I–1 Path of integration in the s-plane.

From this it follows that when these conditions are met, the inversion integral of the Laplace transform can be evaluated as

$$f(t) = \frac{1}{2\pi j} \int_{c-j\,\infty}^{c+j\,\infty} F_1(s)e^{st}\,ds = \frac{1}{2\pi j} \oint_{C_1+C_2} F_1(s)e^{st}\,ds = \sum_j k_j$$

where k_j is the residue at the jth pole to the left of the abscissa of absolute convergence.

The following two examples illustrate these procedures:

Example I-1. Given that a random process has a spectral density of the following form

$$S_X(\omega) = \frac{1}{(\omega^2 + 1)(\omega^2 + 4)}$$

Find the mean-square value of the process. Converting to $S_X(s)$ gives

$$S_X(s) = \frac{1}{(-s^2 + 1)(-s^2 + 4)} = \frac{1}{(s^2 - 1)(s^2 - 4)}$$

and the mean-square value is

$$\overline{X^2} = \frac{1}{2\pi j} \int_{-j\,\infty}^{j\,\infty} \frac{ds}{(s^2 - 1)(s^2 - 4)} = k_{-1} + k_{-2}$$

From the partial fraction expansion for $S_X(s)$, the residues are found to be

$$k_{-1} = \frac{1}{(-1 - 1)(1 - 4)} = \frac{1}{6}$$

$$k_{-2} = \frac{1}{(4 - 1)(-2 - 2)} = -\frac{1}{12}$$

Therefore,

$$\overline{X^2} = \frac{1}{6} - \frac{1}{12} = \frac{1}{12}$$

Example I-2. Find the inverse Laplace transform of $F(s) = \dfrac{1}{s(s + 2)}$.

$$f(t) = \frac{1}{2\pi j} \int_{c-j\,\infty}^{c+j\,\infty} \frac{e^{st}}{s(s + 2)}\,ds = k_0 + k_{-2}$$

From (I-7)

$$k_0 = \left. \frac{e^{st}}{s + 2} \right|_{s=0} = \frac{1}{2}$$

$$k_{-2} = \left. \frac{e^{st}}{s} \right|_{s=-2} = \frac{e^{-2t}}{-2}$$

therefore

$$f(t) = \frac{1}{2}(1 - e^{-2t}) \qquad t > 0$$

References

Churchill, R., *Operational Mathematics*, 2nd Ed., New York, N.Y.: McGraw-Hill Book Co., Inc., 1958.

A thorough treatment of the mathematics relating particularly to the Laplace transform is given including an introduction to complex variable theory and contour integration.

Papoulis, A., *The Fourier Integral and its Applications*, New York, N.Y.: McGraw-Hill Book Co., Inc., 1962.

In Chapter 9 and Appendix II, a readily understandable treatment of evaluation of transforms by contour integration is given.

APPENDIX J

Selected Answers

CHAPTER 1

1–1.1   **b)** 1.633; 2.237
1–4.1   **a)** $\frac{1}{6}$ **b)** $\frac{1}{2}$ **c)** $\frac{1}{2}$
1–4.2   **a)** $\frac{1}{18}$ **b)** $\frac{1}{6}$ **c)** $\frac{1}{2}$
1–4.3   **a)** $\frac{1}{8}$ **b)** $\frac{1}{2}$ **c)** 0.0602
1–4.4   **a)** $\frac{1}{20}$ **b)** $\frac{41}{50}$ **c)** $\frac{20}{346}$
1–4.5   **a)** $\frac{4}{15}$ **b)** $\frac{11}{30}$ **c)** $\frac{1}{5}$ **d)** $\frac{3}{10}$
1–4.6   **a)** $\frac{7}{120}$ **b)** $\frac{1}{20}$ **c)** $\frac{1}{150}$
1–6.1   **a)** $\frac{1}{2}$ **b)** $\frac{1}{2}$ **c)** $\frac{2}{3}$ **d)** 1 **e)** $\frac{5}{6}$ **f)** $\frac{2}{3}$
1–6.2   **a)** $\frac{4}{13}$ **b)** $\frac{1}{52}$ **c)** $\frac{10}{13}$ **d)** $\frac{5}{52}$ **e)** $\frac{1}{4}$ **f)** 0 **g)** $\frac{1}{52}$ **h)** $\frac{1}{13}$ **i)** 0
1–6.3   **a)** 0.0723 **b)** 0.1493 **c)** 0.9955 **d)** 0.7826 **e)** 0.0769 **f)** 0.2174
1–6.5   0.8075
1–7.1   **a)** 0.9246 **b)** 0.9468 **c)** 0.062
1–7.2   **a)** 0.0501 **b)** 0.0729
1–7.3   **a)** 0.2 **b)** $\frac{1}{2}$ **c)** $\frac{1}{8}$
1–7.4 $\frac{3}{4}$
1–7.5   0.99639
1–7.6   **a)** 0.1 **b)** $\frac{1}{2}$
1–7.7   **a)** $\frac{1}{2}$ **b)** $\frac{1}{3}$ **c)** $\frac{2}{3}$
1–8.1   Dependent
1–8.2   **a)** Independent **b)** Independent **c)** Independent
1–8.4   **a)** Independent **b)** Independent
1–9.1   **b)** $\frac{1}{8}$
1–9.2   **b)** 0.7298 **c)** 0.3001 **d)** 0.0111
1–10.1   **a)** 0.1406 **b)** 0.0156
1–10.4   **a)** 0.0123 **b)** 0.5926
1–10.5   **a)** 350 **b)** 150 **c)** 105
1–10.6   **a)** 0.6811 **b)** 0.2701 **c)** 0.04877 **d)** 0.3189
1–10.7   **a)** 10 **b)** 4 **c)** 5
1–10.8   **a)** 0.1268 **b)** 4.32×10^{-5} **c)** 4.36×10^{-4}

CHAPTER 2

2–2.1   **b)** 0.37597 **c)** 0.05469
2–2.2   **a)** 0 **b)** 0.125 **c)** 0.5
2–2.3   **a)** 1 **b)** 0.6321 **c)** 0.3679 **d)** 0.8647

2–2.4
    a) $A = \frac{1}{2}; b = \frac{\pi}{4}$ **b)** $\frac{1}{2} - \frac{\sqrt{2}}{4}$ **c)** $\frac{1}{2}$

2–3.1   **b)** 0.7734 **c)** 0.1718

2–3.2   **b)** 0.2325 **c)** 0.6321

2–3.4   **b)** 0.9653 **c)** 9.158×10^{-3}

2–4.1   **a)** 2 **b)** 5 **c)** 1

2–4.2   **a)** 0 **b)** 0.758 **c)** 0 **d)** 0.758

2–4.3   **a)** $\frac{1}{18}$ **b)** 4 **c)** 18 **d)** 2 **e)** -1.6 **f)** $\frac{2}{n+2} \cdot 6^n$

2–4.4   **a)** 12.5 **b)** 93.75; 9.68 **c)** 0.2373

2–5.1   **a)** 0.9772 **b)** 0.4772 **c)** 0.0228

2–5.2   **a)** 1875 **b)** 26,875 **c)** 0 **d)** 1750

2–5.3   **a)** 1 **b)** 4 **c)** 0.8413

2–5.4   **a)** 0.2867×10^{-6} **b)** 0.9987

2–6.1   **a)** 12 **b)** 288; 0.0832

2–6.2   **b)** 0 **c)** 15

2–6.3   **a)** 32 **b)** 0.3127 **c)** 0.1054

2–6.4   **a)** $= 3$ **b)** 3.76 **c)** 0.003865

2–6.5   **a)** 4.452×10^4 **b)** 0 **c)** 547.7

2–6.6   **a)** 5 **b)** 10 **c)** 3

2–7.1   **a)** $\dfrac{1}{\pi\sqrt{1-x^2}}$ **b)** 0 **c)** $\frac{1}{2}$ **d)** $\frac{1}{3}$

2–7.2   **a)** 58 **b)** 64 **c)** 6

2–7.3   **a)** 0.632 **b)** 0.1353 **c)** 0.0667

2–7.4   **a)** 4000 **b)** 0.2865 **c)** 0.3935

2–7.5   **b)** 0 **c)** 21

2–8.1   **a)** 0.3679 **b)** 8

2–8.2   **a)** $\dfrac{2}{\pi\sqrt{1-x^2}}$ $0 \le x \le 1$ **b)** $\frac{2}{\pi}$

2–8.3   **a)** 0.208 **b)** 0.944

2–8.4   **a)** 5.1856 **b)** 8.0044

2–9.1   **a)** 7.979 **b)** 12.533

2–9.2   **a)** $r\,e^{-\frac{1}{2}(r^2 - r_0^2)}$ $r \ge r_0$ **b)** 1.376

2–9.3   **b)** 2.55

2–9.4   **a)** $\frac{1}{3}$; $\frac{1}{3}$ **b)** 0 **c)** 25

CHAPTER 3

3–1.1   **c)** $\frac{9}{16}$

3–1.2   **a)** 4 **b)** $x^2 y^2$; $\begin{matrix} 0 \le x \le 1 \\ 0 \le y \le 1 \end{matrix}$ **c)** $\frac{3}{16}$ **d)** $\begin{matrix} 2x & 0 \le x \le 1; \\ 0 & \text{elsewhere} \end{matrix}$

3–1.3   **a)** $\frac{1}{4}$ **b)** $\frac{4}{9}$

3–1.4   **a)** $\frac{1}{4}$ **b)** 12.25 **c)** 1.429

3–2.1   **b)** 7.979

3–2.2   **a)** $2x$ **b)** $= 2y$

3–2.3   **b)** 5 **c)** 2

3–2.4   **a)** $-2y$ **b)** -6

3–3.1   **a)** $\dfrac{2}{3\ln 2}$ **b)** $\frac{3}{8}$ **c)** $\frac{88}{315}$

3–3.2 W and V are statistically independent

3–4.1   **a)** 76 **b)** 28 **c)** 28; 76

3–4.2   **a)** 481; 325 **b)** -0.1644

3–4.3   **a)** 16 **b)** $\frac{1}{6}$ **c)** 421

3–4.4   **a)** 3 **b)** $\dfrac{\sqrt{3}}{2}$ **c)** $\dfrac{\sqrt{3}}{2}$

3–5.1   **b)** 0.25

3–5.2   **b)** $\frac{7}{8}$ **c)** 0.68

3–5.3   **a)** $\frac{2}{3}$ **b)** 0.161

3–6.1 $\frac{3}{2}(e^{-t} - e^{-3t})\mu(t)$

3–6.3   **a)** p **b)** p **c)** $p; p - 3p^2$

CHAPTER 4

4–2.1   **a)** 0.3727 **b)** $\frac{1}{12}$ **c)** 455

4–2.2   **a)** 0.2 **c)** 10^6

4–2.3   **a)** 0.9897 **b)** 10 **c)** 17

4–2.4   **a)** 12.5 **b)** 11.218 **c)** 12

4–2.5   **a)** 0.663 **b)** 0.689

4–3.1   **a)** 0.0617 **b)** 1.0×10^{-3}

4–3.2   5002

4–3.3   322

4–4.1   **a)** 0.064 **b)** 0.0602

4–4.2   **a)** $118.66 \le \hat{\overline{X}} \le 121.34$ **b)** $116.42 \le \hat{\overline{X}} \le 123.58$

4–4.3   **a)** $118.95 \le \hat{\overline{X}} \le \infty$ **b)** $117.21 \le \hat{\overline{X}} \le \infty$

4–5.1   **a)** Claim justified **b)** Claim rejected

4–5.2   **a)** Claim rejected **b)** Claim rejected

4–5.3   **a)** 91% **b)** $\overline{X} = 3.997$ years

4–5.4   **a)** 99.78 **b)** 38.94; 6.24 **c)** Claim valid

4–6.1   **b)** $y = 0.5455 + 0.6344x$

4–6.2   **b)** $y = 326.33 - 15.14z$

CHAPTER 5

5–1.1   **b)** 7776 **c)** $\frac{1}{7776}$ **d)** $\frac{1}{7776}$

5–1.2   **a)** 2.5 **b)** 5 **c)** 5

5–2.2   **a)** $\dfrac{1}{4.28}\, e^{\dfrac{-(X_p - 2)^2}{8}}$ **b)** $\delta(X_n)$ **c)** $\delta(X_p X_n)$

5–3.2   **a)** 3   **b)** $9t^2 + 3$ **c)** 40
5–4.2   **a)** Nonstationary   **b)** Wide-sense stationary.
5–6.1   **a)** 0.0362   **b)** $\frac{1}{21}$
5–6.2   1.054

CHAPTER 6

6–1.1   **a)** 0.606   **b)** 3.16   **c)** 16.07
6–1.2   **a)** Wide-sense stationary   **b)** Wide-sense stationary   **c)** Not wide-sense stationary   **d)** Wide-sense stationary.
6–2.1   **a)** $T_1/2T$

$$\textbf{b)}\ \frac{1}{2}\frac{T_1}{T}\left[1 - \frac{|\tau|}{T_1}\right] \qquad j = k \qquad |\tau| \le T_1$$
$$\frac{1}{4}\frac{T_1}{T}\left[1 - \frac{|\tau - (j - k)T|}{T_1}\right] \qquad j \ne k \qquad |\tau - (j - k)T| \le T_1$$
$$0 \qquad \text{elsewhere}$$

6–2.2
$$\frac{1}{4}\frac{T_1}{T}\left[1 - \frac{|\tau - kT|}{T_1}\right] \qquad |\tau < kT| \le T_1$$
$$0 \qquad \text{elsewhere}$$

6–2.3 $\dfrac{1}{T}\, G(\tau)$

6–3.2   **b)** 9   **c)** $\dfrac{3}{2\tau}\sin 6\tau$

6–3.3   **a)** ± 6; 146; 110   **b)** 0, 1, 3, Hz   **c)** 0.318
6–3.4   Only when $T = 2$
6–4.1   **a)** 0.0362   **b)** -0.0460 **c)** -0.0387
6–4.2   **a)** 0.1740   **b)** 0.1664
6–4.3   **a)** $0.959 - 35.52\,|\tau|$ **b)** 0.0345
6–4.4   2118
6–5.1 $A^2 e^{-\alpha|\tau|}$
6–5.2 $2A^2 e^{-\alpha|\tau|}$
6–5.3   **a)** 10.0   **b)** 10.0   **c)** 10
6–5.4   **a)** $\overline{X} = 0$; $\sigma_x^2 = 5$ **b)** Differentiable
6–7.4 $32(\tau - 1)e^{-(\tau - 1)^2}$
6–8.1   **a)** 10.00005   **b)** 0.146

6–8.2   **a)** $\dfrac{0.1}{2}\sin\left[(\Theta - \phi)\right]$

6–8.3   **b)** $\begin{array}{l}\tau_{\max} = 1\\[2pt]\tau_{\min} = 0.1\end{array}$ **d)** 60 m/sec.

6–8.4   0.281 $n$sec.
6–9.2   3

CHAPTER 7

7–1.1 **a)** 6 **b)** $2M \dfrac{\sin \omega T}{\omega}$ **c)** $12 \dfrac{\sin \omega T}{\omega}$ **d)** $\dfrac{6}{\pi}\sigma\delta(\omega)$

7–2.2 4
7–2.3 **a)** 16 **b)** 1 **c)** 16 **d)** 4
7–3.1 **a)** No **b)** Yes **c)** Yes **d)** No **e)** Yes **f)** No
7–3.2 **a)** 3 **b)** 83.5
7–3.3 **a)** 4 **b)** 40 **c)** 0, ± 6, ± 12
7–3.4 **a)** 1 **b)** 1
7–4.1 **a)** $\dfrac{16(-s^2 + 36)}{s^4 - 13s^2 + 36}$ **b)** Poles: ±2, ±3 zeros: ±6 **c)** 0.573
7–4.2 **c)** 11.39
7–5.2 **a)** 11.2 **b)** 11.2
7–5.3 0.05
7–5.4 7
7–6.1 **a)** 10 **b)** $\dfrac{800 \sin^2(0.025\omega)}{\omega^2}$ **c)** $R_X(\tau_1) = 0; S_X(f_1) = 0$
7–6.2 **a)** 24 **c)** 0.0403
7–6.3 **a)** $50/\pi$ **c)** $50/\pi$
7–6.4 **a)** $40/(25 + \omega^2)$ **b)** $4 e^{-5|\tau|}$
7–7.2 **a)** $10\left[\dfrac{\sin 1000\tau\pi}{1000\tau\pi}\right]$ **b)** 0 **c)** 0; 4.135
7–8.1 **a)** 1 **b)** 0 **c)** $16/(\omega^2 + 16)$
7–8.2 $\dfrac{16 - \omega^2}{16 + \omega^2}$
7–9.1 **c)** 0.00299
7–9.2 Hamming window has lower sidelobes.
7–10.2 **a)** 19.2 **b)** 79.4
7–10.4 **a)** $2B_1\sqrt{\sqrt[n]{2} - 1}$ **b)** $2B_1\sqrt{\sqrt[n]{100} - 1}$

CHAPTER 8

8–2.1 **b)** 5 **c)** 64
8–2.2 **a)** $\dfrac{4}{5} B \cos (20t + \theta) - \dfrac{2}{5} B \sin (20t + \theta)$ **b)** 0 **c)** $0.8B^2$
8–3.1 **a)** 0 **b)** 5.0 **c)** 5
8–3.2 **a)** 0 **b)** $8e^{-1}$ **c)** $8e^{-1}$
8–3.3 **a)** 0 **b)** $1/8$ **c)** $1/8$
8–3.4 6.25
8–4.1 **a)** $0.5e^{-10^4|\tau|}$ **b)** $1/2$

8–4.4   **a)** $\frac{3}{4}$ **b)** $\frac{35}{12}$ **c)** $\frac{35}{12} - 3|\tau| + 4\tau^2 - \frac{5}{3}|\tau|^3$ $: -1 \le \tau \le 1$

                                         $\frac{3}{4}$: otherwise

8–5.1   5

8–5.3 $Rxy(\tau) = \frac{13}{2} - \tau$

         $Ryx(\tau) = \frac{13}{2} + \tau$

8–5.4 $\begin{cases} 50\, e^{-1000\,\tau} & \text{for } \tau \ge 0 \\ -50\, e^{1000\,\tau} & \text{for } \tau \le 0 \end{cases}$

8–6.1   **a)** $\frac{2}{3}A^2$

8–6.2   **a)** $\frac{2}{3}$ **b)** $1.\frac{1}{3} \times 10^2$ **c)** $1.\frac{1}{3} \times 10^{-2}$

8–6.4   **a)** 5.7×10^{-3} **b)** 2.65×10^5 **c)** 1.6×10^3

8–7.1   **a)** $\dfrac{s + 1}{3s + 1}$ **b)** $\dfrac{s^2(s^2 - 1)}{(9s^2 - 1)(s^2 - 16)}$

8–7.2   **c)** $\dfrac{46656}{-s^6 + 46656}$

8–8.1   **a)** $\dfrac{\omega^2(\omega^2 + 1)}{2(9\omega^2 + 1)(\omega^2 + 16)}$ **b)** $\dfrac{\omega^4}{2(9\omega^2 + 1)(\omega^2 + 16)(\omega^2 + 4)}$

8–8.2   **a)** $\dfrac{20 \times 216^2}{(\omega^2 + 1)(\omega^6 + 46656)}$ **b)** 20

8–8.3   **b)** 0.04

8–10.1   **a)** 1   **b)** $1.047B_{1/2}$

8–10.2   **a)** 0.1875   **b)** 12   **c)** 12

8–10.3   **a)** $\dfrac{1}{2s + 4}$ **b)** 0.13   **c)** 2.63

8–10.4 $\dfrac{1}{\pi} \times 10^{-8}$

8–10.6 $\dfrac{W}{2}\sqrt{\dfrac{\pi}{1.386}}$

8–10.8   **a)** $F_1 + \dfrac{F_2 - 1}{G_1}$ **b)** 6.1×10^5 **c)** 1.57

CHAPTER 9

9–2.1   Maximum signal-to-noise ratio: (c), (d)

         Minimum mean-square error: (a), (b), (e), (f)

9–2.2   **a)** $\dfrac{40b^2}{\pi[b^2 + (160\pi)^2]}$ **b)** $\dfrac{4b^2}{\pi[b^2 + (160\pi)^2]}$ **c)** $\dfrac{2b^2}{5\pi[b^2 + (160\pi)^2]}$

9–4.1   **a)** 80   **b)** 1.1

9–4.2   0.648

9–4.3   3   **b)** 9.16

9–4.4   **a)** 1.59   **b)** 1.43

9–5.1   **a)** $s(2 - t)u(t)$ **b)** 46.67   **c)** 20

9–5.2   **a)** 50   **b)** 2.661

9–5.3   a) N b) 100

9–5.5   a) $10^{-4}e^{-10^5(4.6 \times 10^{-5} - t)}u(t)$ b) 0.5×10^{-13}

9–6.1   a) $\dfrac{16}{0.1\omega^2 + 17.6}$ b) 0.6

9–6.2   a) $\dfrac{9.27}{s + \sqrt{176}}$ b) 0.927

9–6.3   a) $\dfrac{4}{\omega^2 + 4}$ b) ½

Index